Intelligent Planning for Mobile Robotics:

Algorithmic Approaches

Ritu Tiwari
ABV - Indian Institute of Information, India

Anupam Shukla
ABV - Indian Institute of Information, India

Rahul Kala
School of Systems Engineering, University of Reading, UK

A volume in the Advances in
Computational Intelligence and Robotics
(ACIR) Book Series

Managing Director: Lindsay Johnston
Editorial Director: Joel Gamon
Book Production Manager: Jennifer Romanchak
Publishing Systems Analyst: Adrienne Freeland
Development Editor: Myla Merkel
Assistant Acquisitions Editor: Kayla Wolfe
Typesetter: Travis Gundrum
Cover Design: Nick Newcomer

Published in the United States of America by
Information Science Reference (an imprint of IGI Global)
701 E. Chocolate Avenue
Hershey PA 17033
Tel: 717-533-8845
Fax: 717-533-8661
E-mail: cust@igi-global.com
Web site: http://www.igi-global.com

Library of Congress Cataloging-in-Publication Data

Tiwari, Ritu, 1977-
Intelligent planning for mobile robotics : algorithmic approaches / by Ritu Tiwari, Anupam Shukla, and Rahul Kala.
p. cm.
Includes bibliographical references and index.
Summary: "This book presents content coverage on the basics of artificial intelligence, search problems and soft computing approaches, providing insight on robotics and basic algorithms, special topics in AI, planning, applied soft computing, applied AI, applied evolutionary computing"--Provided by publisher.
ISBN 978-1-4666-2074-2 (hardcover) -- ISBN 978-1-4666-2075-9 (ebook) -- ISBN 978-1-4666-2076-6 (print & perpetual access) 1. Robotics--Mathematical models. 2. Robots--Mathematical models. 3. Artificial intelligence--Mathematical models. I. Shukla, Anupam, 1965- II. Kala, Rahul. III. Title.
TJ211.T596 2012
629.8'932028563--dc23
2012013146

This book is published in the IGI Global book series Advances in Computational Intelligence and Robotics (ACIR) Book Series (ISSN: 2327-0411; eISSN: 2327-042X)

British Cataloguing in Publication Data
A Cataloguing in Publication record for this book is available from the British Library.

All work contributed to this book is new, previously-unpublished material. The views expressed in this book are those of the authors, but not necessarily of the publisher.

Advances in Computational Intelligence and Robotics (ACIR) Book Series

ISSN: 2327-0411
EISSN: 2327-042X

Mission

While intelligence is traditionally a term applied to humans and human cognition, technology has progressed in such a way to allow for the development of intelligent systems able to simulate many human traits. With this new era of simulated and artificial intelligence, much research is needed in order to continue to advance the field and also to evaluate the ethical and societal concerns of the existence of artificial life and machine learning.

The **Advances in Computational Intelligence and Robotics (ACIR) Book Series** encourages scholarly discourse on all topics pertaining to evolutionary computing, artificial life, computational intelligence, machine learning, and robotics. ACIR presents the latest research being conducted on diverse topics in intelligence technologies with the goal of advancing knowledge and applications in this rapidly evolving field.

Coverage

- Adaptive & Complex Systems
- Agent Technologies
- Artificial Intelligence
- Cognitive Informatics
- Computational Intelligence
- Natural Language Processing
- Neural Networks
- Pattern Recognition
- Robotics
- Synthetic Emotions

IGI Global is currently accepting manuscripts for publication within this series. To submit a proposal for a volume in this series, please contact our Acquisition Editors at Acquisitions@igi-global.com or visit: http://www.igi-global.com/publish/.

The Advances in Computational Intelligence and Robotics (ACIR) Book Series (ISSN 2327-0411) is published by IGI Global, 701 E. Chocolate Avenue, Hershey, PA 17033-1240, USA, www.igi-global.com. This series is composed of titles available for purchase individually; each title is edited to be contextually exclusive from any other title within the series. For pricing and ordering information please visit http://www.igi-global.com/book-series/advances-computational-intelligence-robotics-acir/73674. Postmaster: Send all address changes to above address.

Titles in this Series

For a list of additional titles in this series, please visit: www.igi-global.com

Intelligent Technologies and Techniques for Pervasive Computing
Kostas Kolomvatsos (University of Athens, Greece) Christos Anagnostopoulos (Ionian University, Greece) and Stathes Hadjiefthymiades (University of Athens, Greece)
Information Science Reference • copyright 2013 • 349pp • H/C (ISBN: 9781466640382) • US $195.00 (our price)

Mobile Ad Hoc Robots and Wireless Robotic Systems Design and Implementation
Raul Aquino Santos (University of Colima, Mexico) Omar Lengerke (Universidad Autónoma de Bucaramanga, Colombia) and Arthur Edwards-Block (University of Colima, Mexico)
Information Science Reference • copyright 2013 • 347pp • H/C (ISBN: 9781466626584) • US $190.00 (our price)

Intelligent Planning for Mobile Robotics Algorithmic Approaches
Ritu Tiwari (ABV – Indian Institute of Information, India) Anupam Shukla (ABV – Indian Institute of Information, India) and Rahul Kala (School of Systems Engineering, University of Reading, UK)
Information Science Reference • copyright 2013 • 320pp • H/C (ISBN: 9781466620742) • US $195.00 (our price)

Simultaneous Localization and Mapping for Mobile Robots Introduction and Methods
Juan-Antonio Fernández-Madrigal (Universidad de Málaga, Spain) and José Luis Blanco Claraco (Universidad de Málaga, Spain)
Information Science Reference • copyright 2013 • 497pp • H/C (ISBN: 9781466621046) • US $195.00 (our price)

Prototyping of Robotic Systems Applications of Design and Implementation
Tarek Sobh (University of Bridgeport, USA) and Xingguo Xiong (University of Bridgeport, USA)
Information Science Reference • copyright 2012 • 321pp • H/C (ISBN: 9781466601765) • US $195.00 (our price)

Cross-Disciplinary Applications of Artificial Intelligence and Pattern Recognition Advancing Technologies
Vijay Kumar Mago (Simon Fraser University, Canada) and Nitin Bhatia (DAV College, India)
Information Science Reference • copyright 2012 • 784pp • H/C (ISBN: 9781613504291) • US $195.00 (our price)

Handbook of Research on Ambient Intelligence and Smart Environments Trends and Perspectives
Nak-Young Chong (Japan Advanced Institute of Science and Technology, Japan) and Fulvio Mastrogiovanni (University of Genova, Italy)
Information Science Reference • copyright 2011 • 770pp • H/C (ISBN: 9781616928575) • US $265.00 (our price)

Particle Swarm Optimization and Intelligence Advances and Applications
Konstantinos E. Parsopoulos (University of Ioannina, Greece) and Michael N. Vrahatis (University of Patras, Greece)
Information Science Reference • copyright 2010 • 328pp • H/C (ISBN: 9781615206667) • US $180.00 (our price)

www.igi-global.com

701 E. Chocolate Ave., Hershey, PA 17033
Order online at www.igi-global.com or call 717-533-8845 x100
To place a standing order for titles released in this series, contact: cust@igi-global.com
Mon-Fri 8:00 am - 5:00 pm (est) or fax 24 hours a day 717-533-8661

Table of Contents

Preface

Robotics is an ever-growing field, which presents numerous issues, problems, challenges, and solutions at the various levels. The problem of planning a robot deserves a special mention that is the primary enabler of making mobile autonomous robots that can carry out tasks independently without human aid. This calls for making the robots intelligent enough to plan and make any decisions based on inputs from various sensors, which may be later implemented by actuators. Planning is the basic activity that induces this intelligence into the robots for carrying out the various tasks. In such a context, this book is our approach to cover the various aspects of planning of a mobile robot.

At one end, the book deals with the basics of artificial intelligence, search problems, and soft computing approaches. Here, we attempt to explain the theoretical basics of the various paradigms of intelligent robotics. The stress is on conceptualization and to understand and appreciate the presence of numerous algorithms and the means by which these carry out the same task in different manners. The numerous models available in any domain and their widespread use are a natural consequence of active research in these areas. It is, hence, of vast importance to judiciously decide the algorithms, realize their working, and appreciate the methods even before we use them. This calls for a conceptual understanding of all of them and discussion of tradeoffs between them, which this book presents.

Robotic path planning is a specific problem that has an extremely wide range of algorithms applied to it. These algorithms vary from the graph search methods to statistical methods; from neural and fuzzy approaches to evolutionary approaches; from simple algorithms to a combination of algorithms or hybrid algorithms; and from simple scenarios and maps to multi-resolution and hybrid maps. All these have different modeling scenarios in terms of robotic motion, maps, and robotic environment. All of these ultimately solve the same problem of effective planning of the path of the mobile robot and enabling it to reach the goal from the source without any collision with the obstacles.

The application areas usually demand a customized application of the base algorithms that we study in the theoretical aspects. In other words, the problem needs to be well modeled before any standard algorithm may be applied over it. This calls for a lot of novelty in the visualization of the same problem in multiple dimensions. The application of the various algorithms further open before us a variety of issues that at one end enable a clear understanding of the underlying algorithms as well as the problem domain, and at the other end practically expose us towards the concepts and tradeoffs between the algorithms which is of paramount importance.

The inability of any single algorithm to solve the problem has resulted in a paradigm of hybrid algorithms in the soft computing domain. The same applies to the robotic planning domain as well. The numerous algorithms impose restrictions that may not necessarily hold well in real life scenarios. This brings us towards the notion of hybrid algorithms in this field. The manner in which the multiple algorithms fuse with each other to form hybrid algorithms and the manner in which these individual algorithms collaboratively work with each other under the hybrid approach is a unique site. The hybridization may be at the level of the robotic map, problem scenario, or computation. This further adds a new dimension towards the entire outlook at the problem and algorithms. It is again important to conceptualize and understand the reasons for hybridizing algorithms before actually doing so. This calls for a clear understanding of the problems with the individual algorithms and the manner in which they complement each other for an overall good performance.

As per the authors, this is an interesting problem that not only enables learning about this rapidly moving domain but also enables learning algorithms and problem solving techniques that are difficult to learn by other means. It represents a unique amalgamation of theory with applications where we analyze and appreciate how beautifully the theoretical algorithms can be easily cast into the application domain and how the difficult looking application problems can be easily modeled for standard and hybrid customized algorithms.

Robotics is again a very vast field that has numerous burning issues at the intelligent planning front as well. Many specific problems like robot soccer, maze solving, formation control, etc. have emerged. While motion planning is taken as a central issue of the book, an attempt is drawn towards presenting a variety of algorithms for the same; for completeness, other planning problems extensively studied in the domain of robotics are also discussed. An increasing interest is seen towards the use of multi-robotics in problems of all domains. This brings us to the problem of planning and coordination amongst multiple robots, which is another exciting problem of work and can be easily solved knowing the basics of planning for mobile robots.

Chapter 1 covers the basics of the problem of planning a mobile robot. This is done in light of the entire domain of robotics. The chapter aims to broadly introduce the fundamentals of robotics and to highlight the problem of path planning. Further insights into the problem of path planning are presented. The various issues that algorithms must address in terms of optimality, completeness, execution time, etc. are presented. The focus is further to understand the modeling and nature of the problem. The chapter gives an overall insight, which builds the foundation for the rest of the book.

Henceforth starts the task of understanding of the base algorithms, to model the problem so as to solve by these base algorithms, and to present the corresponding results. This is the focus of chapters 2-6. Chapter 2 in particular deals with modeling the problem as a graph search problem. This modeling allows all the various graph search and similar algorithms to be used to solve the problem. Every such algorithm is separately dealt with. The algorithms may be easily seen as intelligent and non-intelligent. Every algorithm has its own advantages and limitations. Chapter 2 explains the concept, working methodology, and issues associated with some of these algorithms. The algorithms discussed in the chapter are breadth first search, depth first search, A* algorithm, multi-neuron heuristic search, Dijkstra's algorithm, D* algorithm, etc. While graphs may be able to solve the problem, it is important to understand the pros and cons of this class of algorithm, which enables us to better understand the problem domain.

Chapter 3 furthers the concepts put forth by chapter 1 and discusses other supplementary algorithms for a similar kind of problem modeling. The algorithms put forth for understanding, modeling, discussion, and experimentation include dynamic programming, Bellman Ford algorithm, Rapidly exploring Random Trees (RRT), artificial potential fields, embedded sensor planning, reinforcement learning planning, and Voronoi graph-based planning. In particular, Bellman Ford is a graph search algorithm used to find the shortest path from source to goal. RRTs carry on the search operation by rapidly growing their search data structure, which by architecture is a tree. Whenever the search operation finds the goal, the search is stopped. The artificial potential field method is inspired by the potential fields and use similar modeling for planning of the robot. Embedded planning takes the problem to an entirely new level by distributing the computation on the entire map rather than sticking to a single robot. Again, emphasis is on highlighting the advantages and disadvantages of individual algorithms.

Chapter 4 opens a new domain of the problem by using evolutionary computing techniques for solving the problem. This paradigm of computing is inspired by the evolution process of the natural world and uses a multi-individual optimization scheme for solving the problem. The chapter presents an in-depth discussion over genetic algorithms, which is extensively used to solve a variety of problems. The same is used for solving the problem of motion planning, which requires a unique technique to model the solution or a robot path. Non-holonomicity is a major issue associated with mobile robots, which emphasizes the individual robotic paths to be smooth enough for a car-like robot to traverse. The B-Splines and Bezzier curves are discussed as curve smoothening agents for the paths optimized in the evolutionary process. The chapter further adds the notion of adaptability of parameters by presenting the evolutionary strategy algorithm for the problem. This algorithm is self-adaptive in nature. Covariance matrix adaptation evolutionary strategy is the studied method of implementation.

Chapter 5 furthers the discussion of chapter 4 by presenting more evolutionary concepts and their application in robot path planning. The chapter explores the use of the swarm techniques for solving the problem. While the modeling technique of the problem of robot path planning is similar to the use of chapter 4, it is interesting to observe how the various techniques differ from each other as an algorithm. All these techniques belong to the evolutionary domain and use similar concepts, but the manner used to tackle the problem is different. Discussed algorithms include particle swarm optimization, ant colony optimization, artificial bee colonies, and probability-based incremental learning. Further, another major pillar of evolutionary computation or genetic programming is presented. Here, the task is optimization of a program that solves the problem. We look into a technique of linear representation of the program called grammatical evolution. The technique is experimented into the problem of planning of the robot.

Chapter 6 opens the last major type of algorithm that may be used for the problem. This is the behavioral planning of the mobile robot. In this technique, in place of planning the robot path, we try to generate a robot behavior that solves the problem of planning. This is similar to the means by which we move and act in everyday life governed by our behaviors. The chapter presents two different algorithms, fuzzy inference systems, and artificial neural networks. The fuzzy method of solving the problem is to use a set of rules, which take the current scenario in which the robot is the input and produces as output the indicative next move of the robot. Based on the same, inputs and outputs are worked over. Rules are generated to produce best moves in any scenario. The neural network method of solving the problem is also introduced. In this technique, the map is embedded with computational units called neurons. These neurons compute the path of the robot, which the robot simply follows.

While use of some base algorithms for solving the problem gives decent results, it is important to get rid of the limitations that they carry. Chapters 7-9 present the notion of hybrid planning of the mobile robot where multiple planning algorithms are merged together to produce a more effective algorithm that to some extent removes the limitations of the individual algorithms. These hybrid algorithms fuse the individual participating algorithms in a characteristic manner so that the resultant algorithm is able to solve the problem more effectively. To effectively hybridize, the advantages and limitations of the individual algorithms must be well known.

Chapter 7 presents the hybrid algorithms whose base algorithm is a graph search. Two algorithms are presented. In the first algorithm, a Multi Neuron Heuristic Search (MNHS) algorithm is merged with itself, that is, it is implemented in a hierarchical manner. A probabilistic map representation scheme is used where we measure the probability of collision with obstacle while traveling inside the cell. The algorithm is iterative and the granularity of the path increases with the increase in iterations. In the second approach discussed in the chapter, a MNHS algorithm is hybridized with an Evolutionary Algorithm (EA). The limitation of MNHS is that it is slow, but the algorithm is complete. On the other hand, EA is not complete but is fairly optimal. The MNHS improves the performance of the algorithm while the EA does the task of time optimization especially in case of complex graphs in the fused algorithm. The EA is used to generate key points in the robotic map, which are further converted into a graph. The A* algorithm uses this graph and finds a solution from source to goal.

Chapter 8 furthers the discussion over the use of hybrid algorithms for the task of planning of mobile robots. In this chapter, the hybrid algorithms are based on evolutionary techniques. Three approaches are presented. In the first approach, EA solves the planning problem in a static map. An additional factor momentum is introduced that controls the granularity with which a robotic path is traversed to compute its fitness. Higher granularity means lesser computation time and a smaller chance of getting the correct result in terms of feasibility and other measures. The fitness function takes into account path length, number of turns, and distance from closest obstacle. Algorithm parameters are kept adaptive and slowly increased with increase in generations. This algorithm is further extended to the use of Hybrid Genetic Algorithm Particle Swarm Optimization (HGAPSO) algorithm in the second approach. The algorithm is a hybrid of genetic algorithm and particle swarm optimization and represents a more sophisticated computational technique. The third approach deals with hybridization of EA with itself in a hierarchical manner. Two hierarchies are used. The coarser hierarchy finds the path in a static environment, which is capable of quickly generating a lower detail path. The finer hierarchy takes a section of the map and computes the path for both static and dynamic environments. The finer hierarchy assumes all obstacles keep moving at the same speed in the same direction to compute the feasible path.

Chapter 9 presents the hybrid techniques that are primarily behavior based. Three approaches are presented. The first approach is a fusion of A* algorithm and Fuzzy Inference System (FIS)-based planner. Here, A* algorithm works over a coarser level algorithm and is able to come up with a path of low-level details quickly. It, however, does attempt to go away from the obstacles. The lower level planning is done by the FIS. The results of A* algorithm serve as a guide for FIS planner. GA is used to optimize the FIS model. Many times the planning algorithm may require a map breakup such that the coarser level graph also has a high degree of resolution and A* algorithm may not be able to work. Hence, in the second approach, we replace the A* algorithm by genetic algorithm. The last approach presented is based on dynamic programming, which in this implementation is similar to use of neurons embedded in the map being planned. In this approach, we talk about the use of extra nodes in the planning framework called accelerating nodes. These nodes are less in number and interconnected. These nodes transmit information about any map change and blockages to each other for sudden re-planning to be initiated. These nodes further guide the robot for path to follow until re-planning completes.

Chapter 10 is not primarily about path planning like the other chapters, but is more for the completeness of the book. The chapter introduces the notion of multi-robot systems, wherein multiple robots are used to solve a problem rather than a single robot. With the addition of robots, it can be clearly guessed that the network would be done much more effectively in whichever task the robots are engaged in. However, this further opens a number of issues and concerns that must be addressed, especially at the planning level. First, a number of specific problems are introduced, which include maze solving, complete coverage, map building, and pursuit evasion. The problem of planning with multiple robots results in a number of issues, which are addressed explicitly. These include optimality in terms of computational time and solution generated, completeness of planning, reaching a consensus, cooperation amongst multiple robots, and means of communication between robots for effective cooperation. The issues are well understood by benchmark problems. In the chapter, we present the problems of multi-robot task allocation, robotic swarms, formation control with multiple robots, RoboCup, multi-robot path planning, and multi-robot area coverage and mapping. The problem of multi-robot path planning is of a greater concern and is taken with more depth. Solutions are classified as centralized and decentralized. We discuss means by which algorithms for single robot path planning may be extended to the use of multiple robots. This is specifically done for the graph search, evolutionary, and behavioral approaches discussed in the earlier chapters of the book.

Chapter 11 gives the overall views and the concluding remarks on the book as a whole. While the problem of planning is challenging, a significantly large number of algorithms exist to solve the problem. The different algorithms differ in optimality, completeness, and modeling. Given a wide variety of solutions, it is important to know the kind of map and use the appropriate algorithm for similar kinds of maps. A tradeoff between algorithms needs to be well understood before using any algorithm for practical problem solving. Even the hybrid algorithms do not fully escape the addressed limitations. Not only is that the advantages are added, but some limitations and assumptions also get added, which limit the entire algorithm. Hence, for characteristic maps, the hybrid algorithms may even be more inferior to the base algorithms. While the book presents numerous algorithms, the quest for a perfect algorithm may still be a quest.

The book is primarily meant for research groups, research students, research scholars, professors, and technicians working in the field of planning in robotics. The book is designed to provide background information before advancing to advanced text. Hence, this book is an ideal choice for entering into research in this domain. Advanced readers having a fair idea of basic planning algorithms may prefer to glance through the introductory material and proceed to the advanced topics. Each chapter is devoted to some specific topic and may be used for reference to individual algorithms for planning.

The book can also be used as a reference book for courses related to robotics, special topics in AI, planning, applied soft computing, applied AI, applied evolutionary computing, etc. The book covers a background of both robotics and basic algorithms, which makes it easier for readers from multiple domains to study the topics presented. Overall, the book provides a rich flavor to the course, covering a wide variety of topics.

Major features of the book are:

- Coverage of all the pre-requisites from AI and Robotics in the book itself
- Uniformity of content and presentation for the specific problem
- In-depth analysis of a large number of approaches to planning in an interdependent manner
- State-of-the-art content coverage
- Coverage of various related topics

Acknowledgment

The book was reasonably easy to conceive. We were engaged for just over two years in the domain within which we already had a lot of literature studied, papers published, and understanding. It was evident that all the knowledge could be converted into a book, which seemed not that difficult a task. Considering ourselves masters of the domain (strictly reference to the degree work reported resulted in for the third author), we went forward towards writing with great zeal and vigor. However, the process was not that simple and during the writing of the book, the affiliation of third author changed, events cropped up at unwilling times, and related stuff—some intentionally unquoted. The deadline changed uncounted times during the writing process, and the work became mix of impulsive and sinusoidal overtime. A book on planning itself not wisely planned is certainly an irony, but we believe such tales in management are common. With all these developments, there were numerous people who became handy, whom we would like to acknowledge.

The authors acknowledge the valuable inputs of the anonymous reviewers whose comments enabled us to improve the quality of the book by a large amount. Not only were the review comments encouraging, but they helped us improve the presentation of the book by a large amount. The authors further thank the editorial team at IGI-Global for all their help throughout the process.

The authors wish to express sincere thanks to Anshika Pal for her contributions to chapter 10, which played a great role in creating its present form. The authors express their gratitude towards Mohit Maru for the help extended in production of many of the line art figures. While it was easy to scribble something on paper that might result in better understanding of the concepts, the production of the same as professional figures was certainly not. We further thank Sanjeev Sharma for the help extended in formatting the entire draft and putting all the files into proper shape for submission. The authors extend their thanks to Anubhav Kakkar and Shivam Chandra for their inputs to various sections on applied and industrial robotics, which are widespread in the final book. The authors further thank Ritesh Maurya for his inputs on robot path planning using ant colony optimization.

The authors thank the support extended by Prof. Kevin Warwick towards the production of the book. Authors acknowledge the facilities and support extended by the host institutions, Indian Institute of Information Technology and Management Gwalior and School of Systems Engineering, University of Reading, UK. The authors thank all the members of the host institutions who contributed towards the book—directly or indirectly.

PERMISSIONS

Excerpts from text, tables, and figures published in this book are taken from a number of sources with kind permissions from their respective publishers. Details are indicated in individual chapters. All publishers retain the copyright for their material reprinted in this book. For details or permission requests, please refer to the original papers. The authors thank the publishers for reprint permissions.

- Chapter 7, parts of "Introduction," "Conclusion," and the Sections "Hybrid MNHS and Evolutionary Algorithms" to "Results" are reprinted with few modifications. © 2009 IEEE. Reprinted with permission from IEEE. Kala, R., Shukla, A., & Tiwari, R. (2009). Fusion of evolutionary algorithms and multi-neuron heuristic search for robotic path planning. In *Proceedings of the IEEE Conference on Nature & Biologically Inspired Computing,* (pp. 684-689). IEEE Press.
- Chapter 6, Section "Path Planning using Neural Networks" has excerpts of text and figures © 2009 IEEE. Reprinted, with permission, from Kala, R., Shukla, A., Tiwari, R., Roongta, S., & Janghel, R. R. (2009). Mobile robot navigation control in moving obstacle environment using genetic algorithm, artificial neural networks and A* algorithm. In *Proceedings of the IEEE World Congress on Computer Science and Information Engineering,* (pp. 705-713). Los Angeles, CA: IEEE Press.
- Chapter 6, Section "Motion Planning by Fuzzy Inference System" and Chapter 9, parts of "Introduction," "Conclusion," and Section "Fusion of A* and Fuzzy Planner" to Section "Simulation and Results" are reprinted with few modifications with kind permission from Springer Science+Business Media. Kala, R., Shukla, A., & Tiwari, R. (2010). Fusion of probabilistic A* algorithm and fuzzy inference system for robotic path planning. *Artificial Intelligence Review, 33,* 275-306.
- Chapter 8, parts of "Introduction," "Conclusion," and Section "Hierarchical Evolutionary Algorithm: Hierarchy 1" to Section "Results" are reprinted with few modifications from Kala, R., Shukla, A., & Tiwari, R. (2010). Dynamic environment robot path planning using hierarchical evolutionary algorithms. *Cybernetics and Systems, 41*(6), 435-454. Reprinted by permission of Taylor & Francis Ltd. (http://www.tandfonline.com).
- Chapter 9, parts of "Introduction," "Conclusion," and Section "Evolving Robotic Path" to Section "Simulations and Results" are reprinted with few modifications from Kala, R., Shukla, A., & Tiwari, R. (2010). Evolving robotic path with genetically optimized fuzzy planner. *International Journal of Computational Vision and Robotics, 1*(4), 415-429. With permission from Inderscience Publishers.
- Chapter 7, parts of "Introduction," "Conclusion," and Sections "Hierarchical MNHS Algorithm" to Section "Simulation and Results" are reprinted with few modifications from Kala, R., Shukla, A., & Tiwari, R. (2011). Robotic path planning in static environment using hierarchical multi-neuron heuristic search and probability based fitness. *Neurocomputing, 74*(14-15), 2314-2335. Reprint is done on the basis of retained author rights.
- Chapter 8, parts of "Introduction," "Conclusion," and Section "Evolutionary Algorithm with Momentum" to Section "Results" are reprinted with few modifications from Kala, R., Shukla, A., & Tiwari, R. (2011). Robotic path planning using evolutionary momentum-based exploration. *Journal of Experimental & Theoretical Artificial Intelligence, 23*(4), 469-495. Reprinted by permission of Taylor & Francis Ltd. (http://www.tandfonline.com).

- Chapter 8, parts of "Introduction," "Conclusion," and Section "HGAPSO with Momentum" to Section "Results" are reprinted with few modifications from Kala, R., Shukla, A., & Tiwari, R. (2012). Robotic path planning using hybrid genetic algorithm particle swarm optimization. *International Journal of Information and Communication Technology* [Accepted, In Press]. With permission from Inderscience Publishers.
- Chapter 5, Section "Swarm Intelligence in Path Planning," Sub-Section "Ant Colony Optimization" has excerpts of text, tables, and figures © 2010 IEEE. Reprinted, with permission, from Maurya, R., & Shukla, A. (2010). Generalized and modified ant algorithm for solving robot path planning problem. In *Proceedings of the 3rd IEEE International Conference on Computer Science and Information Technology,* (vol. 1, pp. 643-646). IEEE Press.

Chapter 1
Introduction

ABSTRACT

The basic purpose of the book is to specifically look into the problem of robot planning, which is an essential problem in robotics. Before the authors discuss the simple and complex algorithms, analyze the same, present the various modeling scenarios, and present some results and associated limitations, tradeoffs, and issues, it is wise to first understand the basics of robotics. The adoption of Artificial Intelligence in robotics is a never-ending effort, a part of which is the crux of the book. It is important to understand the notion of problem of path planning and where and how it fits into the entire domain of robotics. The chapter is an introduction to the entire book and gives an overview of all the background needed, including the underlying problem of motion planning of mobile robotics. In other words, this chapter gives a broader view of the complete system, a part of which, will be looked at in intensive details in this book.

INTRODUCTION

Robotics is a very exciting field, which continues to be of great interest to people. The wide variety of applications of robotics is a natural motivator for people to study this domain and contribute to the same. To build highly sophisticated and intelligent robots that would change the face of society is the aim for everyone working in this domain. The complete domain is highly multi-disciplinary and involves people from various disciples and backgrounds. Building a complete robot and enabling it to work in real life scenarios would involve expertise in a large number of fields of study. The problem includes the design of good robot hardware, as per the terrain and domain of work. This calls for expertise in mechanical disciples along with design skills.

DOI: 10.4018/978-1-4666-2074-2.ch001

The communication of the robot with sensors calls for expertise in the disciples of electronics and communications, along with wireless technologies, if the robot has wireless components. The motor manipulation may further need a lot of electrical and electronic stuff that enables the robot to act to situations. Proper planning and understanding of the situation, of course, requires expertise in pattern recognition, artificial intelligence, knowledge modeling, etc. The robots may be specific to a field of work that forms the grounds of their applications. This puts further constraints of expertise as per the task that they are asked to perform.

The desire to have someone do specific tasks for us has always been a need of humanity. This is why machines have started doing many jobs that the humans once did. The use of robots is not new to society. A lot of effort has always been made, which has become a constant move to automation and industrialization. These robots could always do simple tasks. The desire to do a lot more than just having simple robots as machines kept motivating people to make the field more sophisticated.

A parallel development was in the field of Artificial Intelligence (AI) (Brunette, et al., 2009; Konar, 2000). The scientists made continuous efforts to make the machines or systems as intelligent as possible. The effort was to enable the machines to have the same intelligence as humans. A large part of it was considered impossible in the beginning and attributed to less computational power, less memory, no networking, etc. However, the scenario continuously changed as computation started becoming inexpensive and available in abundance. This made a revolution that bore fruits in every domain. Soon the systems started getting more intelligent and sophisticated. These systems that we use today can perform a variety of tasks that include biometric recognition and verification, automatic biomedical diagnosis, data mining, time series prediction, etc.

AI merges beautifully with the conventional field of robotics (Kaelbling, et al., 1988; Mital, et al., 1988; Murphy, 2000). This gives rise to what we know as intelligent robotics. These robots are able to act on their own without intervention from humans. They are smart and make their own decisions. These decisions may be overseen or assisted by some human expert, if desired. The combination of AI and robotics is seen as a beautiful thing. All robots, to some extent, are fitted with some intelligence that has its basis in AI. AI-based robots started from basic architectures or basic intelligence, which could just move around and perform a small set of tasks. Today, these robots carry out all sorts of tasks. These robots outperform humans in various kinds of tasks in terms of efficiency and accuracy. This book emphasizes, specifically, these intelligent robots, and it deals with the issue of how these robots may be intelligently planned.

The chapter is organized as follows. First, we study some of the commonly used terminologies and concepts in robotics, which explain our precise ground work. We also discuss some of the application areas of robotics. We then deal with the general principle of problem solving in robotics. We then migrate to what is the chief point of concern of the book. Here, we would explain the problem of motion planning in robotics. We further advance the text with a discussion to types of planning, which are followed by some general AI concepts. Towards the end, we present some related topics for better understanding of the complete domain and problems. This includes expert systems and machine learning. This is followed by concluding remarks.

PRELIMINARIES AND DEFINITIONS

Robotics spans across multiple disciples, multiple domains, and levels of work. It is natural that we cannot complete everything at one place with the minutest of details and discussion over the underlying issues. The basic motive of this section (or rather this chapter on the whole) is to showcase the domain of robotics, which forms the basis of the discussion for rest of the book. The complete problem of robotics is usually said to span both hardware and software, which interact at various levels. The level or the degree of complexity of the software may further depend upon the type of robot and the specific task that is being performed (Ge & Lewis, 2006). We study all the various associated terms and concepts in the following sub-sections.

Robot

The basic domain of our study is robot, and it is, hence, essential to formally define it. Robots are machines that can sense their environment, plan their actions, and act as per their plans. Hence, a robot carries out cycles of sense-plan-act (Stentz, 1990), which makes everything that they do possible. This further clarifies that the robot may not always be like a humanoid, a car-like mobile robot, or of any particular shape. A device fitted inside a human heart may also be a robot if it can sense, plan, and act. It may be easily seen that this is more or less the manner in which any human works. Consider any task that you perform in everyday life, no matter how sophisticated it is. Say the task is playing a football match. Now we sense the environment. In this case, it is through our eyesight, where we see the various players of our team and the opponent's team, the ball, locations of the players, etc. It may be additionally in the form of sounds made by the various players or referee.

The other task that we carry out is to plan what needs to be done. Here, we do some type of decision making to decide whether we need to run, and if yes, then in what direction and speed, kick the ball, catch the ball, halt, etc. There may be multiple levels of planning that are done, which result in an instantaneous decision or plan. The last step is to physically execute the plan or to act. For this, we need to move our hands, feet, etc. This results in some sort of manipulation in the physical world. This is being done by all objects in the environment. The world is now manipulated, and this cycle repeats for the other instance of play.

Figure 1. Sense-plan-act cycle of robotics

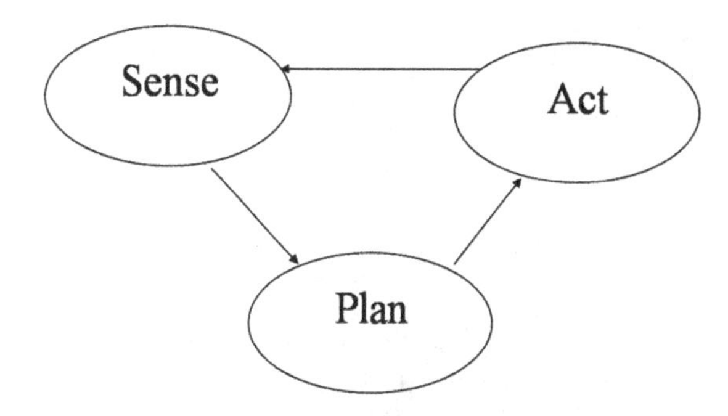

This cycle is shown in Figure 1. It may be further understood that we mainly deal with planning in this book, which is the second step of the cycle.

Hardware

The first task to be done in the robot is to have the hardware as per the requirements, which again depends upon the problem and area of work. We first describe some of the generally used hardware that gives the basic architecture of the robot. Many of the robots built are car-like mobile robots. These robots have almost all the general mobility components that are available in cars. There are a number of wheels that make the base of the robot. These wheels are mounted over the robot base, such that they fit in well and are able to move the complete architecture. The top is reserved for the various sensors of the robot. The commonly used sensors include radars, lidars, infrareds, GPS, lasers, video and audio cameras, stereo vision cameras, etc. (Buehler, et al., 2007; Seetharaman, et al., 2006). The control of the wheels is done by the actuators. These may make the robot move in any direction.

Every robot has motors to drive them. These motors are connected by some power supply that may reside over the robot platform. Wireless communication devices may be embedded in, if the robot is to perform some communication for its operation. The robot may be controlled by an off-board controller, or by some intelligent planning algorithm fused as an electronic circuit

Figure 2. A mobile robot

over a Printed Circuit Board (PCB) at the mobile robot platform, or programmed at the robot platform itself in case the robot has an onboard programming feature. The shape and size of all the components may vary as per the requirement. Robots needed for exploration may be bigger so that they may travel faster. The wheels may again be as per the terrain. Robots moving in deserts may have specialized wheels. Often, they have insect-like legs in place of wheels. The robots for many spying applications may be much smaller. A mobile robot is shown in Figure 2. The robot is called Miabot from Merlin Systems.

The other commonly used robots are the humanoids (Thorisson, 1996). These are human-like robots. These robots have a general shape just like humans. They have face, body, and legs. The motion of these robots is by their legs. The legs may be wheel operated, making them more like the car-like robot, or may be like the conventional human gait. In such a case, balancing is a major issue with these robots. However, the flexibility of their legs may make them reach many places where the car-like mobile robots cannot. Examples include walking up the stairs, moving in sandy terrain, motion in narrow passages, etc. These robots can move in non-smooth paths, which the car-like robots may not be able to do. These are called the non-holonomic constraints of the robot. Here as well, the robot may be equipped with any type of sensor. These may be placed at locations convenient enough as per the processing requirements. For a car-like robot, the directional sensors may be in all directions to enable a 360-degree view of the world. For non-directional sensors, like GPS, placement of sensors would not be that big an issue, as long as they are physically safe and may be able to operate at the location even in hostile environmental conditions.

We focus upon two major aspects from the hardware level, as per our understanding of the sense-plan-act cycle. These are sensors and manipulators that belong to the first and the last step, respectively. The sensors enable the robots to know about the world around. Various sensors have different types of uses. Accordingly, they may be chosen. Many times multiple sensors are chosen to supply redundant information about the robotic world. This helps in cases of high noise, as we shall see later. The video camera is used for taking a pictorial snapshot of the world around. The information provided by these need to be additionally processed by a pattern recognition system to get the understanding of the ways and the obstacles. The video cameras are, however, sensitive to the lightning conditions and sometimes even the weather. The other sensors include a GPS device, which can tell the global position of the robot. These devices convey the global information, which may be very useful for global planning, but not local planning. Similarly, radars and other devices may have very good information about the obstacles in front. Accordingly, they may be used for the planning purposes.

The other important concept is the presence of manipulators in the robots. These are tools used by robots to physically carry forward the task that they are supposed to carry out. These may be in the form of wheels that enable the robot to move in any of the desired directions. Many times the robots have hand-like manipulators used for grasping operations. These are needed in case the robot needs to pick up something and bring it back. Other manipulators include the specialized devices to operate specific tools needed for industrial applications.

Software

The other part of the robot is the software that enables the robot to perform all kinds of intelligent tasks (Holland, 2004; Patnaik, et al., 2005). The software of a complete robot is usually extensive and may be seen at multiple levels from the point of abstraction starting from the hardware level. The complete robot, from the software point of view, may be easily seen as a computer system that we all use in our everyday computing. The robot, hence, needs to have an operating system over which the various modules would be coded. This further necessitates the use of all the computing peripheral into the system with a processing unit or simply a processor and a power unit. Many times the robot may only be for simple purposes, the software to which may be a simple logic coded into a PCB.

The software of most high-end robots would involve a large number of software modules, which all run over the operating system of the robot. All sensors and manipulators are devices that are plugged into the robotic computing system. These may have their own drivers that need to be written/installed for the operating systems to be able to use them. Further, there may be procedures to be written for them that make the operating system call for the input/output to come from them or go to them. All this is done primarily at the device level. The complete software does the task of taking the signals from the sensors, processing them, and producing the output at the manipulators as per our sense-plan-act cycle. A brief of how the planning is done is presented at the next section. The planning may itself have a large number of data structures and other representational techniques that are needed for the processing.

The planning may be in a multi-level manner where the various levels do the different types of planning. Some lower end planners may be needed for finer planning, while the higher end ones do the coarser planning. It is known that due to the presence of multiple inputs as sensors and multiple outputs as manipulators, the complete task of making the software is very complex. The different inputs are processed in different manners. The various levels of planning act in different ways. Further, each manipulator behaves in a characteristic manner, having its own complexity and issues. The planning is usually done on the basis of a knowledge base that helps in decision making. This knowledge base may itself be complex in nature, having complex queries and updates.

Mobile Robotics

We have spoken multiple times of the robots being mobile or about the mobile robotics (Holland, 2004). Mobile robotics refers to the domain where the robots can move from one place to the other. This raises the issues of motion planning and control. Here, the robots are fitted with wheels, legs, or other equipment that can make the robot move physically from one place to another. Mobile robots may be used for problems where the robot has to travel to a distant place, get work done, and return. Because of the issues of mobility, these robots are flexible, but these are usually not big, heavy, and bulky, which restricts motion. These robots are extensively used for multiple operations (Sicard & Levine, 1988). In most of the discussion in the previous sections, we talked about the mobile robots. However, it may never be assumed that all robots are mobile. Some of the highly sophisticated robots in use in industries may not be mobile.

Autonomous Robotics

Autonomous robotics is the domain of study where the robots are autonomous and all the decisions are made by only them (Ge & Lewis, 2006; Holland, 2004). This calls for elements of decision making at multiple stages. These decisions may vary from very simple decisions to a fairly complex decision making. Examples include the decision of how to maneuver around a small obstacle in front to how to react to a blockage in the road of travel. Most of the discussions throughout the chapter are on autonomous robots. In fact, the book assumes the robots to be mobile and autonomous and deals with the planning of these kinds of robots.

The robots may, however, not be entirely autonomous in nature. Many times the robots are semi-autonomous in nature. Here the coarser level of planning may be done by some human expert with the robots being given the autonomy to do smaller, finer level planning and acting accordingly. In many other situations, the humans may be needed to watch out for the robots and intervene in their decision making whenever they see any clash of interests, sub-optimality, deadlock, etc. In all these situations, the humans and the robots work together to get the common goal met. Many other robots may have some autonomous and some non-autonomous components. In these robots, it may be easy to consider the autonomous components and use them as fully autonomous robots. The other category is for the robots that are *non-autonomous,* and their decision making is ultimately done by some human expert. These robots always require a driver or a human to use them.

Unmanned Vehicles

The other concept we cover in this section is the Unmanned Aerial Vehicles (UAVs) (Stipanovic, et al., 2004; Washburn & Kress, 2009). These are aerial vehicles that can fly without any human sitting in them. These vehicles are extensively used for explorations where they are sent over unknown areas for aerial survey. They may be fairly compact in size and are hence very economical. Further, they may be made to fly at reasonably lower altitudes to return the needed information. These vehicles work in a 3-dimensional sky environment, and hence, the task of planning becomes more complex to solve. These vehicles have to further take into account the air current and all the turbulence caused by it. This has its effect on the planning and control of these vehicles. As these are autonomous in nature, these factors become very important to handle.

A related concept is the Unmanned Ground Vehicles (UGVs) (Matthies, et al., 1995; Shimoda, et al., 2005). These are normally car-like vehicles that travel on the ground without any human beings. The planning of these vehicles is in a two-dimensional plane. The effect of the various terrains and the motion kinematics due to the used motion equipments, however, needs to be considered. These vehicles operate without humans and are used for similar purposes.

Robots are not only found in the air or on land. These are found in the water as well. Unmanned Ocean Robotics deals with the use of robots in the water without human beings. Here as well, the planning is on a 3-dimensional scale, and the presence of water currents need to be considered in planning and control. The water currents normally give turbulences that deviate these vehicles from their intended path.

APPLICATIONS

In the past few years, we find an extensive use of robotics in multiple domains. While robots are widespread in numerous domains where they contribute significantly, we list some of the general applications of robots. We further give a historical perspective into robotics.

General Application Domains

Robotics has changed a large part of the industrial world (or industrial robotics). Its adoption in industries is specially worth mentioning. A large number of robots are used in the production lines of multiple products that have completely replaced the humans. The robots are usually more accurate and are capable of performing tasks with more precision, as compared to the humans. Further, the robots are highly efficient and can perform tasks quicker. This may mean an increased production along with saving a large number of resources.

Other interesting uses of robots come from their use in space, known as space robotics. Many satellites orbiting around the earth are intelligent robots that are able to perform a wide variety of tasks (Ellery, 2000; Hirzinger, et al., 1993). It is further useful to have robots carry a large variety of tasks that are prone to be dangerous. For the same reasons we find a very high use of these agents in rescue operations. Here the robots are sent into fires, mines, and all dangerous places to carry operations (even to save the lives of human beings). It is evident that it would be very unsafe to send a human in such conditions. The robots have performed well in such conditions (Carbone, et al., 2008; Hirose & Fukushima, 2002).

In the modern world, where human labor is not only becoming excessively expensive, but also the health consciousness amongst the people prohibits them from doing any task that can involve risk of lives, the robots become very useful tools in such conditions, which take all the risk to themselves. They are engineered in a way that they can survive very hostile environment and still give great results. Further, it is better to lose a robot, rather than to lose the life of an individual. The other similar areas where robots find use include chasing of thieves, safeguarding people, surveillance, etc.

Accuracy is a major advantage of robotics that the humans may not always have. Often some operations need to be carried out at a micro or nano levels (Cavalcanti & Freitas, 2005; Sitti & Hashimotto, 1998). It is naturally impossible for the humans to do these operations. These robots (called micro-robotics and nano-robotics) are fitted with technologies that can enable them to view the area of manipulation, move to the same area, and perform the task with the greatest accuracies. The robotic devices are further designed to operate over the dynamics of that level, which is different from the dynamics of the macro world. In such a manner, these robotic devices are able to carry out even more complicated tasks.

The robots are not only restricted to the industries. They are further used in homes and for entertainment purposes (Turban & Schaeffer, 1988; Wyrobek, et al., 2008) where the domain is known as entertainment robotics or personal robotics. There is an excessive use of robots for cleaning of the house. A constant attempt to make such robots more robust and efficient is still being done. These robots, just like an intelligent maid, do the complete cleaning in structured or un-structured environments, without human aid. Further, these robots may be programmed to carry forward the task of cleaning at any particular point of day, and to return to their charging points afterwards. This eliminates any human intervention in the complete process not even in initiation and completion.

The robots are used for the care of the elderly. These robots can assist the elder people in all their daily tasks, what they might not be able to do because of the old age (Feil-Seifer & Mataric, 2005). These robots (called assistive robotics) can even sense any calamity and report the same to emergency services. An associated field is the use of robots in entertainment industry called entertainment robotics (Fujita, 2000; Veloso, 2002). Here, robots are used to meet people, speak to them, answer their queries, etc. These robots may further perform entertaining tasks for them, as they may desire.

The applications of robots are vast. They extend along disciples, industries, and have touched various aspects of both household and industries. The examples of use of robots are very wide. Numerous interesting applications of robots are still in the laboratory and may be into the public very soon. Examples include the understanding of human gestures for robotic control (Waldherr, et al., 2000). Here, scientists are trying to train robots to understand the hidden meanings inside the various gestures of the humans. These include motion of the hands, fingers, eyes, etc. This enables humans to do their work naturally, while the robots understand the humans and assists them as desired. The robots may be further trained to understand the human voices and the voice commands.

The use of robots for the physically disabled is another key point of research. These robots can be used as artificial limbs to the humans, as hands, feet, etc. To the people who are physically handicapped and have lost some of their body, these robots act as lifesavers. They enable the person to carry out all the tasks by controlling these limbs. The people may be trained to use these artificial limbs. Alternatively, the limb may sense the person's movements and actions and accordingly act in the correct manner. This is similar to how natural limbs work in everyday operations.

Understanding the emotions produced at the brain, and trying to move a robot using these signals is again an exciting field of application of robotics (Millan, et al., 2004). Here, robotic devices may be implanted into the humans for sensing actions, which may be detected by the receivers at the other end. These may be analyzed and used for any kind of manipulation task. Another exciting domain of work is in haptics. Here, the humans operate devices which capture their movements or touch to produce similar or related manipulation at some other end.

Many experiments and applications of tele-robotics are also coming up. Here the humans operate machines at some part of the world to manipulate or operate systems at the other part of the world. An exciting application to this is the tele-robotic surgery, where a doctor at some end operates on patients at another end. This naturally saves a lot of time and money. Doctors to certain rare diseases may be very busy and would not be able to devote enough time to travel to different countries. Further, the patients may themselves be unable to travel for operation or afford the travel. In such a case, tele-robotic surgery has a lot to offer.

Some Robots in Industry

While most of the discussions made previously emphasized the general applications of robots in various domains, for the completeness of the discussion in this sub-section we discuss some specific robots here. All these robots are in use in industries (or industrial robotics), which marks one of the most interesting application areas of robotics. Industrial applications of robots are covered regularly by numerous science and technology magazines, robotics magazines, and robotic association bulletins and other published materials. Readers may take pleasure in subscribing to these.

Robots can be used in a variety of industrial applications ranging from simple pick and place tasks to the complex tasks such as product inspection and testing. The typical industrial application of robotics may include painting, welding, assembly operations, material handling, work cell operations, laboratory applications, dispensing operations, etc. Most of the popular modern industrial robots present the factual marvels of engineering with incredible benefits of handling hundreds of pounds of weight, working repeatedly for 24 hours a day for several years. Although these robots can be reprogrammed, in most of their applications they are programmed once and their repeated tasks are applied in the industry related tasks.

A commonly used robot is ABB IRB 6400 (ABB, 2011), which can perform most of the tasks like spot welding, palletizing, welding, pick and place, woodworking, cutting, spraying, painting, etc. The unique control of ABB IRB 6400 allows the robot to optimize its acceleration and retardation as per the actual load that results in shortest possible cycle time. Standing for Selective Compliant Assembly Robot, the SCARA (Wisegeek, 2011) robot is an assembly machine, which is designed to imitate the actions of the human arm by installing the parts or carrying the items that can be used in several types of jobs from manufacturing factories to underwater constructions. This robot has the full range of motion in its X and Y-axis which can be used for performing specific jobs repetitively like carrying the items from one place to another, installing the pins, etc. It highly imitates the construction of the human arm, wrist, and elbow to perform various actions in a fraction of the time compared to the average human arm.

Robocaster from the KUKA Roboter (KUKA Robotics, 2011) provides practically unrestricted freedom in terms of motion and dynamic performance. It can be used for carrying workers from one place to the other through difficult terrains and restricted locomotion for humans. The robot offers a wide range of ride profiles, speeds, motion sequences, etc. ARC Mate 50iC/5L (Fanucrobotics, 2011) from FANUC Robotics is used for welding applications in industries. This robot is equipped with slim arms, which improves the part and tooling accessibility. This robot supports a variety of intelligent functions with its internal Programmable Machine Controller (PMC).

The delta robots (Bonev, 2011) come into the category of parallel robot and basically consist of three arms connected to a universal joint. The high speed of the Delta robot allows its deployment in the packaging, medical, and pharmaceutical industry. In addition, the popular usages of delta robots have been in picking and packaging works with the speed sometimes reaching to 300 picks per minute. The key feature in the design of the delta robot is the utilization of parallelograms in its arms that maintain the required orientation of end effectors. As all the actuators are located in the base region, the arms are often made with the help of light composite material allowing very small inertia for the moving parts of delta robots, which gives allows for smooth acceleration.

History of Robotics

While the history of robotics is deep and widespread, we discuss some of the major and interesting episodes here (Bellis, 2011) to stimulate the readers towards the domain. The robotics community over the years has developed a wide variety of Web resources besides academic resources, which contain detailed coverage from a historical perspective.

The history of robots takes us far back to ancient myths and legends when an ancient Greek engineer named Ctesibus used to make organs and water clocks equipped with movable parts around 27 BC. However, it was not until the period of industrial revolution that the modern concepts related to robotics began to develop when the more complex mechanics and introduction of electricity made the use of small compact motors and other entities used in modern robotics possible. The informal but relevant literature regarding the creation of manmade machines capable of performing human-like tasks started with the publication of *Frankenstein* by Mary Shelly in 1818, which depicted a startling artificial life-form.

Most of the historic developments in the field of robotics started taking place in 20th century. The term ‘robot’ was used for the first time in “Rossum’s Universal Robot,” a play written by Karel Capek. Another science fiction writer, Isaac Asimov, made the first use of the word “robotics” for describing the technology of robots in addition to predicting the rise of a powerful robotic industry. The same writer gave the three famous laws of robotics in his story “Runaround” in the year 1942. Another sound literature was *Cybernetics*, published in 1942 by Norbert Wiener about artificial intelligence. In the same year, William Grey Walter invented the robots named “Elmer” and “Elsie” that imitated the lifelike behavior by making use of simple electronics.

Soon after this, George Devol and Joe Engleberger developed the first programmable robotic arm along with coining the term “Universal Automation" for the first time. The same scientists formed the first robot company in 1956 along with the invention of Squee, an electronic squirrel. In the year 1961, the first industrial robot, named Unimate, began working in a General Motors automobile factory located in New Jersey. The next remarkable innovation was Dendral, which was the first expert system designed for executing the accumulated knowledge of the subject experts. In 1968, octopus-like Tentacle Arm was invented by Marvin Minsky. However, it was in 1969, when the first electrically powered computer controlled robot arm named "Stanford Arm" was built.

Soon after this invention, SRI International introduced the Shakey, which was the first mobile robot organized with the help of artificial intelligence. After this, there has been no dearth of innovations, with the subsequent inventions coming up in the history of robotics such as Silver Arm designed to perform the assembly of small parts by utilizing the feedback from touch and pressure sensors. The first robot with successful area exploration and path planning was achieved with the demonstration of Standford Cart, which crossed a char-filled room without any sort of human assistance with the help of TV cameras mounted on it.

PROBLEM SOLVING IN ROBOTICS

In the previous section, we talked about the simple sense-plan-act cycle that makes it possible for any robot to carry out all the wonderful tasks in robotics. In this section, we discuss this complete procedure of the robot and try to throw some more light over the various layers of planning. The various steps involved in the process are given in Figure 3. The processing starts with the sensors giving the sensed information. This represents the raw information that is further processed. Based on the information available, we try to localize the robot into the world coordinate system using localization algorithms. At the same time, we try to build the global world map of the robot by mapping algorithms.

The next task is planning, where we decide the motion of the robot. The planning results in a set of high-level instructions that once taken would make the robot achieve its goal. The low-level implementation is facilitated by the robot control. The control is responsible for ensuring that the path tracked physically by the robot is as close as possible to the planned path. The physical motion is done by the physical manipulators available with the robot. There are extra steps carried out if there are multiple robots (Ge &

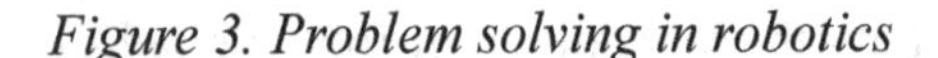

Figure 3. Problem solving in robotics

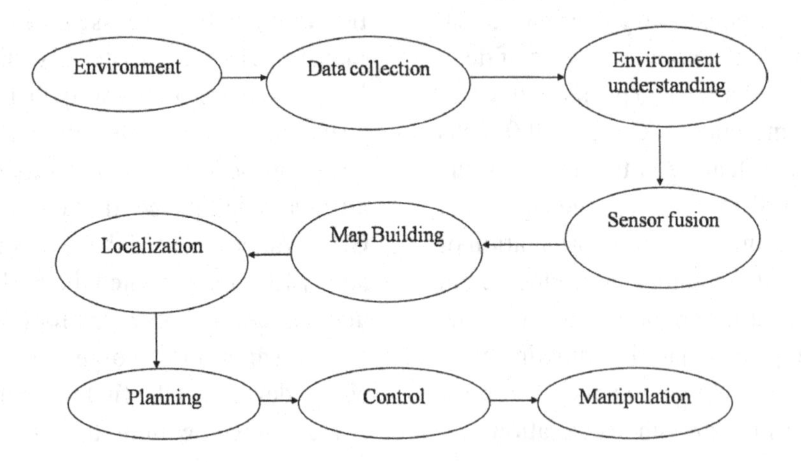

Lewis, 2006; Shukla, et al., 2010a). We discuss each one of these steps in the next subsections.

Data Collection

The robot has numerous sensors of multiple kinds. The first step is to collect the data from all of these sensors. Each sensor may have its own frequency by which it gives the information about the world as it sees. The data provided by the sensors is always raw data. This needs to be further processed into information or a form by which some inference may be made. In addition, the different types of sensors may give different types of data. Hence, for each sensor, we carry forward a sensor specific processor to convert the data into some kind of information or knowledge. For a GPS device as a sensor, we may need to parse the data it sends to get the idea regarding the position and the speed of the robot. This would give us the current location in a global coordinate system, which, when used along with the robotic map (if available), gives us an idea of the current robot position in the map and its direction and speed of motion.

Similarly, the radars or lidars may give core numeric values, which we may need to further process to get an idea of what obstacles surround the robot. Many times video cameras may be used for recording the environment around the robots. The use of multiple cameras may record each and every detail all around the robot. The video cameras especially give high dimensional information about the robotic environment. This helps a lot in moving. A natural analogy can be drawn that all the human planning and actions are chiefly drawn by whatever we see around by our eyes. For related discussion refer to Buehler et al. (2007) and Seetharaman et al. (2006).

Understanding the Environment

The information given by the various sensors usually represents raw numbers. This raw data needs to be processed to know what exactly the scenario or the robotic environment is. This processing generates useful information that may be processed further for complete understanding of the robotic environment. The understanding of the sensor outputs is specific to every sensor. Most sensor outputs may be easily understood by simple mathematical analysis. This may give information about the path to be traversed, obstacles, obstacle dynamics, speed, etc. (Florczyk, 2005).

The case of video (and similarly audio) cameras is different. Here, we get a snapshot of the surroundings, which requires intensive processing as discussed above. Since these cameras are widely used, we briefly discuss this cycle of processing. The aim here is to analyze the image and to obtain the obstacles and their locations, despite the noise that may be prevalent in the system. These may later be used for the next stage of processing where we make use of the information about these obstacles. The complete system (Shukla, et al., 2010b) is given in Figure 4. The video taken by the sensors is split across image frames. The image comes for further processing. The image may normally be converted into a grayscale image, as color seldom has useful information to aid the process of object recognition. The image is then further passed through filters for noise removal. The noise removal results in undoing the effects of noise that always creeps into such systems. The image is then processed further by the application of *edge detection*.

The edge detection technique separates all edges from the non-edges. This helps in separating any object from the rest of the contents of the image. It may however be noted that simple edge detection does not work for overlapping obstacles, which would interpret the two objects as one obstacle. A very common technique of edge detection is to take a first order derivative of the image. The edges show a non-zero value of the derivative. The non-edges have almost constant value across neighboring pixels, showing a zero derivative. The second order derivatives may be used for more information about the edges.

Figure 4. Object detection

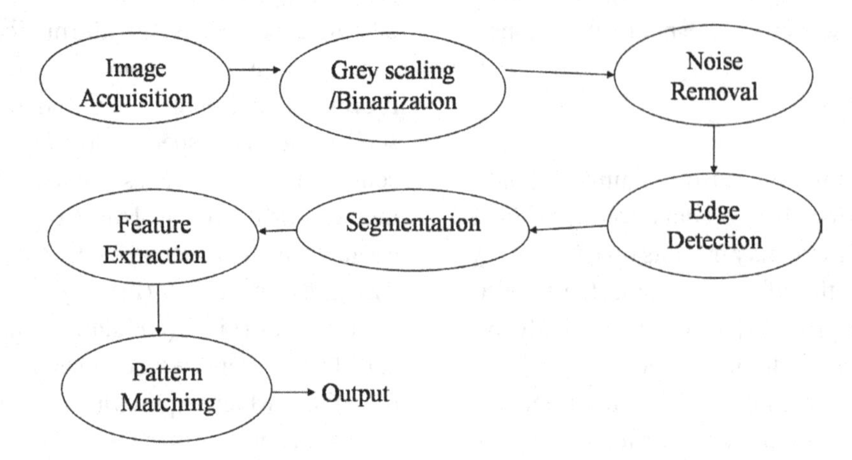

Canny edge detection is a commonly used technique of edge detection. The other step involved is segmentation. The image so far has multiple objects whose edges have been identified. In the next stage, we separate the various objects from one another. This gives us a collection of objects present in the input image. Each of these objects is analyzed separately.

The next task is of pattern matching, where we try to identify the isolated objects. The isolated object itself represents a fairly high magnitude of input. Any classifier may not be able to deal with such a large magnitude of data. Hence, we carry forward the step of dimensionality reduction. Here, we perform feature extraction where the complete set of image input is converted to a set of features. Features represent the interesting facts about the original image that aid in classification or identification of the object. They represent the numerals that may uniquely identify any object. The features must ideally remain constant for any recording of the same object and must change drastically as we change the object. This helps in easy classification.

In this problem, we may use a few top coefficients of the Fourier transform of the pixels representing the object boundary as the features. In this manner, we assume that only the obstacle boundary can enable effective recognition and any information about the rest of the object is not required. This is a valid assumption to make considering the fact that we need to mine down the dimensionality to a great extent and can hence retain only very important features. Alternatively, we may use statistical techniques like Principle Component Analysis (PCA), or Linear Discriminant Analysis (LDA) for the dimensionality reduction.

The last step is to use the set of recorded features and use them for pattern matching. This may be done by any classifier. The classifiers may be neural, statistical, or based on any other technique. The classifiers usually take a historical multiple recording of all the objects along with their feature values. This enables them to know what types of features are associated with what type of objects. These classifiers may be learnable or non-learnable. The learnable classifiers learn the patterns from the historic data and try to use the same learning for the new input that is applied to the system.

The final output of the complete system is the objects present in the system along with their locations. Each of these objects may be an obstacle that we need to avoid in the motion. Many times the object may be the goal that we need to attain in the motion.

In many scenarios, object tracking is additionally performed. The recordings of the sensors may be useful in moving robots in the physical world. The sense-plan-act cycle would demand the robot to sense again and again, after it has moved by some amount. Since the problem may have a small sense-plan-act cycle, it is evident that the robot may have similar sensor values, or that the obstacle positions may only change by small amounts, unless there happens to be sudden emergence of some obstacle. In such scenarios, it is possible to track the manner in which the various obstacles or landmarks are changing their positions with time and use them in planning and control of the robot. This is done by passing the sensor inputs through particle filters. The use of the Extended Kalman filter is common. Tracking helps in minimizing the effect of sensor errors. They are further tools for localization of the robot.

Sensor Fusion

Working over the specific sensors gives us a lot of sensor-specific information about the environment of the robot, which we may analyze using the stated means. In a practical scenario, a single sensor may not be able to give the complete information about the entire robotic environment. Consider that the robot was only to be moved based on a front camera. Now the video camera used may not be able to capture everything. Many times the vision algorithms, which have their duty to analyze the captured images, may fail to detect all the obstacles. This would result in a wrong interpretation of the robotic world and may further result in collisions. This is highly undesirable. The failure is usually due to noise inherent in the system. Many times the obstacle boundaries, from which the classifiers work, are not clearly captured. Many other times the camera calibrations are wrong.

Dependency of lightning and other weather conditions further worsen the scenario. Even a single camera may not convey complete information about all directions as its visibility is limited. Effective planning may require knowledge of what is happening at every side of the robot. An effective short-vision camera would not provide information about far off areas, needed for higher level of planning and control. Similarly, a far-vision camera may not provide detailed information about nearby obstacles, making planning and control a difficult task.

The other important aspect is that the sensors may not necessarily be placed over the robots. Sometimes there may be external sensors placed at the place of work of the robot, which might record and analyze valuable information. These recordings may then be communicated to the robot to use in its own planning and control tasks. In such a manner, a combination of sensors on the robot and outside the robot may be used for carrying out all the intelligent tasks that the robot does.

It is usually not possible to work using a single sensor in most practical scenarios. We need multiple sensors to record different aspects of the robotic environment. The other problem comes in using the individual readings of the various sensors and to use them in order to make the complete robotic environment map. The complete environment of the robot needs to be represented in a manner that can be worked by the higher levels of robot problem solving. This final representation of the world of the robot is known as the robotic map. Chiefly, the planner needs to query the robotic map to carry the task of computation of the trajectory that the robot should follow to complete the given task. The robotic map is a knowledge base that can be easily understood by the artificial intelligence algorithms that work at the later stages.

One of the commonly used methods to carry the task of conversion of sensor outputs to the robotic map is to use Occupancy Grids. Here the complete map is represented in forms of grids. Each sensor is taken and is analyzed for the presence of obstacles and waypoints, over which the robot may move. These are casted into the occupancy grids. One of the common tasks to be done before casting is to convert the local frame of reference of the sensor to a global frame of reference to which all the sensors agree. This frame of reference may be any central point on the robot, or sometimes even static points in the robotic environment outside the robot. For each sensor, we mark its findings onto the occupancy grids.

The problem arises in sensor fusion when the various sensors do not agree with each other. One sensor may report an obstacle at some location, whereas the reporting of the different sensor for the same point may be a clear way. In such a case, the decision lies in how to decide the reality of the point. This is a common problem while working with occupancy grids wherein different sensors try to fill different values in the grids. The task is handled by an integration scheme that uses an integrator for decision-making. The input to the integrator depends upon the level at which fusion is performed. Hence, every sensor reports its findings to the integrator. The integrator analyzes all the decisions to decide the presence or absence of obstacle, or likewise whatever the conflicting decision is. This is given in Figure 5.

There are four different levels in which the fusion can take place. The first level is the decision level in which every sensor informs the integrator about its decision regarding the presence or absence of obstacles. The integrator analyzes these decisions and makes the final decision. A simple technique is voting, where every sensor casts a vote. The second level of fusion is score level. Here, every sensor outputs a confidence value, a score, or probability of presence of obstacle. All these values from multiple sensors may be integrated in a variety of ways. In this technique, the confidence of a sensor in its decision may be assessed looking at the value it generates, which may be helpful in decision-making by integrator.

The third level of fusion is feature level. In this technique, the features from the different sensors are all pooled up to generate a common pool of features. All these features are used for collective decision-making. Here, there is a single decision-making module, which takes as input all features from all sensors. The last level is the data level. In this level, raw observations or data from various levels are pooled and used for feature extraction and further decision-making.

A common integration technique is to take a weighted mean of all the individual sensor decisions for a score-based fusion technique. In this technique, every sensor is assigned a weight.

Figure 5. Multi-sensor fusion

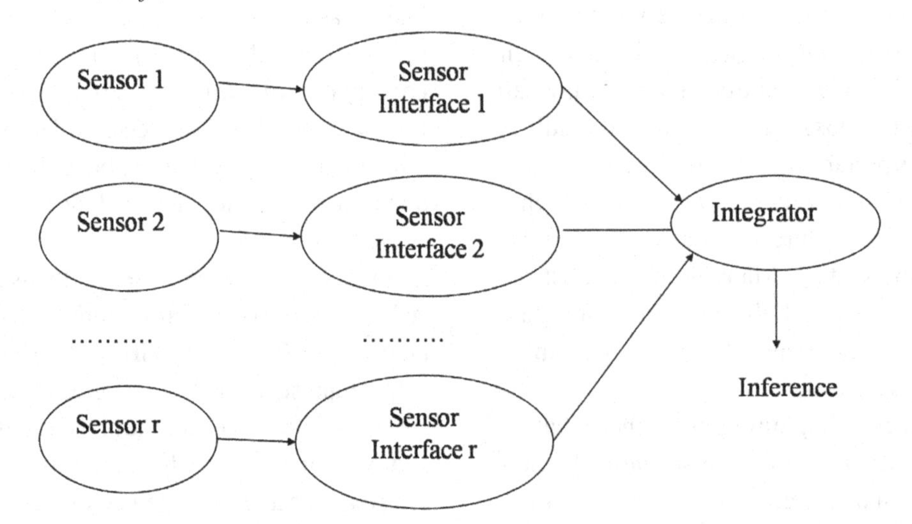

The higher weighted sensors have a greater say towards the final decision made. The various sensors may be weighted as per their sensing capabilities towards the particular conflicting decision, which may be the presence or absence of obstacles. These weights may be dynamic in nature, where the change in weather, lighting conditions, dusts, and other interferences are taken into account dynamically. The weights may be decided based on the accuracies recorded by the sensors to known obstacles and other metrics that they are supposed to measure.

Many other times, however, the system needs to be very secure, and it is a requirement to have no collisions at all. In such cases, the decision by the integrator may be made differently. The recording of even a single sensor as an obstacle may be taken as an obstacle. The commonly used integration techniques include polling, where the different sensors vote in favor or against the decision; maximum, where the maximum confidence values to a decision are considered; minimum, where minimum confidence values towards a decision are considered; product, where we multiply the confidences in place of adding them up; median, where the median confidence value is taken; fuzzy integration, which uses fuzzy arithmetic for decision making; etc.

Map Building

As already stated in the previous section, our intermediate goal is to make a robotic map that completely represents the robotic world. Numerous representations of the robotic map may be possible. The type of representation to be used for solving any problem largely depends upon the problem, depth of information desired, time constraints, planning algorithm used, etc. One of the commonly used maps is a topographic representation. Here, we represent the complete maps as grids. This gives a high degree of information ready accessibility to the planner or other robotic modules. The other technique of map representation includes the Voronoi Maps. This is essentially a graph-based representation. Every node of this map represents an accessible region. The different accessible regions may be reachable from one another or not. In case a region is accessible by the other region, the two nodes are connected by an edge. In this manner, the complete map is built. A sample Voronoi map is as shown in Figure 6.

Figure 6. Voronoi map

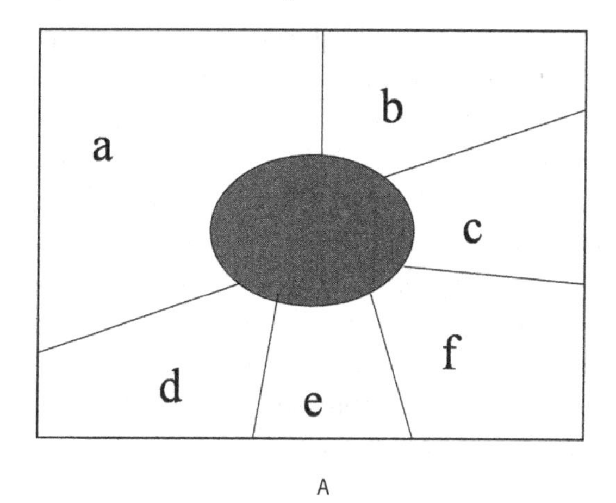

A

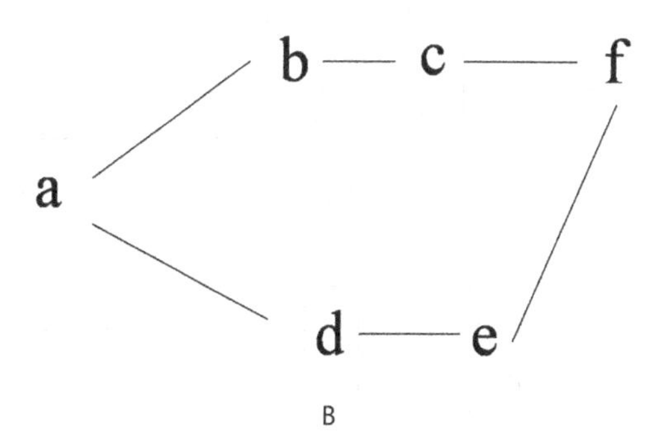

B

The different map representations may additionally be used in a hybrid mechanism, where two different representations are used at different levels. This results in hybrid robotic maps. This representation tries to combine the advantages of the individual representations, and tries to customize the resultant representation as per the problem and algorithm requirements.

In the entire robotic literature, even though we attempt to clearly demarcate and separate the different functions in various modules, the various modules have a large dependency with each other. A higher-level module has a number of assumptions from the preceding modules, and in turn caters to the needs of a number of requirements of the next module. It is not essential that all modules be present in all problems. Many simpler problems may be solved by a lesser number of modules. Many other problems, on the other hand, would demand a very heavy dependency between the various modules.

Localization

One of the major problems while working with maps is to be able to find out the location of the robot. Maps may not be dynamically made at every iteration of the robotic cycle. Instead, it is possible that some part of the workplace of the robot is already known. This is the case where robots operate inside the office. The basic map showing the numerous rooms and their interconnectivity is known prior. In such a case, prior to the start of the journey and at any time during the motion of the robot, it would be vital to know the position of the robot in this map as accurately as possible. This information would be useful at multiple times in the planning, mostly in decision-making regarding whether to turn or go straight, whenever options are available. This task of locating the robot in the map is called localization.

Another context where localization is important is in obstacle tracking. Sensors may indicate the presence of an obstacle and accordingly the robot may have moved to avoid the obstacle. In this process, it is possible that the obstacle is out of the coverage area of the robot's sensors. Now, since the robot has already seen the obstacle once, its subsequent motion must be such that it does not move at the place occupied by the sensed obstacle. At later stages, it would be fair to assume the obstacle might be cleared by some manual mechanism. Hence, at every stage it is important to localize the robot and the obstacle, as the robot moves towards its goal.

The task of localization may essentially be done in two ways. The first way is using a globalized localization scheme. The Global Positioning Systems, or GPS, gives the location of the place on earth where the GPS receiver is to be found. A GPS receiver can always be embedded into the robot to compute its position at any time. This, however, assumes that the entire zone of operation of the robot would be where GPS is usable. In case the robot moves into a region where GPS accessibility is poor, the subsequent operations might not be possible. Further, the accuracy of GPS systems is restricted to a few meters, which means that uncertainty is high. For the same reasons, global localization systems may be usable to compute approximate positions in world map for strategic decision making, but may not be used for smaller tasks like obstacle avoidance.

The other manner is to localize the robot within a smaller frame of reference. Here, we select a single or few features in the robotic world. These features may be entry point door, tables, etc. in an office environment. The position of features are assumed to be stable and not change during the motion of the robot. These features may be given arbitrary coordinates in the world map. The position of the robot from these features is computed. This helps us to know the position of the robot, under the assumed positions of the features in the map. As the robot moves, the various distances to features change. All this gives an idea of the location of the robot. Tracking algorithms are also used for the purpose. Given the motion model of the robot, and the distances of the robot from the various features, the tracking algorithms determine the actual position of the robot. These take into account the uncertainties associated with the individual measurements. Particle filters and Extended Kalman filters are widely used.

Motion Planning

The next step to carry forward is planning of the robot. The planning decides the trajectory that should be followed by the robot for its motion. The planning algorithm only decides the path of the robot, it does not physically move the robot, which is the task performed by the next layer. The planning algorithm may sometimes be classified under two separate heads. These are planning in static environment and planning in dynamic environment.

Planning in static environment assumes that the position of the various obstacles do not change along with time. In such a case, the robot may take its own time in planning the optimal path as far as the task desired can wait. In such a case, we may experiment with computationally expensive algorithms that may tend to offer better results, meeting the stated goals in a much better manner. The static environment is a very likely case for a robotic manipulator that has to align itself to some orientation, or to move to a specific goal.

The motion of the robotic manipulator in such a case may be easily planned. The environment is assumed not to change within the time frame. When the entire trajectory has been planned, it may be given to the control for physically implementing the trajectory. In such a case, there may be only a single or sometimes a few cycles of sense-plan-act. If the robotic path has been planned with utmost care considering all the kinematic constraints, the path may be easily implemented with no necessary feedback to the planning algorithm again. However, if the computed path is found infeasible, or the environment changes somehow, a feedback to planner would be there to invoke re-planning. This planning is also known as offline planning. A sample problem of robotic manipulation is given in Figure 7.

The case with dynamic environment planning is different. Here the planner needs to give the desired path within a small time frame. As a result, there are many sense-plan-act cycles of very small duration each. The planning is performed multiple times, and necessary instructions are given to the controller for physically executing the plan. This changes the environment as well as the robotic scenario, which is catered by the next cycle of sense-plan-act. The environment may be dynamic in nature. These algorithms can save collisions from any moving obstacle, or even obstacles that happen to suddenly enter into the map. These algorithms are highly time-efficient and designed for real-time planning. These may however not work for situations where a high dimensional map with complex object structuring is prevalent. This planning is also known as online planning. It is important for online planners to return the results within the time span that causes significant change in environment. Or else, it would be possible by the time the planner returns results and the controller acts over them, for the map to have changed and what was supposed to be a free position as per the map given to the planner is now occupied by some obstacle.

Often, planning by a single algorithm may not be able to cater to the needs of the algorithm. It is then that we plan the robot in multiple layers using multiple algorithms. Many of them, the higher ones, may be offline, while the lower ones are online. The highest level of planning may be

Figure 7. Planning of a mobile manipulator

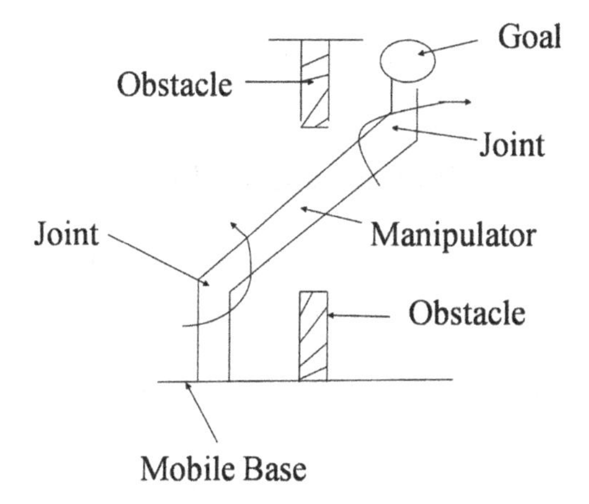

Figure 8. Levels of robotic planning

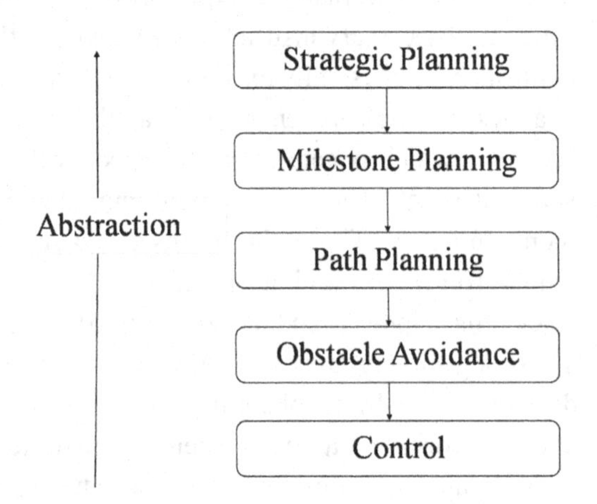

strategic planning, guiding the robots through what goals are to be achieved and in what order. A lower level may involve a milestone planning, which tells the robots how the current goal can be achieved. A still lower planning may be planning the specific path to be followed. This tells the robot how to reach the various landmarks. A still lower planning may be for obstacle avoidance. This tells the robot how to avoid the static and dynamic obstacles it meets on the way. Still lower may be the robotic control that moves the robots as per all its kinematic constraints. This hierarchical planning concept is given in Figure 8

It may be easily seen from Figure 8 that the complete concept of planning is rather generalized in nature. At the higher levels, it touches the actual problem of robotics being solved and the general strategy by which it needs to be solved. At the lower level, it touches the smallest issues of ensuring a collision free path.

Robotic Manipulation and Control

The robot path planner gives as its output a plan of action or a robotic trajectory. The next step to carry forward is to physically move the robot in this path. This is done by a control mechanism. The controller acts as an interface between the planner and the actuators that physically move the robot. The controller analyzes the desired trajectory of the robot path. It then generates a series of commands to all the actuators such that the desired trajectory is followed as closely as possible. It may not be possible to exactly follow the desirable trajectories as per the kinematic constraints and other constraints prevalent in the controller and the actuators. The controller usually undertakes a feedback mechanism, where it sends some control signals to the actuators for motion of the robot. The motion is noted by a set of observations. Accordingly, the errors are noted. These errors are fed back into the system to be used for the next stage of planning. The cycle is shown in Figure 9.

Figure 9. Closed loop robotic control

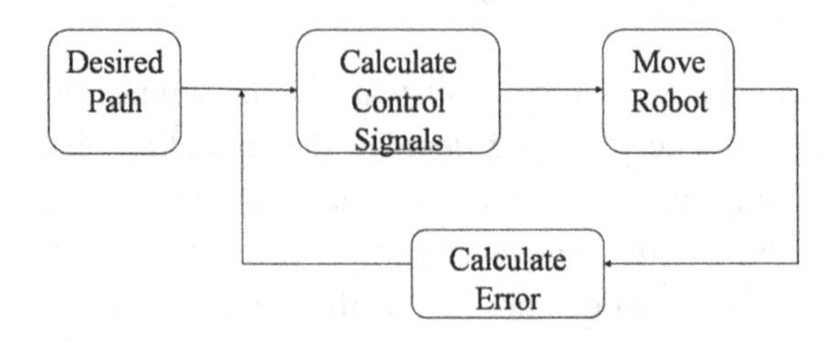

Again, like the planning, the control may be hierarchical in nature. The higher level of hierarchies of control generates the higher-level control signals and so on. It may be noted that the control may not always be to move a robotic wheel for motion. It may as well be to move some manipulator like a robotic arm, leg, etc.

The degree of freedom is a very important concept while dealing with manipulators. The degrees of freedom denote the independent motion (linear or rotational) of the robot. Usually, every degree of freedom would be manipulated by a single motor or actuator. This needs a separate control signal for its output. A single degree of freedom manipulator has less options of movement and hence can be easily planned and controlled. However, as we increase the degrees of freedom, the dynamics increase drastically. Every combination of motion of the independent degrees of freedom might mean a different motion. Hence, the planner as well as the controller needs to consider the complete dynamics and work accordingly. This makes the problem complex and planning a difficult task. Any position in a 3-dimensional environment may be reached by a 3-degree of freedom manipulator. Higher degrees of freedom manipulators mean a redundant way of solving the problem or reaching the point, for which intense planning is needed to compute the optimal one.

Multi-Robot Coordination

The current demands of the tasks to be carried down by the robots are immense and may not be possible by a single robot. Hence, in place of a single robot, multiple robots are now being used for problem solving. The entire task may be divided into the various robots, which may do their part of it and contribute towards the entire problem solving. The division of the task amongst the robots is problem specific. Consider the problem of exploration of an unknown land. Here, we may intend to use multiple robots for faster exploration, and we may employ multiple robots to explore different directions. It is essential to have a communication amongst the robots so that the different robots do not end up exploring a piece of land already explored by another robot, and further, to know which areas are yet to be explored.

The entire problem of use of multiple robots for problem solving distributes the work but creates a set of new issues as well. The first issue is the division of task itself. The approach followed may be centralized or decentralized. In a centralized approach, a designated authority, usually one amongst the participating robots, carries forward the task. The division is based on some predefined heuristics. The leader, or the central authority, may be predefined, or may be made dynamically. The latter is more towards the decentralized approach. Here, the various robots may vote in favor of a leader, which may have some time frame for which it assumes the authority.

The other approach is a decentralized approach. In this approach, there is no centralized authority to carry forward the task, but all the robots collectively come up with a plan of action. The plan decides which robot needs to carry forward what type of task. Here the various robots may vote for a particular task or plan. The major issue associated with these systems is that the complete task needs to be carried out efficiently. Inefficient division may lead to the entire task being even slower than a single robot, due to the excessive communication between the robots.

The next major issue associated is fault tolerance. Any robot may break down at any point of time. In such a case, its work needs to be distributed by the other robots with it not being assigned further work. It may additionally recover at any other point of time. It particularly becomes a problem when the leader falls, which requires appointment of a new leader to carry forward the task of coordination. A proper synchronization amongst the robots is further desirable. Additional means to monitor synchronization may be required. Imagine multiple robots are working over the same component in a manufacturing production line, with two robots required to simultaneously strike a component. Any lack of synchronization may lead to wrong outputs, which may not be desirable.

Communication is another major problem. The communication is bound to be costly and all the robots may not be able to share constant information with all others. The transfer of information from one robot to another requires the making and breaking of connections, along with the communications, which have their own overheads. Further, the protocol to use and the choice between uni-cast or broadcast is always important. In uni-cast the robot selects which robot it needs to communicate the message and then communicates to it by making a new connection. In broadcast, the messages are sent to all the robots, which receive them and act accordingly.

ROBOT MOTION PLANNING

In the previous section, we gave a broad overview of the entire methodology of problem solving in robotics. We give some more details towards the problem of robot motion planning or path planning in this section. This forms the basic problem of the entire book, where we would solve the problem using a variety of methods. Further, in this section we stress the specific modeling scenario we use in all the applied algorithms. Hence, the discussion would be more specific and forms the basis of our further discussions in the rest of the chapters. While we deal with the problem of motion planning, it would be assumed that the other modules have been made and are present in the system. Hence, the focus would only be on the planning part.

Inputs and Outputs

The input to the problem of path planning is the representation of the robotic world or a robotic map. It is assumed that this map has already been formed. We carry forward the planning for the case of 2-dimensional maps. This is like the scenario where we are given the map of a ground and a vehicle has to traverse the same. The robotic map consists of specifications of the way the robot can move and the obstacles. Hence, we assume that we have an *map function* available map(x,y) that returns whether any pixel (x, y) is free or is occupied by some obstacle. The map function is given by (1)

$$map(x,y) = \begin{cases} 0 & \text{pixel (x,y) is not occupied by obstacle} \\ 1 & \text{pixel (x,y) is occupied by obstacle} \end{cases} \quad (1)$$

The robotic environment may be of numerous types. The major categorization of the environment is in terms of structured and unstructured environments. The structured environment has obstacles of structured shape and nicely placed within the map. This is similar to a perfect office environment. In such a case, the planning becomes a highly easy affair. An example of structured environment is given in Figure 10. It may be easily seen that the complete map may be decomposed to a set of interconnected paths, and the planning algorithm is simply supposed to figure out the correct route to take. The complexity of the graph may be easily decomposed.

The other type of environment is the unstructured environment. In this environment, the obstacles may be of a particular structure and placed at tilted orientation (obfuscated) or the obstacles may be of different shapes and sizes. In such a case, the planner has to find out the optimal path. This is a very complex problem to solve as the algorithm needs to compute the feasibility and the optimality of the path. The robot should be able to reach the goal without collisions. An example of unstructured environment is given in Figure 11.

Further, we assume that the source and the destination are perfectly known. Hence, we assume that the given high-level task has been decomposed to a task of navigation of a robot from a pre-specified source to a pre-specified goal.

There may be three types of obstacles in the path of the robot. The first type of obstacles are the static obstacles. These obstacles do not

Figure 10. Structured robotic environment

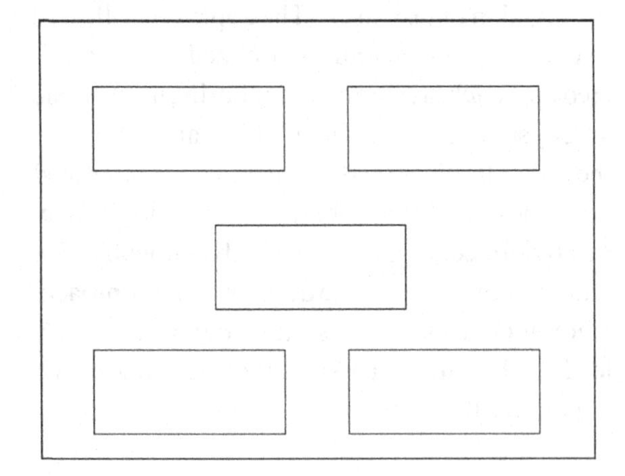

Figure 11. Unstructured robotic environment

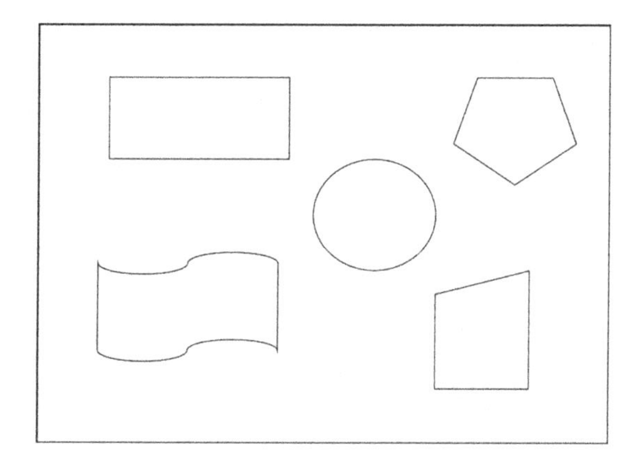

change their position with time and are constant throughout the duration of problem solving. Examples include the furniture in office environment, etc. The other type of obstacles that we deal are moving obstacles. These obstacles change their position with time, but the dynamics of motion is known or can be computed. Examples include moving vehicles in outdoor environments, as the vehicles usually tend to move in constant velocities for small durations of time. The last type of obstacles are the sudden obstacles. These appear suddenly into the robotic maps. These obstacles are the hardest to deal with and require real time planning to be done by the algorithm. The dynamic obstacles may further cause blockages in the robotic path. Hence, if the robot moves as per some plan of action, and the path of motion later gets blocked, it needs to re-plan its journey and compute a feasible path to reach its goal.

The output of the algorithm is always a trajectory or the robotic path. It is assured that the robot would not collide with any obstacle, if it follows this trajectory, as per the information available with the planner. This trajectory is given to the robotic control for the physical movement. Sometimes, cycles of sense-plan-act are used, where the process is carried multiple times. The planners are modeled accordingly. It may not be possible for the controller to track the precise trajectory. Some planners may additionally leave scope for errors by the control.

Planning Algorithm

The task that we deal with is the planning algorithm that takes the inputs and processes them to produce the output. The planning algorithm is our crux of study in this book. A huge number of planning algorithms have been employed to solve these problems. We broadly group these algorithms into a number of categories for easier understanding and organized discussion.

The first category includes the graph-based algorithms. These algorithms model the complete problem as a graph and then use graph search algorithms to compute the path from source to destination. The other category includes the branch and bound algorithms. These algorithms try to generate some random solutions to the problem or robotic paths. The intent is to keep improvising these paths as we proceed with the algorithm. The last group is the behavioral planning algorithms. These approaches try to model the robot behavior and make the robot move similar to the mechanism we use to move in our daily lives. The behavior is modeled such that the robot is encouraged to avoid obstacles and follow a straight-line path towards the goal. We study each of these classes of algorithms in detail towards the later chapters of the book.

Issues

The complete problem of robot motion planning has a number of issues that make the problem challenging to solve. The first issue associated is the completeness. Completeness states that a planning algorithm must return the results, if there exists one. This property emphasizes that the algorithm returns a solution before it terminates, if there exists any solution to the problem. The other issue is of computational time. Every problem has a designated time within which the algorithm must return the solution. For real-time algorithms this time would be much less as per the stated issues of dynamic planning.

Figure 12. Smoothness of path: (a) unsmooth path, (b) slightly smooth path, (c) very smooth path

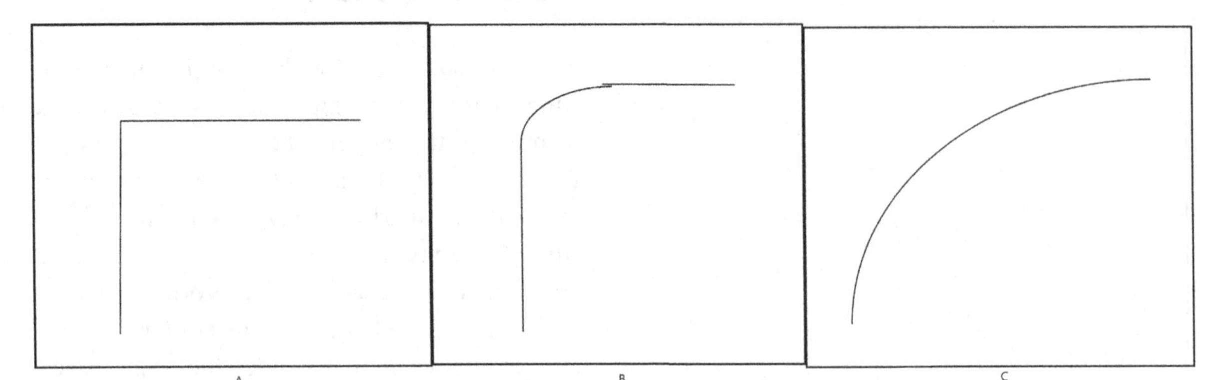

The optimality of path is another major issue. There are a number of factors over which the quality of the path or its optimality may be defined. These are the total length of the path, the total time needed for traversal of the path, and straightness of the path. All these more or less point to the same notion, but the optimal path these compute may be different. A short length is desirable if one needs the robot to travel less, especially in cases of fuel economy. The shortest path may not give the smallest travel time, as it may be going through regions where high speeds cannot be maintained or too many turns prohibit maintenance of high speeds. For the same reasons, many times straightness of a path is measured and straight paths are preferred. The straighter paths mean that the robot can navigate at higher speeds with lesser time. They further indicate straightness and hence may lead to a smaller path length.

The other important issue associated with the planning is the validity of the non-holonomic constraints. It is natural that the car-like mobile robots cannot make very sharp turns. They are only capable of making smooth turns in their paths. The smoothness defines the maximum velocity that the robot can acquire. In other words, a very steep path would have the robot to slow down to a large degree before taking the turn. The maximum speed that a robot can acquire is proportional to the radius of the curve over which it is moving. Hence, if a planner returns a path shown in Figure 12a, it cannot be tracked by the controller. Hence, every path shown in Figure 12a needs to be smoothened. This may be done by application of a smoothening operator, which returns smoothened paths as shown in Figure 12b. This may be traversed by the controller as it obeys the non-holonomic constraints. If we further smoothen the path, as shown in Figure 12c, the robot would be able to traverse the path at a still higher speed.

CONCLUSION

This chapter was intended to give a general introduction to the domain of robotics. While the complete domain is large and extensive, we gave a general overview of the various terms, concepts, and underlying technology. The focus was upon general problem solving in robotics as per the sense-act-plan cycle. This cycle forms the basis of the intelligence in robots, to enable them to perform all the wonderful tasks that were highlighted from multiple domains. A detailed insight into the cycle gives us the various modules that are present in an intelligent autonomous robot. This forms the general software architecture of the robot.

The modules include data gathering, visualization, or understanding of the environment, sensor fusion, map building, localization, mapping, motion planning, and robot control. The discussion over each and every module highlights the tasks they perform, how they integrate with the other modules, and the various issues and concepts related to them. The discussion then pointed over the specific problem of robot motion planning. Here, we clearly understood the role of inputs, outputs, and the planning algorithms. The crux here was the various issues associated with the planning algorithms that all need to be addressed and studied while we develop any planner for robots.

REFERENCES

ABB. (2011). *IRB 6400RF*. Retrieved on November, 2011, from http://www.abb.com/product/seitp327/6331470730f0261fc12570e70046bb3a.aspx

Bellis, M. (2011). *Timeline of robots*. Retrieved November, 2011, from http://inventors.about.com/od/roboticsrobots/a/RoboTimeline.htm

Bonev, I. (2011). *Delta parallel robot — The story of success*. Retrieved on November, 2011, from http://www.parallemic.org/Reviews/Review002.html

Brunette, E. S., Flemmer, R. C., & Flemmer, C. L. (2009). A review of artificial intelligence. In *Proceedings of the IEEE Conference on Autonomous Robots and Agents*, (pp. 385-390). IEEE Press.

Buehler, M., Iagnemma, K., & Singh, S. (2007). *The 2005 DARPA grand challenge: The great robot race*. Berlin, Germany: Springer-Verlag. doi:10.1007/978-3-540-73429-1

Carbone, A., Finzi, A., Orlandini, A., & Pirri, F. (2008). Model-based control architecture for attentive robots in rescue scenarios. *Autonomous Robots*, *24*(1), 87–120. doi:10.1007/s10514-007-9055-6

Cavalcanti, A., & Freitas, R. A. Jr. (2005). Nanorobotics control design: A collective behavior approach for medicine. *IEEE Transactions on Nanobioscience*, *4*(2), 133–140. doi:10.1109/TNB.2005.850469

Ellery, A. (2000). *An introduction to space robotics*. New York, NY: Springer-Verlag.

Fanucrobotics. (2011). *ARC Mate 50iC/5L*. Retrieved on November, 2011, from http://www.fanucrobotics.com/file-repository/DataSheets/Robots/ARC-Mate-50iC-5L.pdf

Feil-Seifer, D., & Mataric, M. J. (2005). Defining socially assistive robotics. In *Proceedings of the 9th International Conference on Rehabilitation Robotics*, (pp. 465- 468). IEEE.

Florczyk, S. (2005). *Video-based indoor exploration with autonomous and mobile robots*. Berlin, Germany: Wiley. doi:10.1007/s10846-005-3508-y

Fujita, M. (2000). Digital creatures for future entertainment robotics. In *Proceedings of the IEEE International Conference on Robotics and Automation,* (Vol. 1), (pp. 801-806). IEEE Press.

Ge, S. S., & Lewis, F. L. (2006). *Autonomous mobile robot*. Boca Raton, FL: CRC Press.

Hirose, S., & Fukushima, E. F. (2002). Development of mobile robots for rescue operations. *Advanced Robotics*, *16*(6), 509–512. doi:10.1163/156855302320535845

Hirzinger, G., Brunner, B., Dietrich, J., & Heindl, J. (1993). Sensor-based space robotics-ROTEX and its telerobotic features. *IEEE Transactions on Robotics and Automation*, *9*(5), 649–663. doi:10.1109/70.258056

Holland, J. M. (2004). *Designing autonomous mobile robots*. Boston, MA: Elsevier.

Kaelbling, L. P. (1988). Artificial intelligence and robotics. In *Proceedings of the IEEE Conference on Computer Society International Conference,* (pp. 59-61). IEEE Press.

Konar, A. (2000). *Artificial intelligence and soft computing: Behavioral and cognitive modeling of the human*. Boca Raton, FL: CRC Press.

Kuka Robotics. (2011). *Robocoaster*. Retrieved on November, 2011, from http://www.kuka-robotics.com/usa/en/products/systems/robocoaster/start.htm

Matthies, L., Kelly, A., Litwin, T., & Tharp, G. (1995). Obstacle detection for unmanned ground vehicles: A progress report. In *Proceedings of the IEEE Intelligent Vehicles Symposium*, (pp. 66-71). IEEE Press.

Millan, J. R., Renkens, F., Mourino, J., & Gerstner, W. (2004). Noninvasive brain-actuated control of a mobile robot by human EEG. *IEEE Transactions on Bio-Medical Engineering, 51*(6), 1026–1033. doi:10.1109/TBME.2004.827086

Mital, D. P., Teoh, E. K., & Yong, I. N. (1988). A robotic vision system with artificial intelligence for automatic wafer inspection. In *Proceedings of the IEEE Conference on Intelligent Robots,* (pp. 289-295). IEEE Press.

Murphy, R. (2000). *Introduction to AI robotics*. Cambridge, MA: MIT Press.

Patnaik, S., Jain, L. C., Tzafestas, S. G., Resconi, G., & Konar, A. (2005). *Innovations in robot mobility and control*. Berlin, Germany: Springer-Verlag.

Sariff, N., & Buniyamin, N. (2006). An overview of autonomous mobile robot path planning algorithms. In *Proceedings of the IEEE Conference on Research and Development,* (pp. 183-188). IEEE Press.

Seetharaman, G., Lakhotia, A., & Blasch, E. P. (2006). Unmanned vehicles come of age: The DARPA grand challenge. *Computer*, *39*(12), 26–29. doi:10.1109/MC.2006.447

Shimoda, S., Kuroda, Y., & Iagnemma, K. (2005). Potential field navigation of high speed unmanned ground vehicles on uneven terrain. In *Proceedings of the 2005 IEEE International Conference on Robotics and Automation*, (pp. 2828-2833). IEEE Press.

Shukla, A., Tiwari, R., & Kala, R. (2010a). *Real life applications of soft computing*. Boca Raton, FL: CRC Press. doi:10.1201/EBK1439822876

Shukla, A., Tiwari, R., & Kala, R. (2010b). *Towards hybrid and adaptive computing: A perspective*. Berlin, Germany: Springer-Verlag.

Sicard, P., & Levine, M. D. (1988). An approach to an expert robot welding system. *IEEE Transactions on Systems, Man, and Cybernetics*, *18*(2), 204–222. doi:10.1109/21.3461

Sitti, M., & Hashimoto, H. (1998). Tele-nanorobotics using atomic force microscope. In *Proceedings of the IEEE/RSJ International Conference on Intelligent Robots and Systems,* (vol. 3), (pp. 1739-1746). IEEE Press.

Stentz, A. (1990). *The navlab system for mobile robot navigation*. (Ph.D. Thesis). Carnegie Mellon University School of Computer Science. Pittsburgh, PA.

Stipanovic, D. M., Inalhan, G., Teo, R., & Tomlin, C. J. (2004). Decentralized overlapping control of a formation of unmanned aerial vehicles. *Automatica*, *40*(8), 1285–1296. doi:10.1016/j.automatica.2004.02.017

Thorisson, K. R. (1996). *Communicative humanoids: A computational model of psychosocial dialogue skills*. (PhD Thesis). Massachusetts Institute of Technology, Program in Media Arts & Sciences. Cambridge, MA.

Turban, E., & Schaeffer, D. M. (1988). Expert system-based robot technology: A systems management approach. In *Proceedings of the IEEE Conference on System Sciences,* (Vol. 4), (pp. 227-235). IEEE Press.

Veloso, M. M. (2002). Robots: Intelligence, versatility, adaptivity. *Communications of the ACM, 45*(3), 59–63.

Waldherr, S., Romero, R., & Thrun, S. (2000). A gesture based interface for human-robot interaction. *Autonomous Robots*, *9*(2), 151–173. doi:10.1023/A:1008918401478

Washburn, A., & Kress, M. (2009). Unmanned aerial vehicles. In *Combat Modeling, International Series in Operations Research & Management Science* (pp. 185–210). Berlin, Germany: Springer.

Wisegeek. (2011). *What is a SCARA robot?* Retrieved on November, 2011, from http://www.wisegeek.com/what-is-a-scara-robot.htm

Wyrobek, K. A., Berger, E. H., Van der Loos, H. F. M., & Salisbury, J. K. (2008). Towards a personal robotics development platform: Rationale and design of an intrinsically safe personal robot. In *Proceedings of the IEEE International Conference on Robotics and Automation*, (pp. 2165-2170). IEEE Press.

KEY TERMS AND DEFINITIONS

Map: An interpretation of the robotic world as seen from multiple sensors embedded in the robot or elsewhere. The map states the regions that robot may use for travel and regions that may not be used for travel as they have some objects or obstacles.

Mobile Robot: A robot that can move around in its workplace and is not stuck at one place. The robot can, hence, go to a place, get task done, and return back. Motion may be restricted as a result of the robot's design, called kinematic or non-holonomic constraints.

Motion Control: The task of physically moving the robot. Each actuator is given a control signal that results in some actuation in the physical world. The resultant motion of the robot as a result of all actuations produced should carry the task as planned by the planning algorithm. In case of availability of a desired trajectory, it must be tracked as closely as possible.

Planning: The task of deciding the actions that the robot must take based upon the available inputs. Actions may be long-termed or short-termed. A planned action may have sub-plans. At the lowest level of planning, actions must be realizable, meaning that they can be executed by the robot.

Sense-Plan-Act: A typical cycle governing intelligent robotics, wherein a robot first senses the environment so as to get an idea of the situation, makes plans so as to best solve the problem it is intended to solve, acts over the plans so as to get the task done. Cycle may be repeated multiple times. Frequency of repeats depends upon the abstraction of task being dealt with. Many tasks are broken down into multiple abstractions resulting in different components working with different frequencies.

Trajectory Generation: A lower level planning where the task is to decide the way that the robot should follow in order to reach the goal in the best possible way, given its current location and the robotic map. The way must ensure no collision of the robot, fast motion, small path lengths, and other application specific objectives. The path should be realizable as per robot constraints.

Chapter 2
Graph Based Path Planning

ABSTRACT

Graphs are keenly studied by people of numerous domains as most of the applications we encounter in our daily lives can be easily given a graph-based representation. All the problems may then be easily studied as grap-based problems. In this chapter, the authors study the problem of robot motion planning as a graph search problem. The key steps involve the representation of the problem as a graph and solving the problem as a standard graph search problem. A number of graph search algorithms exist, each having its own advantages and disadvantages. In this chapter, the authors explain the concept, working methodology, and issues associated with some of these algorithms. The key algorithms under discussion include Breadth First Search, Depth First Search, A Algorithm, Multi Neuron Heuristic Search, Dijkstra's Algorithm, D* Algorithm, etc. Experimental results of some of these algorithms are also discussed. The chapter further presents the advantages and disadvantages of graph-based motion planning.*

INTRODUCTION

Graphs have been of great interest to mathematicians for a long time. Graphs are able to model a large number of real life scenarios. The scenarios extend from game playing, where computers are used to play games like chess, to general everyday problems like the Hannibal and cannibals problem. The domains that the graphs touch are very wide. It is extremely exciting to see how the graphs model the various scenarios and further solve all the problems associated with them. Hence, graphs are constantly used by multiple people from multiple domains. A large number of graph algorithms have come up as well that do a variety of tasks and return the solutions. Graphs are commonly used to solve problems. These include social network analysis. Here, scientists used graph theory to

DOI: 10.4018/978-1-4666-2074-2.ch002

analyze the social networks that are prevalent in all social networking sites. These may be further extended to research groups by figuring relations existing in citations or co-authors.

The discussion forums are further potential sources of networks that may be analyzed by using graph modeling techniques. Another major application of these algorithms is in the domain of planning and scheduling. Here, the scientists are interested in scheduling the various tasks in a project. This gives rise to automated project management and optimal resource utilization. Planning and scheduling may be easily converted into a graph by pointing relations and dependencies amongst the various sub-tasks. These may be then analyzed by graph algorithms for coming up with a strong plan of action regarding the project execution. The graphs have become useful tools for problem solving.

One of the classic benchmarks of graph search algorithms is the IBM Deep Blue project. Chess playing is seen as a challenging AI problem. Much like human players, the computer programs play chess intelligently by trying to forward compute all possible moves of them and the opponents, and try to figure out good moves. In 1997, IBM Deep Blue defeated Gary Kasparov, a world chess champion. Chess playing represents a highly complex graph with too many vertices and edges. It requires significant computing to search the graph in order to find out a move, which may be near optimal. This needs to be done within the timeframe as allowed in the game. The system used intensive parallel processing using high-end systems from IBM. A chess specific processor chip was developed by the team, which was proficient of evaluating and examining 2 to 3 thousand positions per second. It generated thirty billion positions for every move with an average depth of fourteen, routinely (IBM, 2011).

The numerous graph algorithms that are already extensively researched and designed play an important role in motivating conversion of all scenarios and problems into graphs. Hence, the entire task of problem solving gets divided into two parts. The first part, which is more application specific, is to concert the problem as a standard graph. This would require expertise and understanding of the problem. One further need to understand what exactly is the intended output of the system. Hence, the end deliverables must be precisely known. Once this step has been done, we may just be required to pick up a standard graph algorithm that can give us the desired output. The existence of a variety of standard graph problems and a variety of algorithms for each of these problems should usually mean that the algorithm for the task being desired is already available in the library.

Robot motion-planning is a problem that may be easily solved by using graphs. The problem here is twofold. The first part, or the major part deals with how we convert the motion-planning problem as a graph problem, deals with the representation aspects of the problem. The other half of the task is to simply use a standard graph search algorithm for working over the graph generated and to return the solution. We, hence, fundamentally divide the chapter into two parts. The first part deals with the problem of robot motion planning and its representation as a graph. The second part illustrates numerous graph search algorithms. The experimentation, where we apply some of these algorithms to get pictorial results, would be presented at the very end of the chapter. Before everything, it would be wise to brush up the concepts of graphs.

GRAPHS

In this section, we describe the general graph theory. In the entire chapter, we will be making numerous references to the concepts of graphs, which are all summarized in this section. We start by some definitions and terms, and later describe the graph representation and the graph search algorithms. For more detailed discussion over graph and graph search techniques, readers are referred to more general texts (Cormen, et al., 2001; Coppin, 2004; Horowitz & Sahni, 1978; Thornton & du Boulay, 2005; Tutte, 2001).

Preliminaries

Graphs are defined as a set of vertices and edges. The vertices denote the nodes that may have their own significance based upon the problem being dealt. The various vertices are connected to each other by means of edges. An edge always connects two vertices. Hence, let us denote a graph by G(V, E). Here V is the set of vertices (V= $\{V_1, V_2, V_3, \dots V_n\}$) and E is the set of edges (E= $\{E_1, E_2, E_3, \dots E_m\}$). Here V_i is any vertex in the graph and E_i is any edge. Any edge E_i connects two vertices, say a and b. Hence, we may write V_i: a → b. Figure 1 shows a graph with some vertices and edges.

Figure 1. A sample graph with 4 vertices and 5 edges

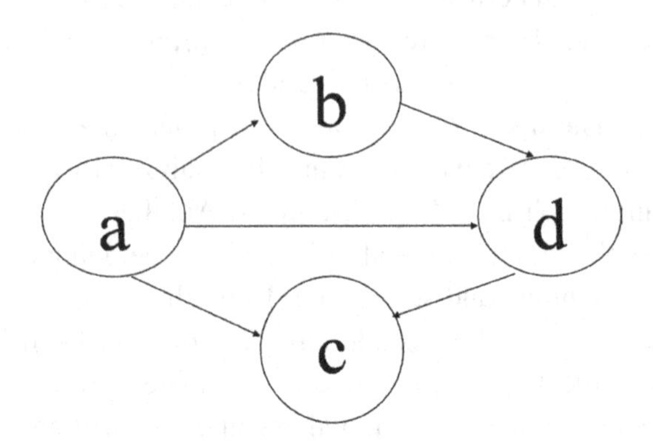

A graph is called an undirected graph if an edge between any vertex a to any vertex b also implies an edge from vertex b to vertex a. Hence a→b ≈ b→a. This means that if we can move from vertex a to vertex b using an edge, then we can use the same edge to move from vertex b to vertex a. If a graph is undirected, we may drop of the arrows shown in Figure 1. Technically, it is equivalent to drawing two arrows denoting the possibility of traversal from both ends in the graph as shown in Figure 2. In case the edge from one vertex to another vertex does not imply an edge the other way round, the graph is knows as a directed graph. A common example of directed graph is the graph of railway lines. If a line joins two stations, it can be used for going to and from between the two stations. The graph is undirected. An example of the directed graph would be the road graph of a region where there are multiple one-way lanes. We may use the lane for going from one place to the other, but we may not use the same lane for coming back.

The edges may further be weighted edges or un-weighted edges. The weighted edges have some weight that denotes the weight to travel between two vertices using that edge. Let the weight of edge e_j: a→b be given by w_{ab}. The weight or the cost may again have a different meaning as per the problem representation. Un-weighted graphs or un-weighted edges have no representation of weights. In other words, it is assumed that all the edges have a weight of unity. An example of the weighted graph is the road graph, where the weights may correspond to the road length. An example of the un-weighted graph

Figure 2. Directed and undirected edges

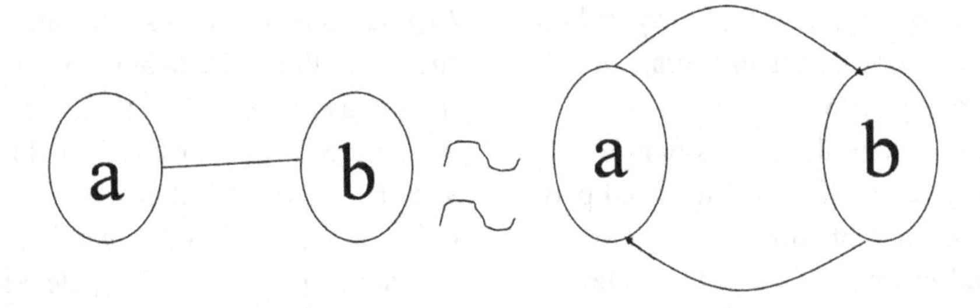

is the game playing graph, where every state of the game is a vertex and edges denote possible transitions amongst states. Here, every move denotes a constant weight and the entire graph is un-weighted. Graphically, a weighted graph may be represented by writing the weight over the edge.

Representation

The graphs discussed in previous sub-sections are easy to understand and manually work over. However, we need a suitable technique to represent in our computer systems, so that the various graph algorithms can be designed and worked over. In this section, we deal with the data structure to be used for storing and working over a graph. This would enable us easy and suitable initialization of the graph, modification of the graph architecture whenever required, and to further query the graph as per the algorithmic requirements. We discuss two major representation schemes. These are adjacency matrices and adjacency lists.

The adjacency matrices store the complete information about the graphs as a square matrix or a 2-dimensional array. Each of the number of rows and the number of columns of this matrix correspond to the number of number of vertices in the graph. Every row i in the matrix denotes the vertex V_i. Similarly, every column i denotes the vertex V_i. Hence, let the matrix be given by $X=[x]_{nxn}$. Here, n is the number of vertices in the graph. We store 1 at a location (i, j), if there exists an edge from vertex V_i to vertex V_j. If there is no edge connecting the two vertices, we simply store NIL. There is no edge joining a vertex to itself. Hence, X_{ii} may always be NIL for all i. Hence, we may represent X as given in Equation 1.

$$X_{ij} = \begin{cases} 1 & \text{if } V_i \rightarrow V_j \in E \\ NIL & \text{if } V_i \rightarrow V_j \notin E \end{cases} \quad (1)$$

Figure 3. Adjacency matrix representation of directed graph

	a	b	c	d
a	0	1	1	1
b	0	0	0	1
c	0	0	0	0
d	0	0	1	0

In case the graph is a weighted graph, we may additionally have the weights of the edges stored at the adjacency matrix in place of a unity. Hence, the same Equation 1 changes to 2.

$$X_{ij} = \begin{cases} w_{ij} & \text{if } V_i \rightarrow V_j \in E \\ NIL & \text{if } V_i \rightarrow V_j \notin E \end{cases} \quad (2)$$

Here w_{ij} denotes the weight of the edge connecting V_i to V_j.

Figure 1 shows a small graph. The adjacency matrix of the graph is given in Figure 3.

In case the graph is undirected, it is evident that the existence of an edge from i to j is the same as the edge from j to i. In case of the adjacency matrix, we have $X_{ij}=X_{ji}$. This is nothing but the condition for a matrix being symmetric. Hence, the adjacency matrix representation of an undirected graph is symmetric. In other words, the transposition of the matrix is equivalent to the matrix. This may be represented as given in Equation 3.

$$X^T=X \quad (3)$$

We may easily store any half of the adjacency matrix for such graphs. This would save half of the storage space. The adjacency matrix representation for an undirected graph is given in Figure 4.

Figure 4. Adjacency matrix for undirected graph

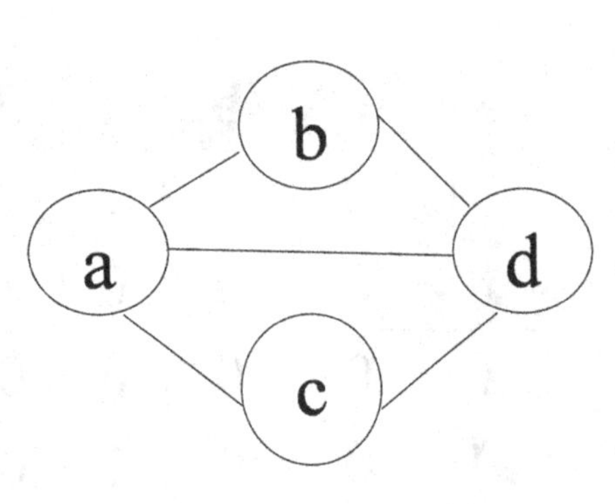

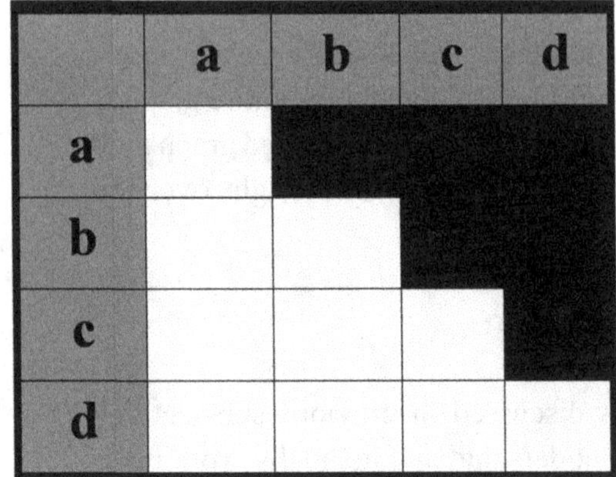

The other representation used is the adjacency list representation. In this representation, we take an array of lined lists. The size of the array is equal to the number of vertices in the graph. Each cell of the array X_i corresponds to a linked list. This linked list stores all the vertices to which the vertex V_i is connected. If the vertex V_i is not connected to any of the vertices, the linked list X_i is empty and simply points to NULL. A vertex that has connections from other vertices to it but does not point to any vertex is known as a sink. Figure 1 shows a graph and Figure 5 shows the corresponding adjacency list representation. As per norms of the lined list, all the linked lists are terminated by pointing to NULL.

The two techniques of representation have their own advantages and disadvantages.

Figure 5. Adjacency list representation of graph

a	b ⟶ d ⟶ c ⟶ Nil
b	d ⟶ Nil
c	Nil
d	c ⟶ Nil

- **Edge between 2 Vertices Query:** The adjacency matrix representation is swiftly able to handle any query (produced by the algorithm) to check the presence of edge between any two vertices. This may take time in an adjacency list, where a complete traversal of the linked list is desired for the query to be answered.
- **Memory:** Adjacency matrix representation takes a lot of space. This is especially true in case the graph is sparsely connected, or that the edges in actual are very less than the maximum permissible number of n(n-1). In such a case, maximum elements of the matrix would be NIL. This, however, is not the scenario in the case of the adjacency list, where only the needed connections are stored.
- **Edges from a Vertex Query:** Any query to get all edges emerging from a vertex is a problem in sparse graphs in adjacency matrix representation. The algorithm would have to traverse all the elements in the row and return all non-NIL elements. In the case of the adjacency list, only the linked list of the corresponding cell needs to be returned.

Searching

Searching is one of the major algorithms used in graphs. A graph search algorithm tries to find a set of vertices (or equivalently edges) that enable reaching a predefined vertex called source, to a predefined vertex called goal. The graph search algorithms take as input the source vertex (S) and the goal vertex (G). They produce as their output the series of vertices (or edges) in order that enable reaching the goal from the source ($S \rightarrow V_1 \rightarrow V_2 \rightarrow V_3 \ldots \rightarrow G$), along with the distance from source to goal. Some graph search algorithms may further work towards returning the best or the shortest path from source to the goal. It is possible that no path exists between the source and the goal, for which the graph search algorithms may simply return an empty set or NIL as their output. Graph search is a very common problem and is used as it is for most of the applications, or may additionally be used to provide details for more complicated problems. The problem of robot motion planning, as we shall see later, is simply a graph search problem.

ROBOT MOTION PLANNING

We intent to solve the problem of robot motion planning by a number of graph-based algorithms. As discussed earlier, for this we first need to convert the problem as a graph, before we can use the standard graph search algorithms for solving the problem. Hence, let us first clearly define the problem along with the assumptions involved. Then we proceed with the modeling of the problem as graph. We use the same general notion of the problem that was introduced in chapter 1. For related works, please refer to Fernandez and Gonzalez (1998), Hui-Ying et al. (2010), Kala et al. (2009), Liu and Arimoto (1990), Lozano-Perez (1987), and Shukla et al. (2008).

Problem Description and Assumptions

The problem is to find a robotic path that enables the robot to reach the goal from a source. We assume that the complete map is already known. Further, we know the start point as well as the goal point. The output to be produced is the points that the robot must travel to reach the goal from the source. The solutions generated by the graph-based algorithms may take time, and hence, the problem may not always be suitable for real-time path planning. We assume that the environment is completely static in nature and there are no dynamic obstacles involved in the process. Further, we assume that the non-holonomic constraints are not of concern here. Hence, the paths generated may have very sharp edges as well. For the validity of the non-holonomic constraints an additional smoothening algorithm may be used, or maybe the robot is holonomoic and is capable enough to take sharp turns. We further assume that the robot may be taken as a unit point. Obstacles in map may be swollen up by robot size or alternatively the robot may be too small compared to the complexity of the map.

We further assume that the entire map is broken down into grids, with clear understanding of presence or absence of the obstacle in every grid. The graph search algorithms, as we shall see later, are very sensitive to the dimensionality of the problem, which is reflected by the resolution of the map. Resolution here is refers to the quality of the map or the image, which is a measure of the number of points in the map. Since we have a grid-based representation of the map, where a grid is of size m x n, the resolution may simply be taken as m.n.

The algorithms based on some heuristics may perform fine in simpler paths, but as the complexity of the path from source to goal becomes large, these algorithms starting taking a reasonably long time. The other algorithms not based on heuristics may only be able to work over significantly low resolution maps. If for any of these maps, we increase the resolution, we may observe an increase in computational time. For any robotic task, there is always a maximum time within which the algorithm must return the solution. In such a case there is a heavy threshold on the resolution that can be used in the graph-based algorithms. We assume that the map given to the planner for execution is of reasonably small resolution.

Since the map is produced by the map-building algorithm, the complete map may be of high resolution. In such a case, we may perform a resolution reduction to convert the map into a lower resolution counterpart, before using it as an input to the graph search problem. This indeed reduces the effective information content in the map, but it enables to use graph search algorithm over high dimensional maps, which may not have been possible otherwise. In the later chapters, we will see how we can iteratively use a technique to increase the resolution with every iteration of a master algorithm to get more precise results.

Consider that the high-resolution map is given by the function given in Equation 4.

$$map(x, y) = \begin{cases} 0 & \text{pixel (x,y) is not occupied by obstacle} \\ 1 & \text{pixel (x,y) is occupied by obstacle} \end{cases} \tag{4}$$

Now we need to reduce the map from the original dimension m x n to a lower dimension by reduction in resolution by a factor of α. This may be done by passing a map through an averaging window. This window computes the average of all the points lying inside it. The size of the window is equal to the size of the reduction factor. Hence, the size of the window is α x α. The resultant map map' may be given by Equation 5.

$$map'(x, y) = \frac{\sum_{i=1}^{\alpha}\sum_{j=1}^{\alpha} map(x\alpha + i, y\alpha + j)}{\alpha^2} \tag{5}$$

However in our assumptions, we stated that the algorithms would require a binary map representing the locations of both the obstacles as well as the movable paths. Hence, we need to further work over the map given by Equation 5 and convert it into a perfect binary map. The value of any cell in the averaged map map'(x,y) basically represents the probability of the cell being occupied by an obstacle. Hence, we set a threshold. Below the threshold all points are classified as free and above the threshold all the points are classified as with an obstacle. This is given by Equation 6.

$$map''(x, y) = \begin{cases} 0 & \text{map' (x,y)} \leq \text{Th} \\ 1 & \text{map'(x,y)} > \text{Th} \end{cases} \tag{6}$$

Here Th is the threshold. It may be conveniently be set as 0.5, or be adjusted as per the safety needed for the robot motion or the complexity of the environment. In the complete discussion in the chapter, we would be referring the input as map(x, y) and the resolution as m x n. They may be conveniently taken to be the modified map with the reduced resolution for ease of discussion.

The concept of drop of resolution is illustrated in Figure 6. Figure 6a shows the original map and Figure 6b shows the reduced map after averaging.

Graph Representation

Once we have worked over the inputs, the other major task to be carried out is to work over a strategy that the complete problem can be represented as a graph. This may be followed by a simple graph search algorithm to compute the path to be traversed by the robot.

The graph consists of vertices and edges. In this approach, we assume each of the free cells in the input map to be a vertex. A cell is called free if there is no obstacle on the cell (map(x, y)=0). Since there are m x n cells in the map, there can be a maximum of m x n vertices in the problem graph. Each vertex may be conveniently labeled

Figure 6. Effect of drop of resolution on map: (a) original map, (b) map after drop of resolution

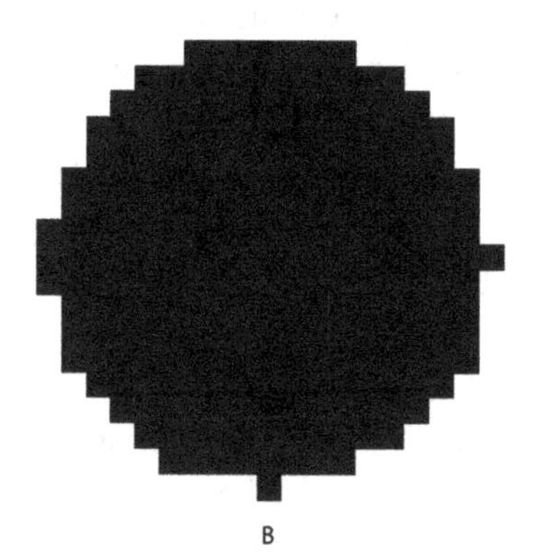

by the position it has on the map. Hence, we denote any vertex by V(x, y) which represents the location (x, y) in the input map.

The other part of the graph representation is edges. We state that the robot is allowed a set of moves, and at every point of time, it has to make one of these moves. This is unlike the real world, where we may move any number of steps (even fractional) at any point of time. Further, we may turn by any degree in our entire movement. Hence, motion in the real world is all about working in a continuous domain where any step may be taken by any degree and at any orientation. However, in the case of motion planning of robots, we have restricted the motion of the robot to a set of moves. In other words, we restrict the motion to a discrete domain or finite move set. It may be noted that a similar task was performed on the map, where we divided the entire map into discrete grids, which were finite in number, rather than working over with possibilities of infinite number of grids. This restriction is due to the graphical representation of the problem. The total number of vertices and edges needs to be highly limited in a graph for getting results in finite time.

Let us define the set of possible moves as a small collection of commands as given in Equation 7:

move = {north, north-east, east, south-east, south, south-west, west, north-west} (7)

This means that at every point the robot can either make a step north, or a step east, or a step west, or a step in the south direction. Hence, at every point there are a maximum of 8 possible moves. The possible moves are graphically shown in Figure 7.

We say that an edge exists between two vertices, $V_1(x_1, y_1)$ and $V_2(x_2, y_2)$, if the robot can reach point (x_2, y_2) from the point (x_1, y_1) by making motion from one of the moves in the set move given by Equation 7. It needs to be additionally checked for an edge to exist between vertices V_1 and V_2; the robot straight line path from V_1 to V_2 must be collision free. All motions in Equation 7 lead the robot from one cell to some adjacent cell.

Figure 7. Maximum possible moves for robot when placed at any map grid

NW	N	NE
W	R	E
SW	S	SE

In physical robot terms, they indeed constitute a unit move. In such a case, the path between these vertices may indeed be collision free if the two vertices are collision free. Many times the set move may be rather extensive such that two connected edges V_1 and V_2 may be little far apart. In such a case, this additional collision check must be performed.

It may be clearly seen that if the robot can move from $V_1(x_1, y_1)$ to $V_2(x_2, y_2)$ by making one of the allowed moves, it may also be able to move from $V_2(x_2, y_2)$ to $V_1(x_1, y_1)$ by taking opposite move. In place of north, it should take south and so on. Hence, the generated graph is undirected one.

We further define the weight of the edges. The weight of any edge from $V_1(x_1, y_1)$ to $V_2(x_2, y_2)$ may be easily taken as the Euclidian norm between $V_1(x_1, y_1)$ and $V_2(x_2, y_2)$. This means that the distance between two points denotes the cost of motion between these points. In other words, the algorithm that we built for finding the optimal path from source to goal would try to reduce the net distance of travel. If the robot moves in a constant speed, this would mean minimization of the time of travel as well. Hence, we may define distance by Equation 8.

$$w_{V_1(x_1,y_1)V_2(x_2,y_2)} = \left\|V_1(x_1,y_1) - V_2(x_2,y_2)\right\| = \sqrt{(x_1 - x_2)^2 + (y_1 - y_2)^2} \quad (8)$$

Now consider any small map as shown in Figure 8a. We may easily convert the map to a graph as given in Figure 8b.

Hence, in this manner we may easily convert any map into a graph. The source and the goal are already known. Hence, in the converted graph we may easily mark out the source vertex and the destination vertex. It is then that the problem of robot motion planning reduces into a standard problem of graph search.

It may again be noted that in the problem the moves were restricted to 8. This means that every vertex can be connected to a maximum of 8 vertices, or any vertex can have maximum 8 edges. The number would be less than 8 if some movement results in collision with an obstacle, i.e. the corresponding vertex does not exist. An increase of this number would mean that the vertices would start becoming connected to all the other vertices. The density of edges would increase and the graph would start taking an all-connected form. Many graph search algorithms are unable to handle large number of edges or such high con-

Figure 8. Conversion of map to graph: (a) sample map, (b) equivalent graph

a	b	c	
d		e	f
g			h

A

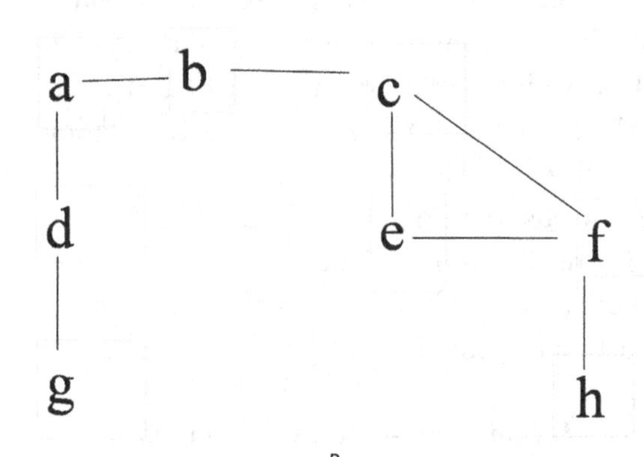

nected graph forms, as the complexity increases with the increase in edges. As we shall see later, in such a case, every vertex exploration would result in many vertices being uncovered that the algorithm would have to explore. In such a case, computational time is likely to increase.

On the other hand, keeping the number of possible moves too small would again be poor. Larger number of moves means having more flexibility in motion. This allows robots to make movements that result in better position towards the goal. This may be conceived as a step forward towards continuity from discreteness. Hence, the net quality of path increases as we increase the number of allowed moves. If we analyze the graph, we would find traversal from lesser number of vertices from source to goal enable us to reach the goal as we increase the number of possible moves. However, if the optimal path has a lesser number of vertices in its path from source to goal, the computation time is bound to be low. This is because if the exploration of the search algorithm is in correct direction, it has to visit a lesser number of vertices.

The computation time must decrease as we increase the number of moves. This contradicts our above statement where larger edges resulted in increase in complexity and higher computation time. In totality, contradictory forces of both these effects play a role in decision to computation time. The increased edges warrant an increase in time, whereas reduction in path length warrants a decrease in the same. We speak more about the computation time and the associated complexity after discussing the specific algorithms.

Let us understand the relation between the path, length of path, and the maximum allowable moves by a simple example. We take three scenarios. The first assumes the total number of moves to be only 4, i.e. move={north, east, south, west}. The second scenario is when the algorithm may have 8 moves as discussed in Equation 7. The third scenario is when the algorithm may have 16 possible moves. These are move = {north, north-north-east, north-east, east-east-north, east, east-east-south, south-east, south-south-east, south, south-south-west, south-west, west-west-south, west, west-west-north, north-west, north-north-west}.

All these three cases of moves are shown in Figure 9. It would be interesting to note that we did not opt for cases like north-north in the third scenario, even though it was valid and represented the same notion as the other marked distances. It may be noted that the movement north-north is

Figure 9. Possible robot moves for different number of moves allowed: (a) 4 moves, (b) 8 moves, (c) 16 moves

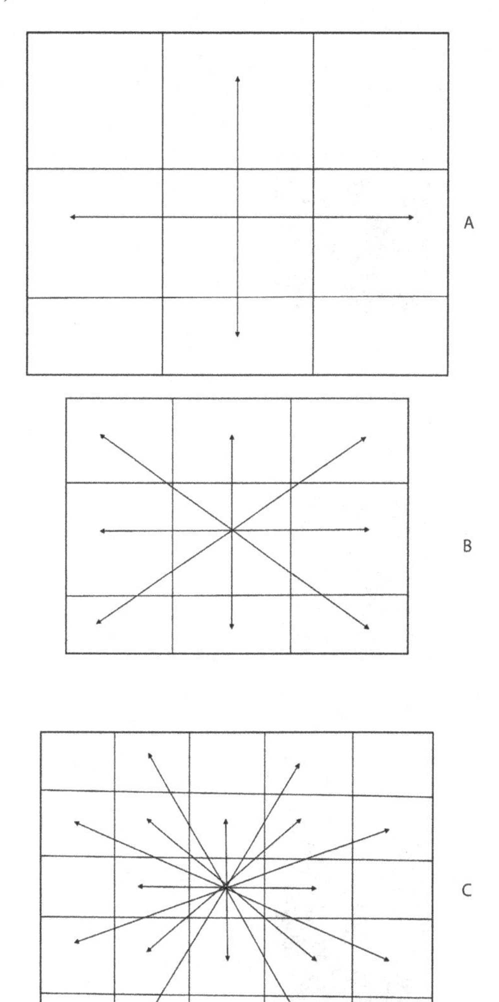

exactly similar to two independent movements followed one after the other that is north and north. Hence, it may be generated by the planner without the need of specifying as an edge.

Further, let us take a sample map as given in Figure 10. We try to manually find the optimal path generated by each of these three scenarios from the source to the goal (which are marked in the map itself). The paths generated are given in Figure 10. It may be easily seen that the flexibility of the path increases as we increase the number of moves, and the net paths generated start getting shorter in length. Note that logic has been used to compute which cells must be obstacle free for the robot to move by any move. For moves which lead to adjacent cells, the only condition is that the present location and the location after the move being expanded must be free. There are 8 moves which take the robot by √5 distance. For all these moves, we see which additional cells the robot would intersect to reach the goal. All these cells must additionally be ensured to be free.

Figure 10. Paths from source to goal with graph search algorithm: (a) 4 possible moves, (b) 8 possible moves, (c) 16 possible moves

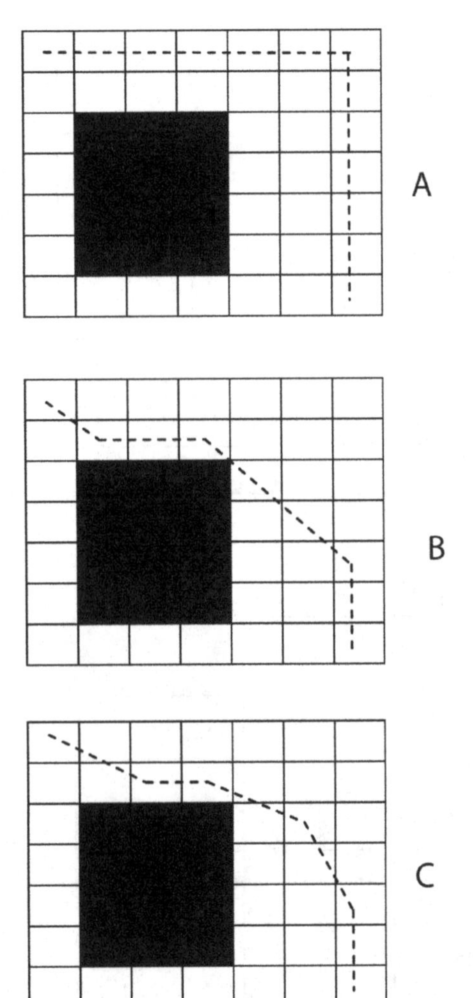

GRAPH SEARCH

There are a variety of graph search algorithms, each running on their own theories, concepts, and notations. Before touching all these algorithms, we build up a general framework for graph search algorithms. This would greatly formalize our approach towards dealing with different algorithms, as well as enable us to easily compare the working of the various algorithms. Hence, in this section, we try to look at the various algorithms from a broader perspective and explain the concept behind the working of graph search algorithms. Please note that the different algorithms discussed in different textbooks might appear different in appearance and discussion, but the general working and concepts would be the same as presented in this chapter.

The search algorithm in case of the graphs is similar to the manner in which we perform search in everyday life, or like a linear search of an array where the search process involves traversal of entire array. The search algorithm has the task to search the entire graph structure traveling from one node to the other. In its travel across the graph structure, the search algorithm compares the node to be searched with the current node. The entire expertise in the search lies in ordering the various nodes or in other words the sequence in which the search would be performed amongst the various nodes. It may be clearly seen that difference in this scheduling would result in difference in performance of the algorithm. The search algorithms need to report not only the goal node that was intended to be searched but also the path from the start node to the goal node. This path need not be shortest path possible for all search algorithms, but may be a factor in certain requirements.

There are a number of factors based on which the performance of the different search algorithms may be studied. The different search algorithms, by their modeling scenarios and working methodologies, differ in performance based on these factors, which again depend upon the problem being tackled. The first major factor is the time complexity. The search must be able to return the results within the time requirements imposed by the problem. The other important factor is the memory complexity. The different search algorithms differ in their memory requirements. The memory required needs to be less than the memory available in the system. The last factor is the optimality of the search path. Some problems demand the search algorithm to return the shortest path from the source to the goal, which the search algorithm has to ensure.

The search algorithms usually maintain a schedule or list containing the vertices of the graph. Usually two lists are prepared which are known as the open list and the closed list. The entire graph at any instance of time would have three kinds of vertices. The first kind of vertices are the ones that are yet to be discovered. The second kind of vertices are the ones that are discovered but yet to be processed by the algorithm. The third kind includes the vertices that have been discovered as well as processed by the search algorithm. The open list contains all the vertices of the second kind, which are discovered but yet to be processed. The closed list contains all the vertices of the third kind, which are discovered as well as processed. The undiscovered vertices are not yet within the knowledge of the system; hence, these vertices are not stored. Initially, the system knows about only one vertex, that is the source. Hence, the entire algorithm has the source in the open list, and the closed list is empty.

As the algorithm proceeds, the nodes are taken from the open list, processed, and added to the closed list. Hence, the closed list grows along with time or iterations of the search algorithm. The processing involves computation of costs, as well as the expansion of the vertex being processed. The expansion means discovering new vertices joined or connected to the vertex being processed. Many of these vertices may have previously been unknown to the system. In other words, these might not be occurring in the open or the closed list. All these vertices are added in the open list. The expansion results in the increase in the size of the open list. During any time of the entire process, if the vertex being processed is found to be the one we were searching for, the search is said to be complete and the vertex is returned. However, if at any moment of time the open list gets empty, the search is said to have been complete with the desired goal node not found.

It may be easily visualized that the emptiness of the open list means we have traversed all the vertices that were connected directly or indirectly to the source. The goal vertex not being found may be due to the vertex not being present in the graph, or that the graph was segmented into disconnected sub-graphs and the source and the goal did not appear in the same sub-graph. In such a case we may further attempt to carry the search operation in another sub-graph, if only the goal node is intended to be found with not necessarily there being any path from source to goal.

Based on these discussions we may easily write a search algorithm as Algorithm 1.

In this algorithm G<V, E> is the graph to be searched,

Select(x) is the algorithm specific operation that returns the node to be processed in the entire open list

Expand(x) returns a set of vertices that are directly connected to x by an edge

Algorithm 1. General search algorithm

```
Search (G<V,E>, source, goal)
source -> parent ← null
closed ← ϕ
open ← {source}
while open is not empty
        x ← Select(open)
        if x=goal return x
        δ ← Expand(x)
        for all y in δ
                if y ∉ open∪ closed
                  compute costs of y
                  y -> parent  ← x
                  open ← open∪{y}
            end
        end
        closed ← closed∪{x}
        open ← open - {x}
end
return null
```

The cost is an optional step that may be needed by certain algorithms. The costs may be used to denote the priority of a node being produced in the open list, or otherwise.

The parent attribute of a vertex is noted to backtrack the path that connects the source to the goal. As discussed certain problems are interested in the roadmap that connects the source to the goal. This is a sequence of edge from which we may reach the goal, while starting from the source. We make this provision in the search algorithm itself. Every vertex is given a parent, which is the vertex through which it was discovered. The parent of the source is taken as null. While the execution of the search algorithm goes on, we keep the parent information update for all vertices. After the execution of the search algorithm is complete and we have the goal node, we may simply trace the path that took us to goal node by the information of the parents. We start from the goal vertex to get the information about which vertex was it discovered through; that is its parent. Then we move to the parent vertex to find out how this vertex was discovered. In this manner, we go on checking for parents of the vertices, unless we reach the source, whose parent is null. This gives us the path from the goal to the source; the reverse of which joins the source to the goal. The algorithm to get the path is given by Algorithm 2.

Algorithm 2. Path retrieval algorithm

```
Path(goal)
P ←ϕ
I ← goal
While I ≠ null
              P ← I∪P
              I ← I -> parent
end
return P
```

This is the general mechanism to deal with a graph problem. We discuss more about the same along with examples when we study the different graph algorithms.

STATE SPACE APPROACH

So far we have assumed that the graph would be available in the form of a collection of vertices and edges. We further assumed that the graph may be represented as an adjacency list or an adjacency matrix. However, in many cases this may not be possible. There might be an infinitely large number of vertices possible as per the problem-modeling scenario. Each of these vertices may be connected to only a smaller number of vertices. Solving such a problem using a graph search method would require initialization of all the vertices, which may mean an excessively large computation time. In most of the problems, it would not be possible to complete the initialization of the graph within finite time implying computational constraints. Other problems may not allow such a large graph to be stored within the memory constraints. On the contrary, the search problem after such a large computation may not take time of that large order.

Consider the problem of motion planning of mobile robot. We know that the resolution of the map may be significantly high. Hence, the image vector is reasonably big. The number of vertices, which as per the same problem-modeling scenario would be equal to the number of pixels, would be reasonably large. In such a context, we would be expected to initialize a large array. This may be quite time consuming.

In such problems, we do not store the complete problem as a conventional graph. Rather, we develop mechanisms to understand the states and deal with all the queries that the graph algorithm would have by this understanding of the states. From our discussion over the problem of robot motion planning and its modeling as a graph, we know that we may not initialize all the possible states at the start. Many of them would not be discovered in the search at all, and only a small fraction of them might be used in the search process. We define a *state* of the algorithm in such a case similar to the definition of the vertex, except that the states may not be stored or initialized. From the basic architecture of the graph search algorithm, we know that the algorithm needs to query the graph to get the next connected vertices and to compute the distance between any two vertices. We may develop the function that takes in a state and returns all the possible states that may be generated from the current state as per the problem modeling. This may be simply done as per our discussion over the edges in the previous section.

Similarly, the weights of the edges or the costs may be simply computed using the norm measure we discussed. In such a case there happens to be no need of making a formal data structure storing the adjacency list or adjacency matrix. We may simply use our understanding of the state and query the state as per the graph algorithm requirements. In case of graph search algorithms, the states are stored in the open list, retrieved from the open list, processed, and added in the closed list. The processing enables discovering newer states that may be added accordingly, if previously unknown. In this manner, the approach differs from the other approaches where the complete information about every vertex was stored in a special data structure. All the vertices were initialized at the start, and the updates were made as the algorithm proceeded.

BREADTH FIRST SEARCH

The fist algorithm in this class of graph search algorithms that we study is the Breadth First Search (BFS). This algorithm is applied over an un-weighted graph, or in other words, it may be assumed that the weight of all the edges is unity. This search algorithm returns the smallest distance between the source and the goal. Hence, the search is very useful in problems that deal with shortest path computation. This search algorithm, however, has a very large memory complexity. A large number of nodes need to be stored in the memory for long time. Further, the time complexity is also large and it would take a long time to get to the goal node from the source node. We look into the concept and working of the algorithm in the coming sub-sections.

Concept

The basic concept of the breadth first search is that the processing of the vertices need to take place such that all the breadths of a vertex are explored first, before exploring the depths of each of the vertices connecting it. Hence, from the source, we first intend to explore all the vertices that are connected directly to the source. We later attempt to explore all vertices that are connected to the vertices directly connected to the source. In this manner, before attempting to explore any vertex, we ensure that all the vertices at a lower level to it, in terms of depth from the source, have been visited. This means a simple level-by-level exploration, where level is defined as the minimal distance from the source vertex, or the depth of the vertex. Since the exploration is in a level-by-level manner, giving more regard to the breadth rather than the depth, we know that if a goal vertex is found, there would be no mechanism in which it would have been found at a lower level, or at a lower distance from source. Hence, this algorithm ensures the path returned from the source to the goal is the smallest possible path if one exists.

The implementation of this search mechanism is to simply implement a queue as a data structure of the open list. In other words, the selection and adding mechanism of the algorithm is similar to a queue implementation. The implementation is first in first out, wherein a vertex that is first to get inside a queue is also the vertex that is processed first and removed from the queue. This plays a role in implementation of the concept that results in finding the smallest path from the source to the goal. When a vertex is inserted in a queue, it naturally means that the entire vertices at a smaller distance from source have already been inserted. Since these are inserted earlier, these would be processed earlier.

The complete algorithm of the search is given in Algorithm 3. Please note that if an adjacency matrix, adjacency list, or a similar approach is used, the vertices would be indexed separately. The parent and other information may be added at the same place. In such a case, the closed list is not needed. However, in case the state space approach is used, we need to use a closed list to store the vertices along with the other information.

In Algorithm 3, we have taken an extra attribute of distance that stores the distance from the source. Every vertex in this case has a distance from

Algorithm 3. BFS algorithm

```
BFS(G<V,E>, source, goal)
source -> parent ← null
source -> distance ← 0
closed ← φ
open ← {source}
while open is not empty
        x ← Select 1st vertex from open
        if x=goal return x
        δ ← Expand(x)
        for all y in δ
                if y ∉ open∪closed
                        y -> distance ← x-> distance +1
                        y -> parent  ← x
                        add y to end of open list
                end
        end
        closed ← closed∪{x}
        open ← open - {x}
end
return null
```

source unity greater than the distance from the source of its parent. This is because it has been assumed that all the edges have a unity weight, since the graph considered is un-weighted.

Example

We further study the algorithm by a small illustrative example. Consider a small graph, as shown in Figure 11. The source node and the goal node are marked in the figure. Initially the source is kept in the open list making the open list Open = {a}. The closed list is empty that is Closed = {}. Now in the first iteration this vertex is extracted from the open list. The vertex is added to the closed list and all the adjoining vertices are added to the open list. Hence the open list becomes Open = {b, c, d} and closed list becomes {a}. At the next iteration, vertex b is taken from open list, expanded, and added to closed list. The two lists now become Open = {c, d, e} and Closed = {a, b}. Next iteration vertex c is processed making the two lists Open = {d, e, f} and Closed = {a, b, c}. In the same manner, the lists are explored until the time goal is the vertex to be processed from open list.

In this manner, it may be verified that the algorithm took a top down approach or a breadth first approach in searching for the goal from the source. In addition, the smallest path from the source was figured out by the algorithm of the goal.

It is clear from the algorithm as well as the results that the breadth first search produces good results when the goal is to be found at shallow regions, or quite close to the source. In case this is not true, the algorithm would expand a large number of nodes as it comes at lower regions. All the expanded vertices would be accumulated in the open list, making the list heavy. Further, many times the graph may have a very high connectivity, where a vertex is connected to a large number of other vertices. As per the working methodology of the algorithm, all the vertices at smaller levels need to be visit before iterating into the levels. In such a case, the algorithm would waste a large amount of time in visiting all the upper nodes.

DEPTH FIRST SEARCH

The other similar algorithm we study is the Depth First Search (DFS). This algorithm is also used in case of an un-weighted graph, or in other words, we assume the weights of the various edges to be unity. The algorithm, unlike the BFS, explores by giving priority to the depth. The algorithm does not return the smallest distance of a source to the

Figure 11. Graph search by breadth first search

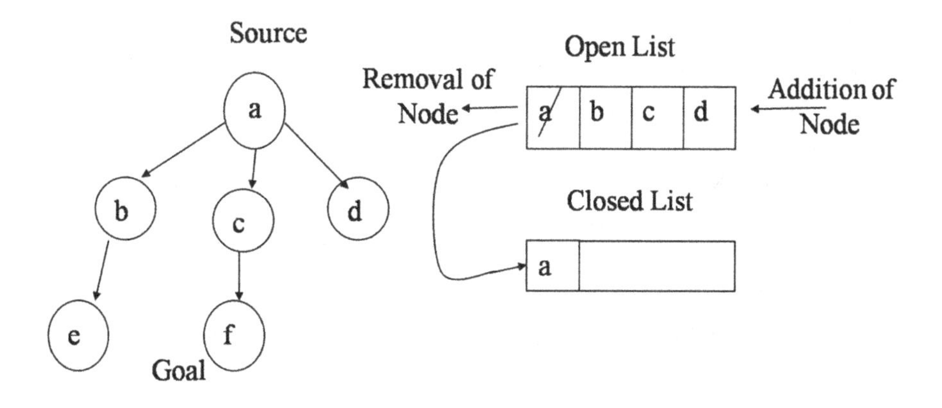

goal, but the memory requirements (or memory complexity) of the algorithm for most of the general cases are reasonably low. The algorithm is very useful when the vertex being searched lies deep inside the search graph. It may be noted, as the path retuned by this algorithm when applied to the problem of robot motion planning can be unreasonably long, the algorithm is not used for the particular problem. The sole motive of discussing this algorithm in this book is for completeness of the topic of graph search.

Concept

The basic concept of this search algorithm is that the depth needs to be explored. Hence, whenever the algorithm selects a vertex for processing, it tries to march deep inside the graph using the edges. This is prioritized over the task of finding and exploring vertices at same level from the source. This naturally does not result in smallest paths from source to the goal. The implementation of the algorithm is same as the BFS, with the major difference that we need to use a stack implementation in place of a queue implementation. The stack has a last in first out implementation, and hence the vertex that is last to be entered into the stack is the one that is processed first of all. This is responsible for the depth first nature of the algorithm. A vertex being processed has all its adjoining vertices being added right at top of the open list. The first vertex being processed is naturally the vertex that was just expanded. Later, when the expanded list is processed, the last added vertex is further processed and its expanded vertices are expanded. In this manner, the algorithm expands the vertices, until there is no possibility of further processing the vertices.

The complete algorithm is given in Algorithm 4.

In Algorithm 4, we maintain time of discovery of the goal in place of maintaining the distance, as in the case of the BFS. This metric is not needed in the search operation, but may be needed for some sorting and other operations needed by the other graph algorithms.

Another common implementation uses the recursive function call, which we do not write because of the uniformity of the methods. It may be noted that recursive functions themselves have a stack implementation and hence the same approach generalizes to the one presented.

Example

We present a small illustrative example to show the working of this algorithm. Consider a small graph as shown in Figure 12. The source node and the goal node are marked in the figure. Initially the source is kept in the open list making the open list as Open = {a}. The closed list is empty or Closed = {}. Now in the first iteration, this vertex is ex-

Algorithm 4. DFS algorithm

```
DFS (G<V,E>, source, goal)
source -> parent ← null
closed ←φ
open ← {source}
time ← 0
while open is not empty
  time ← time+1
  x -> time ←  time
  x ← Select 1st vertex from open
  if x=goal return x
  δ ← Expand(x)
  for all y in δ
          if y ∉ open ∪ closed
                  y -> time ← time
                  y -> parent  ← x
                  add y to top of open list
          end
  end
  closed ← closed ∪ {x}
  open ← open - {x}
end
return null
```

tracted from the open list. The vertex is added to the closed list, and all the adjoining vertices are added to the open list. This makes Open = {b, c, d} and Closed = {a}. The next vertex to be processed is b, which makes the lists Open = {e, c, d} and Closed = {a, b}. At the next iteration, vertex e is processed and still later vertex c is processed making the lists Open = {f, d} and Closed = {a, b, e, c}. At next iteration, f is processed which is the goal and hence the algorithm stops.

In this case, it may be seen that the algorithm explored all the depths of the processed vertices, before going through the breadths.

A* ALGORITHM

The next algorithm we study is A* algorithm. This presents a shift from the conventional search algorithms to the intelligent search algorithms where we can use some pre-defined knowledge of the problem for better search. This may result in earlier discovery of the goal vertex and thus the lower search time of the complete operation. We present the concept and working associated with the algorithm in the following sub-sections. For more details refer to Konar (2000), Russell and Norvig (2003), and Shukla et al. (2010).

Concept

The algorithms studied so far, BFS and DFS, tried to search the goal from the source by their own assumptions about the problem. It is clear that these search algorithms may guide the search near the goal or far from the goal from time to time. Many times, we may have an idea of the *direction of search*, or we may be able to judge the goodness of a vertex in terms of its possibility of lying near the goal. In such a case, it is sometimes wise to use this understanding of the problem to direct the search algorithm towards the goal, as far as possible. This may result in the search algorithm to proceed, as much as possible, towards the goal, which may further result in an early search of the goal.

We define heuristics to be the goodness of a vertex that may lead to early discovery of the goal. This heuristic is not computed by the algorithm, but is rather a pre-conceived goodness defined for every node by a function called as the heuristic function. The heuristic function is problem specific. It takes any vertex or state and tries to guess its goodness, which is returned as the heuristic value. The goodness may generally be taken as the distance of the vertex or the state from the goal state. The heuristic function is just a guess depending upon the understanding of

Figure 12. Graph search by depth first search

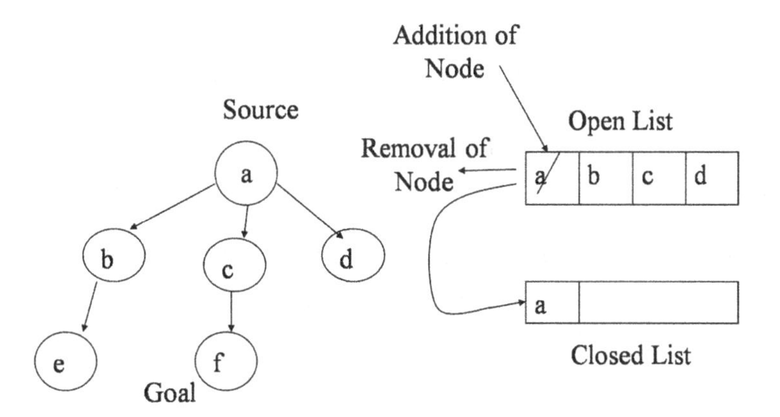

the problem by the algorithm designer. Hence, it may be accurate, over-expect, or under-expect the actual value. In any case, it might be wise to consider this value while making decisions regarding the order in which the vertices are processed from the open list. The general guideline should be that the vertices that are expected to lie closer to the goal to have higher priority in processing.

The paradigm of heuristics makes the search an intelligent search where the intelligence is in the mechanism in which we can model the heuristic function, knowing the problem requirements as well as the graph modeling of the problem. While for most of the problems a trivial selection of the heuristic function should solve the problem, for most of the other problems a clever selection of the heuristic function might be needed. Heuristics decides the strategy of working of the algorithm and hence needs to be carefully chosen. The change in heuristic value may drastically change the performance of the algorithm.

In the problem of robot motion planning the heuristic function may be simply chosen as the distance of the vertex from the goal. Hence, the heuristic cost of any state or vertex at location $V(x_v,y_v)$ with the goal $G(x_g, y_g)$ may be given by Equation 9.

$$h(V(x_v,y_v)) = \left\| V(x_v,y_v) - G(x_g,y_g) \right\| = \sqrt{(x_v - x_g)^2 + (y_v - y_g)^2} \quad (9)$$

This selection of the heuristic means that we are trying to minimize the total distance from the source to the goal, and that the goal is likely to lie at places with a lower heuristic cost. Hence, at any time the algorithm may prefer processing vertices that are likely to be near the goal out of all the available vertices in the open list. Please note that whenever we use the term cost, it would be implied that a lower value is better.

Working

The basic intention of the algorithm is to process vertices from the open list, with regard to their heuristic value. Hence, for implementation of the algorithm, we maintain the open list as a priority queue. This queue has a priority associated with every element in it. The elements are processed strictly in the order of their priorities. The basic motive of the A* algorithm is to try to figure out a path with smallest distance from the source to the goal. Hence, we take two costs associated with every vertex. The first cost is the historical cost g(V) that measures the distance of a vertex from the source. Since the path until the vertex being explored is already figured out, this cost is perfectly known. The other cost is known as the heuristic cost h(V) that is a guess of the distance of the vertex from the goal. The priority (lower values preferable) of any vertex, also its total cost f(V) is taken as a sum of these two costs given in Equation 10:

$$f(V) = g(V) + h(V) \quad (10)$$

The role of the heuristic is to assign a guess about the goodness of a vertex to the historical cost. The total cost may be taken as the historic cost biased by the heuristics. The Equation 10 is sometimes also taken as Equation 11:

$$g(V) = \alpha\, f(V) + \beta\, h(V) \quad (11)$$

Here, α is called as the historical contribution factor to the total cost and β is called as the heuristic contribution factor to the total cost. If α is taken to be 0 ($\beta = 1$), the algorithm is entirely driven by the heuristics. In such a case, it tries to go as near as possible to the goal, without considering the historical factor. In other case with β=0 (α=1), the algorithm is entirely driven by the historical factor. In such a case, it tries to process vertices close to the node and the resultant algorithm is equivalent

Algorithm 5. A algorithm*

```
A*(G<V,E>, source, goal)
source -> parent ← null
closed ← ϕ
g(source) ← 0
f(source) ← g(source) + h(source)
open ← {source} with cost f(source)
while open is not empty
        x ← Select vertex from open with lowest cost
        if x=goal return x
        δ ← Expand(x)
        for all y in δ
                g(y) ← g(x) + w_xy
                f(y) ← g(y) + h(y)
                if (y ∉ open∪closed)
                        y -> parent  ← x
                        add y to open list with cost f(y)
              else if y ∈ open∪closed and y is better in terms of cost f(y)
              update list to make vertex y (and all its decedents if in closed list) point to x
                as parent with newly calculated cost f(y)
                end
        closed ← closed∪{x}
        open ← open - {x}
end
return null
```

Figure 13. Graph search by A algorithm*

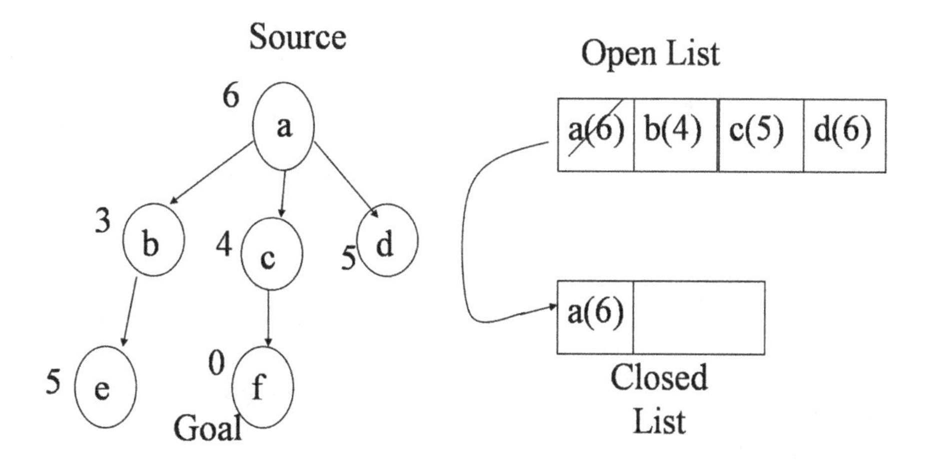

to BFS for weighted graphs, in same case if all weights are positive the algorithm behavior would be similar to Dijkstra's Algorithm (Cormen, et al., 2001) and Uniform Cost Search Algorithm (Konar, 2002). Here the heuristics are not at all considered for decision-making. It may be noted that since cost is a relative measure, once one of α or β is zero, the value of the other is irrelevant.

The algorithm works as per the same working fundamentals of the general search we presented. We take vertices from the open list, expand them, and put them in the closed list. The processing is strictly done on the basis of costs, and the vertex with the least cost in the open list is selected. However, this algorithm is based on heuristics, and it is possible that at a later stage we may find a route to a vertex that is better than the previously known path. Hence, when we re-discover a vertex, it is possible that the re-discovery may be through a shorter path. Hence, we may update the costs and the paths (by changing the parent attribute) to point to the newly discovered path. There is, however, no assurance that the re-discovered path is best and no other smaller path exists to link the vertex with the source. The complete algorithm may hence be given by Algorithm 5.

In Algorithm 5, we have assumed that the heuristic function h(V) is already known for every vertex V. We further assume that the historic function g(V) increases by a factor of weight of the edge w_{xV} while traveling from vertex x to vertex V. In our considered problem of robot motion planning, this would correspond to the physical distance between the two points given in Equation 12.

$$g(y) = g(x) + \|y - x\| \qquad (12)$$

where ||.|| is the norm, which may be taken as the Euclidean norm.

An important issue associated with the A* algorithm is of optimality. Nodes are operated strictly in the order of their total cost values given by Equation 10. The algorithm is guaranteed to give optimal results if the heuristic function is admissible. A heuristic function is called as admissible if for every node it returns a value which is less than or equal to the actual shortest cost (or the actual shortest distance to goal). This means that a heuristic function may overestimate, but it may never underestimate. In the heuristic function of Equation 9, we know that shortest distance from a point to goal can never be less than the Euclidian distance to goal. The robot may not be able to travel in straight line from the point to goal because of presence of obstacles or the path may not be traceable due to limited allowable robot moves. Hence, we can say that the path returned using this heuristic function would be optimal. In general cases, unless admissibility of heuristic function can be determined, A* algorithm is regarded non-optimal.

Example

We further our discussion of the algorithm by taking a small example. Consider a graph as shown in Figure 13 to be searched using A* algorithm. All weights are supposed to be of unity. The heuristic costs for the various vertices are given in figure. Initially, the source vertex is pushed in the open list with historic cost of 0 and heuristic cost of 6 making the total cost as 6. The two lists are hence Open = {a(10)} and Closed={}. The number in parenthesis denotes total cost. The vertex a is processed and added in the closed list. The expanded vertices all have a historic cost of 1 and heuristic cost as indicated in figure. The lists hence become Open = {b(4), c(5), d(6)} and Closed = {a(6)}. The list is always sorted according to total cost and the first element is processed, which in this case is b with a cost of 4. The expansion produces vertex e with historic cost 2 and heuristic cost 5 making total cost as 7. The lists hence become Open = {c(5), d(6), e(7)} and Closed = {a(6), b(4)}. In the same manner, the lists are worked until goal is found.

Algorithm 6. MNHS algorithm

```
MNHS(G<V,E>, source, goal)
source -> parent ← null
closed ← ϕ
g(source) ← 0
f(source) ← g(source) + h(source)
open ← {source} with cost f(source)
while open is not empty
        X ← Select α vertices
        for each x in X
if x=goal return x
                δ ← Expand(x)
                for all y in δ
                        g(y) ← g(x) + w_xy
                        f(y) ← g(y) + h(y)
                        if (y ∉ open∪closed)
                                y -> parent  ← x
                                add y to open list with cost f(y)
              else if y ∈ open∪closed and y is better in terms of cost f(y)
              update list to make vertex y (and all its decedents if in closed list) point to x
              as parent with newly calculated cost f(y)
                        end
                end
                closed ← closed∪{x}
                open ← open - {x}
end
end
return null
```

MULTI-NEURON HEURISTIC SEARCH

The other algorithm we study is Multi-Neuron Heuristic Search (MNHS) (Shukla & Kala, 2008). This algorithm is like the A* algorithm with the change that in place of extracting a single vertex from the open list, and processing it, we extract and process multiple vertices. The basic problem with the A* algorithm is that it does not consider the scenarios where heuristic decision may be poor. For most of the cases we encounter in general problems, heuristics gives a fair idea of the goodness of a vertex. However, in many cases it may drive the algorithm in false directions. In such a case, the algorithm would have to uncover a fairly large number of vertices, only to find that the goal cannot be reached. It would then back track and start the expansion of the vertices it did not do earlier because of the heuristic values. Hence, the A* algorithm may not give a good performance in case uncertainties associated with the heuristic values is high. In such a case, one may prefer not to use the heuristic values at all. However, this may further result in poor performance. It may be wise to take use of the heuristic values to some extent, even if they are uncertain.

Consider the example of the use of A* algorithm for solving a maze. In this case, let every block marked as a vertex and an edge connect the vertices where the corresponding blocks are reachable by a unit distance. Let us take the Manhattan distance between the vertex and the goal as the heuristic function. In this case, if we go by the conventional A* approach, we may reach very near to the goal, only to find there is no way that we can reach the goal. In such a case, we need to backtrack the entire path traced and return to a fairly far away point, which might make the algorithm reach the goal. In this case, a lot of time is wasted because of a poor decision regarding the expansion of vertex that was dominated by the heuristic value.

Concept

The MNHS algorithm advocates the expansion of a number of vertices that range in their costs from low to large. At every iteration, we select α vertices which are produced. For selecting these vertices, we divide the entire sorted open list into α regions, as per their cost. Each region hence has a set of vertices with varying cost. From every region, we select the vertex with the smallest cost and use it for further processing. In this manner, we make selection of a total of α vertices. All these are processed as per the conventional working methodology of the A* algorithm. The resultant algorithm is hence given by Algorithm 6.

The algorithm is especially relevant from the point of view of robot motion planning. Here, it is possible that following a path may take us to an end in a maze like situation. It is hence necessary to have some backup path expanded to a reasonable degree. In case the map path fails due to blockage, we always have multiple backup paths ready and already expanded to some extent. There would be wastage of time in case the main path does lead us to the goal. However, this time is an assurance that the blockage of this path would have a backup path ready to some extent. The algorithm may now just further work over it. In case the backup path also fails, we would have another backup path ready. This concept is shown in Figure 14. For related works in robot path planning using MNHS refer to Kala et al. (2009a, 2009b, 2011).

Example

In this case, we take a simple graph and study the manner of working of the algorithm. Consider the graph shown in Figure 15. Similar to the A* algorithm the open list is initially Open = {a(6)} and closed list is Closed = {}. We assume α as 2. Since initially we have only 1 vertex in open list, we process it, producing open list as Open = {b(4), c(5), d(6)} and Closed = {a(6)}. Now in the next iteration, we divide the sorted open list into 2 parts which are Open1={b(4), c(5)} and Open2={a(6)}. The first element of each of these lists is processed one after the other. The open list hence becomes Open={c(5), e(7)} and Closed = {a(6), b(4), d(6)}. In the next step, the sorted open list is again divided into two parts, and the first elements, which are c and e, are expanded. The process follows in same way until the goal is processed.

Figure 14. Concept behind multiple paths in MNHS (obstacles not shown in the figure)

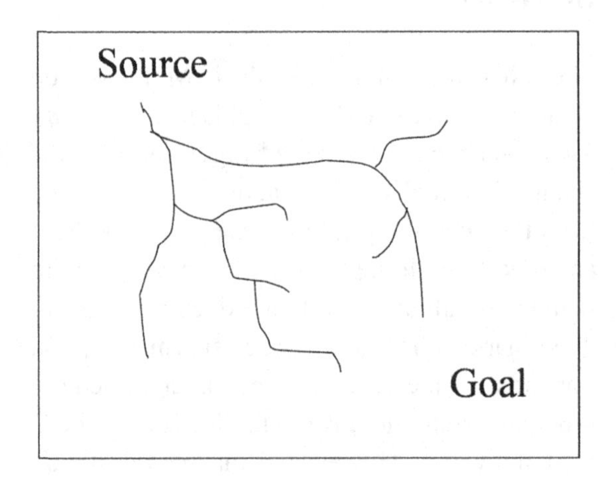

Figure 15. Graph search by multi-neuron heuristic search

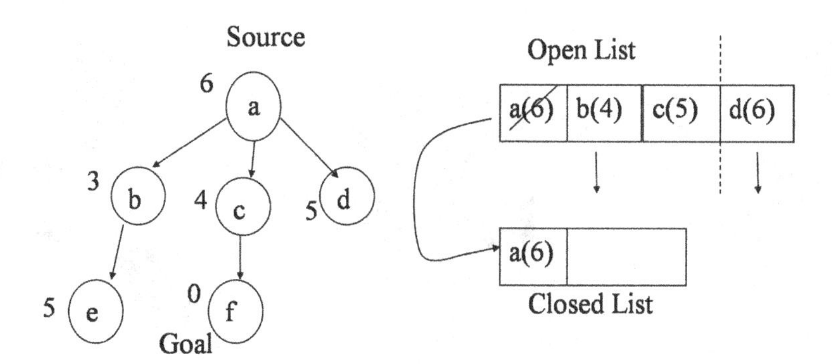

D* ALGORITHM

The next algorithm we study is the D* algorithm (Stenz, 1994). This algorithm is similar to the A* algorithm, with the major difference that it works for the dynamic maps and is hence known as dynamic A* or D*. In this algorithm, it is assumed that the cost of traversal between the various vertices, or the weights of the edges may change along with time. The entire algorithm is designed to make changes in the computed costs as these changes are monitored and updated to the algorithm.

The entire algorithm has two major phases. The first phase is the modify cost phase. This phase deals with monitoring of the graph and reflecting any change necessary to the costs involved in the graph representation. The other phase is the process state phase. In this phase, the algorithm processes the open list to continue the search process. Both these phases take place one after the other, until the final path is computed. The algorithm follows the similar modeling scenario like the A* algorithm where the open and closed lists are prepared to track the various vertices. In this algorithm, however it is possible that a vertex is removed from the open list, and later it is found that the cost affecting this vertex has changed. This means re-computation of the parent and cost of this vertex. Hence, this vertex may be again placed on the open list for further processing as per the changed modeling scenario. The updates done to this vertex would be as per the changes made to the graph in the modify cost phase and communicated to the other vertices, which later propagate these changes and accordingly all the vertices are modified.

The modify cost phase seeks to monitor the change in cost between any vertex X to any vertex Y. In case there is a change in the costs involved, the same must be reflected in the working plan of the panning algorithm. In case of a change the algorithm inserts the vertex X into the open list for re-computation of its costs, in case it is not already in the open list or has not yet been discovered. The insertion of the node in the open list gives an opportunity to the algorithm to carry the computations as per the changed costs. The changed may be an increase or decrease in the total cost. The increase is marked by a raise event and the decrease is marked by a lower event.

The process state phase works over this re-computation of the costs. The re-computation is simply initiated by the insertion of vertices in the open list. The process state phase looks into the minimum cost element and checks its change of cost to capture the raise or lowered event. Accordingly, the processing is done with the vertex. All the neighboring vertices as well may be modified to work in the similar manner as per the new costs, and inserted into the open list if needed. Insertion would make the costs of these vertices updated

Figure 16. Experimental results

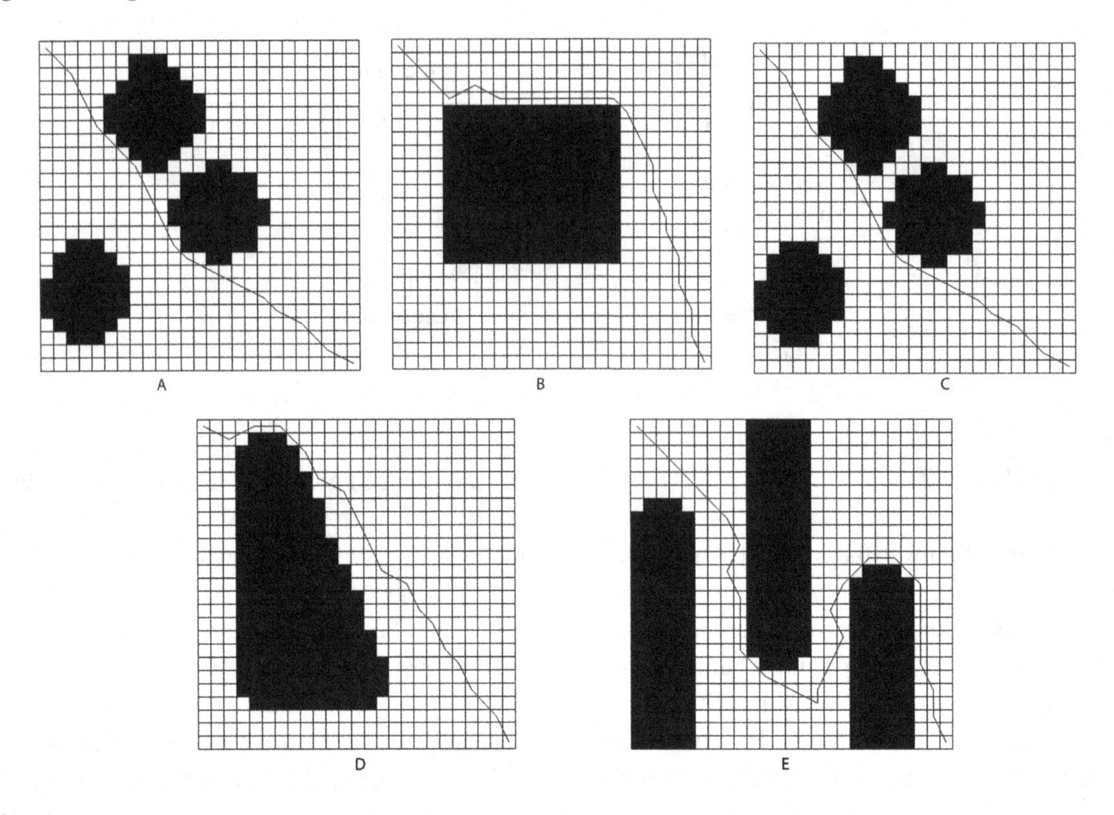

and in turn have their children vertices be placed in open list. In this manner, the costs are updated in all the vertices tracked by the planning algorithm. The update may however demand placing of vertices from the closed list to the open list. For more details, please refer (Ferguson & Stenz, 2005; Guo, et al., 2009; Hu & Yang, 2004; Koenig & Likhachev, 2002; Stenz, 1995).

Figure 17. Path traced by robot: (a) 4 allowable moves, (b) 8 allowable moves, (c) 16 allowable moves

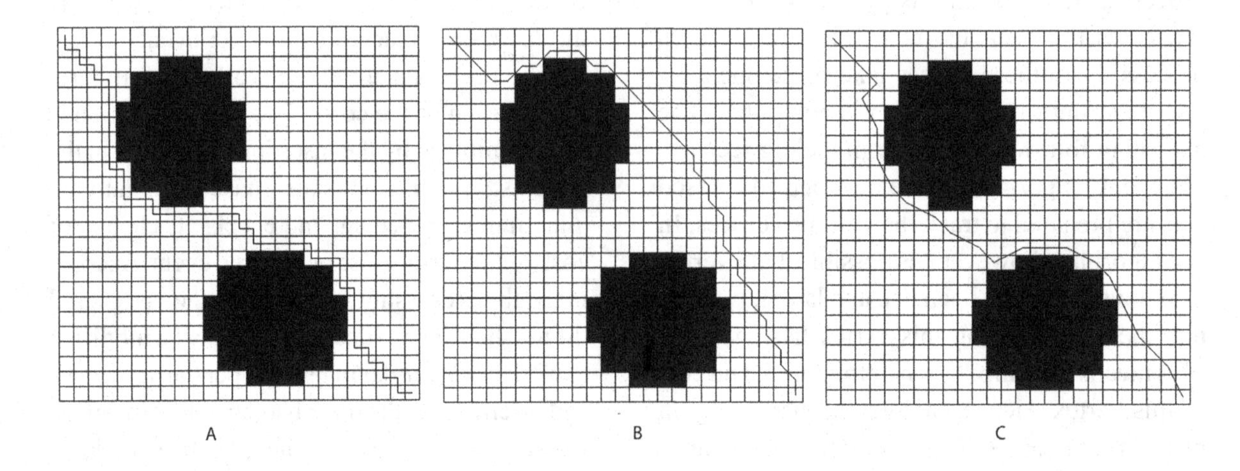

EXPERIMENTAL RESULTS

The graph algorithms presented so far used a similar modeling scenario and the graph representation. We experiment a number of maps and display their experimental results. In all these cases, the simulation was done on a JAVA platform and the map was input to the platform as a bmp file. This map was processed by the various algorithms. The final output computed by these algorithms was plotted as an image showing the map as well as the computed path. In all the cases, the resolution of the map was taken to be 100 x 100 pixels. The results to some of the input maps are given in Figure 16.

We further try to understand the relation between the path optimality and the number of allowable moves of the robot. We earlier stated that increase in the number of allowable moves results in the paths being more flexible and hence the optimality of the path should increase. The corollary was that the increase in number of allowable moves may increase the computational time. The paths traced by the robot for a sample map for various numbers of allowable moves are given in Figure 17.

CONCLUSION

In this chapter, we studied the graph-based motion planning of mobile robots. The basic intention was to understand the concept of graphs and the working of the graph algorithms, and to later use the same for solving the problem of motion planning. The planning problem was modeled as a graph, which included the modeling of the vertices and the edges. This forms a standard input to the graph search algorithms. A number of graph search algorithms were studied. This included Breadth First Search, which tried to explore the breadths of a graph, before touching its depths. The other algorithm included the Depth First Search, where the algorithm tried to go deep inside the depth. From the heuristic domain the algorithms studied included A* algorithm, heuristic search and the Multi-Neuron Heuristic Search. These algorithms were guided by heuristics, which tried to guess the goodness of a vertex. The last algorithm studied was the D* algorithm that modeled the A* algorithm for dynamic environments. Results to various search algorithms were also presented.

REFERENCES

Coppin, B. (2004). *Artificial intelligence illuminated.* Boston, MA: Jones and Barlett Publishers.

Cormen, T. H., Leiserson, C. E., Rivest, R. L., & Stein, C. (2001). *Introduction to algorithms* (2nd ed.). Cambridge, MA: MIT Press.

Ferguson, D., & Stentz, A. (2005). The delayed D* algorithm for efficient path replanning. In *Proceedings of the IEEE International Conference on Robotics and Automation*, (pp. 2045-2050). IEEE Press.

Fernandez, J. A., & Gonzalez, J. (1998). Hierarchical graph search for mobile robot path planning. In *Proceedings of the IEEE International Conference on Robotics and Automation,* (vol. 1), (pp. 656-661). IEEE Press.

Guo, J., Liu, L., Liu, Q., & Qu, Y. (2009). An improvement of D* algorithm for mobile robot path planning in partial unknown environment. In *Proceedings of the IEEE Conference on Intelligent Computation Technology and Automation,* (vol. 3), (pp. 394-397). IEEE Press.

Horowitz, E., & Sahni, S. (1978). *Fundamentals of computer algorithms*. Baltimore, MD: Computer Science Press.

Hu, Y., & Yang, S. X. (2004). A knowledge based genetic algorithm for path planning of a mobile robot. In *Proceedings 2004 IEEE International Conference on Robotics and Automation,* (vol. 5), (pp. 4350-4355). IEEE Press.

Hui-Ying, D., Shuo, D., & Yu, Z. (2010). Delaunay graph based path planning method for mobile robot. In *Proceedings of the IEEE Conference on Communications and Mobile Computing*, (vol. 3), (pp. 528-531). IEEE Press.

IBM. (2011). *Deep blue*. Retrieved on December, 2011, from http://www.research.ibm.com/deepblue/meet/html/d.3.html

Kala, R., Shukla, A., & Tiwari, R. (2009a). Fusion of evolutionary algorithms and multi-neuron heuristic search for robotic path planning. In *Proceedings of the 2009 IEEE World Congress on Nature & Biologically Inspired Computing*, (pp. 684-689). Coimbatote, India: IEEE Press.

Kala, R., Shukla, A., & Tiwari, R. (2009b). Robotic path planning using multi neuron heuristic search. In *Proceedings of the ACM 2nd International Conference on Interaction Sciences: Information Technology, Culture and Human*, (pp. 1318-1323). Seoul, South Korea: ACM Press.

Kala, R., Shukla, A., & Tiwari, R. (2011). Robotic path planning in static environment using hierarchical multi-neuron heuristic search and probability based fitness. *Neurocomputing*, *74*(14-15), 2314–2335. doi:10.1016/j.neucom.2011.03.006

Kala, R., Shukla, A., Tiwari, R., Roongta, S., & Janghel, R. R. (2009). Mobile robot navigation control in moving obstacle environment using genetic algorithm: Artificial neural networks and A* algorithm. In *Proceedings of the IEEE World Congress on Computer Science and Information Engineering*, (pp. 705-713). Los Angeles, CA: IEEE Press.

Koenig, S., & Likhachev, M. (2002). Improved fast replanning for robot navigation in unknown terrain. In *Proceedings of the IEEE International Conference on Robotics and Automation*, (pp. 968 – 975). IEEE Press.

Konar, A. (2000). *Artificial intelligence and soft computing: Behavioral and cognitive modeling of the human*. Boca Raton, FL: CRC Press.

Liu, Y., & Arimoto, S. (1990). A flexible algorithm for planning local shortest path of mobile robots based on reachability graph. In *Proceedings of the IEEE Conference on Intelligent Robots and Systems*, (vol. 2), (pp. 749-756). IEEE Press.

Lozano-Perez, T. (1987). A simple motion-planning algorithm for general robot manipulators. *IEEE Journal on Robotics and Automation*, *3*(3), 224–238. doi:10.1109/JRA.1987.1087095

Russell, S. J., & Norvig, P. (2003). *Artificial intelligence: A modern approach*. Upper Saddle River, NJ: Prentice Hall.

Shukla, A., & Kala, R. (2008). Multi neuron heuristic search. *International Journal of Computer Science and Network Security*, *8*(6), 344–350.

Shukla, A., Tiwari, R., & Kala, R. (2008). Mobile robot navigation control in moving obstacle environment using A* algorithm. In *Proceedings of the International Conference on Artificial Neural Networks in Engineering*, (pp. 113-120). St. Louis, MO: IEEE.

Shukla, A., Tiwari, R., & Kala, R. (2010). *Real life applications of soft computing*. Boca Raton, FL: CRC Press. doi:10.1201/EBK1439822876

Stentz, A. (1994). Optimal and efficient path planning for partially-known environments. In *Proceedings of the International Conference on Robotics and Automation*, (pp. 3310–3317). IEEE.

Stentz, A. (1995). The focussed D* algorithm for real-time replanning. In *Proceedings of the International Joint Conference on Artificial Intelligence*, (pp. 1652–1659). IEEE. Thornton, C., & du Boulay, B. (2005). *Artificial intelligence: Strategies, applications, and models through search*. New Delhi, India: New Age Publishers.

Tutte, W. T. (2001). *Graph theory*. Cambridge, UK: Cambridge University Press.

KEY TERMS AND DEFINITIONS

Computation Time: The total time taken by the algorithm to either return a path (if one exists) or to terminate safely. If this time is large, the algorithm may only work in stationary environments. If this time is small, the algorithm may work in dynamic environments as long as the map does not change appreciably while computations are made.

Graph Search Algorithms: The process of finding out the sequence of vertices, or the sequence of edges, starting from a predefined vertex called source, which ends at a predefined vertex called goal. Some algorithms may additionally ensure that out of all possible sequences or paths, the shortest one is retuned.

Graph Search Strategy: The ordering policy or the order in which the graph search algorithm processes the various nodes available (in the open list). The computation time, chance of finding goal early, path length, etc. depend upon the search strategy.

Heuristics: An indicative measure or a guess regarding the goodness of a node of a graph in the graph search process. For motion planning problem heuristic denotes a guess regarding the distance of the node from the goal.

Path Optimality: The objective being attempted to minimize in the graph search operation. Path length and time to reach goal are commonly used measures. In case the robot travels at constant speed, both measures give the same path.

Robotic Graph: Conversion of the available robotic map into a graph format. All points in the map are taken as vertices. From every point or vertex, we define a finite set of allowable moves. An edge exists between any two vertices if motion from one vertex to other vertex may be carried using these allowable moves, and the entire motion is collision free.

Chapter 3
Common Planning Techniques

ABSTRACT

Motion planning of mobile robots can be done by a variety of mechanisms, which makes it a very interesting field of work. In this chapter, the authors discuss some of the commonly used techniques that resemble or closely follow the graph-based search techniques discussed in chapter 2. They take a detailed study of the use of algorithms, like dynamic programming, Bellman Ford algorithm, Rapidly exploring Random Trees (RRT), artificial potential fields, embedded sensor planning, reinforcement learning planning, and Voronoi graph-based planning. Bellman Ford is a graph search algorithm that finds the shortest path between the source and all the vertices in graph. RRTs are tree-like data structures where every node is a free position in the map. The trees grow to explore areas and thereby find the path from source to goal. The artificial potential field method uses the concept of potential by every obstacle and goal for motion planning of the robot. The planning in embedded sensor networks is distributed in the sensors distributed in the entire map. The discussions enable a multi-dimensional outlook over the problem. The authors note the similarities, differences, advantages, and the disadvantages between the different algorithms.

INTRODUCTION

Motion planning of mobile robots is a very interesting problem, which has a large number of issues associated with it. The choice between online and offline planner, degree of optimality of the path, complexity of the map, etc. play a major role in choosing the search algorithm or designing of a new algorithm. In many algorithms, the internal algorithm parameters may need to be tuned for better performance. The computational time and optimality of a path for most general graphs with no roadmap information may generally be contradictory to each other. The algorithms giving

DOI: 10.4018/978-1-4666-2074-2.ch003

early results may not guarantee the optimality of the paths generated and vice versa.

Graph search methods are extensively used for motion planning, as we discussed in chapter 2. Here the task is to convert the given problem into a graph, and then to use any standard graph search technique for solving the problem. There are, however, many approaches that use similar modeling scenarios or a similar working methodology. The purpose of this chapter is to present these approaches and possibly analyze them for performance in various scenarios.

Motion planning deals with computation of a feasible path from a given source to a given goal. The robot must avoid all the possible static and dynamic obstacles on its way. We saw in chapter 2 that the graph modeling more or less assumed that each pixel be a vertex of the map. This enabled the conversion of a simple maze into a search graph. In general, we may have different nodes kept at different places of the map that help in the computation of the path from source to goal. This node based planning enables us to use the information processed by the various nodes, to get the optimal path. Here each node may denote a pixel, or a still higher area in case the complete map is of a high resolution.

The graph-based planning is a complete algorithm. Completeness of algorithm refers to the fact that a solution must be returned by the algorithm if it exists. The computational time for many cases may however be high. For these reasons graph based approaches are widely used for higher level planning where the complete problem at a very abstract level may be given to the algorithm to figure out a vague work plan. The other real-time planners may be used for making the exact plan to be physically traversed by the robot.

The choice of algorithm between online and offline planners is always hard. The real world with all its uncertainties advocates the use of online planners. This ensures that the robot would be able to escape any possible obstacle that might suddenly crop in. Dynamism is again a very common feature in any practical planning scenario where the multiple obstacles in form of other robots, humans, machines, etc. may be moving around the map. This further necessitates the planners being online.

However, the online planners can only perform limited computation and mostly make decisions based on the current surroundings. This may not ensure the overall path optimality or completeness. In most cases, the robot may not at all reach the goal, or may reach the goal in sub-optimal paths. The offline planners with some heavy computation do attempt to try as many combinations of paths as possible to figure out the most efficient path. For most of the scenarios where the optimal path of the robot is not very simple, or there is a possibility of multiple paths, these planners may perform a lot better. However, the scenario might change in the presence of uncertainties. The robot may make some computations, plan a path, and move as per the generated plan, only to find that the scenario in the meantime has changed and the plan is no longer valid. In such scenarios, this planning fails.

Autonomous vehicles are cars, which are not driven by a human driver, but the vehicle drives itself. These vehicles are a lot safer as the vehicles do not fatigue, and further, the combined view of multiple sensors is less prone to errors as compared to humans. In 2007, DARPA organized the DARPA Urban Challenge competition, which was a race between autonomous vehicles. The vehicle named Talos from MIT stood third in the contest. The vehicle's success was largely due to the planning algorithm, which enabled the vehicle to avoid other vehicles and obstacles, at the same time staying within the road boundaries (Kuwata, 2009). The planning algorithm used by the team was Rapidly Exploring Random Trees, which is one of the topics of this chapter.

In the next sections of the chapter, we take in numerous algorithms for the motion planning of the robots. Each of these algorithms has its own modeling scenario that is helpful in some manner or the other for the problem of motion planning. We broadly study four major algorithms. The first is a graph based search approach, which finds the smallest distance between a source vertex and all the other vertices in the graph. This is the bellman ford algorithm. The next approach of discussion is the rapidly exploring random trees. Here the focus is to analyze how tree-like structure can be grown over the map of work and used for the purpose of motion planning. The next approach we deal in this chapter is the embedded sensor planning. In this approach, the entire map of the robot is embedded with sensors that carry forward the process of planning. The last approach is the artificial potential field. Here every obstacle and goal applies some positive or negative potential. The robot moves according to these potential forces. Further, a number of other methods for planning are discussed. All the different methods are discussed as individuals sections of the chapter.

PLANNING USING BELLMAN FORD ALGORITHM

The first algorithm we study and use in this chapter is Bellman Ford algorithm. This algorithm uses the concepts of Dynamic Programming for solving the robotic map and finding out a solution. Hence, the same algorithm may be called motion planning using Dynamic Programming as well, without much difference in the problem solving methodology. The Bellman Ford is a single source all paths algorithm. Hence, it computes the minimal distance of all the vertices from the given source, in place of computing the minimum distance of only the goal. We study the various characteristics and details of the algorithm in the coming sections.

Dynamic Programming

Dynamic Programming (DP) (Cormen, et al., 2001; Sniedovich, 2011) is a commonly used computing tool that tries to solve problems using a divide and conquer approach. Here, we try to iteratively build up a solution to a bigger problem using the solutions of smaller problems. Once the solutions to the smaller problems are known, we further attempt to solve still larger problems. The solution builds up in a bottom up manner. The most trivial or the simplest problem is assumed to be already known. The entire setup of solving a problem in this manner is done using a recurrence relation. The recurrence relation states the manner in which a single or multiple smaller problems may be used to solve a bigger problem. The algorithm may be solved in the same manner until the solution to the biggest problem, that we intended to solve, is available.

The concept is similar to the use of recursive function calls to solve the problem of divide and conquer. In this approach while trying to solve a problem of higher order, we assume that the problems of the lower order are already solved. The task is hence just to divide the problem and use recursive function calls for each sub-problem. In the next phase or the conquer phase the solutions to each of these sub-problems needs to be integrated as per the problem logic to return a solution to the major problem. While solving the problem in this mechanism, we assume that the solution to the smallest problem is already known and, therefore, can be found out without any division involved. This ensures that the algorithm does not indefinitely divide the problems. The general architecture of recursive solutions may hence be given by Algorithm 1.

The problem with the recursive calls is that it may be possible that a sub-problem is solved multiple times. It may be hence useful to store the solutions in a memory, so that the next time the algorithm needs to just look at the solution from the memory, in place of searching it in the complete problem domain. The dynamic programming technique is motivated form the same concept. Here, we intend to store the results of the sub-problems in an organized memory structure,

Algorithm 1. Divide and conquer algorithm

```
Function(problem)
if problem is smallest problem
return solution
else
        sub-problems ← Divide(problem)
        for every sub-problem s in sub-problem
                sol(s) ← Function(s)
        solution ← Conquer(sol)
        return solution
end
```

so that these results may be directly looked up whenever needed to solve any other problem. For the same reasons we try to build up the solution in a bottom up manner. In such a case, we may easily work over any part of the problem, since the results to all the smaller sub-problems are already known.

The recurrence relation states the manner in which the smaller problems are related to the larger problem. A recurrence relation queries one or more of the smaller problems and forms the solution to the larger problem. The solutions to the smaller problems are available as a direct memory lookup. The solution to the larger problem is in turn stored into the memory. For related works in motion planning please refer to Jihong (1995), LaValle et al. (1997), and Shin and McKay (1986).

Bellman Ford Algorithm

This is a graph search algorithm that uses the concepts of Dynamic Programming (DP) to compute the shortest path from a given source to all the other vertices of the graph. It may be noted that this algorithm for most of the maps with decent resolution is very time consuming as compared to the heuristic algorithms like A* (Konar, 2002; Thornton & du Boulay, 2005). However, this algorithm ensures that it returns the smallest path from the source to the destination. This was not the case with the A* algorithm with general (non-admissible) heuristics. Hence, this algorithm may be better when the graph has a very low number of pixels, or that the entire map has been reduced to some small number of vertices for a coarser planning. In present modeling of the motion planning problem, since we have a good admissible heuristic formulated, there seems to be no betterment of this approach for single source single goal problems.

The Bellman Ford algorithm (Cormen, 2001; Horowitz & Sahni, 1978) also works over a graph which has vertices and edges. The entire graph modeling of the problem may be taken as it is from the previous chapter. The purpose is to compute the shortest path from the source to any other vertex. Hence, let us define a function f(V, k) that returns the shortest distance of a vertex V from the source after k iterations, where k is the value of the iterator of the dynamic programming. Fundamentally, it means that the values stored in f are correct to a maximum of k edges. Any vertex whose smallest path lies within an interval of k edges from the source is correctly stored in the function f. If the smallest path lies at an interval of greater than k edges, it may not be correctly stored. It is natural that any vertex must have its smallest path from the source within an interval of N-1 edges. Here N is the total number of vertices in the graph. In this case, we need to travel through all the vertices to reach a vertex V by the smallest path. This case is possible when all the vertices are arranged in a linear manner. Hence, f(V, N) gives us the correct value of distance of all vertices from the source.

The task is to formulate the recurrence relation f(V, k) assuming the solutions to all the smaller problems are known, that is f(V, t) for all t<k. The minimum distance of any vertex V is always equal to the minimum distance of the neighboring vertices plus the cost to come to this vertex from the neighboring vertices. This assumes that the minimum distances of the neighboring vertices are precisely known. As the value of the iterator k increase, this would eventually happen, when the nearby edges would start taking their smallest values from the source. We further need to check that the smallest part of the problem is known. In this case, we do this by setting a minimum value of 0 to the source, and infinity for all other vertices. These vertices would obtain their minimum values as the algorithm proceeds. The recurrence relation may be given by Equation 1.

$$f(V,k) = \begin{cases} \min\{f(V,k-1), f(X,k-1) + w_{vx}\} & k \neq 0 \\ 0 & k = 0, V = S \\ \infty & k = 0, V \neq S \end{cases} \quad (1)$$

Here, X is a collection of vertices connected to V. w_{vx} is the weight of the edge connecting V to X.

Along with the recurrence relation, we may be further interested in how the path was built from the smaller problems to the larger problems. This is done by storing the information during the algorithm iterations. For implementation we may simply take a memory structure to store all the computed values of f, such that all queries to f(V, k) can be answered by a simple lookup.

The Bellman Ford algorithm uses the same recurrence relation to compute the smallest distance of any vertex from the source. The implementation may be simply to iterate the recurrence relation presented in Equation 1 a number of times from k = 1 to N. f(V,N-1) would then give us the needed values of the minimum path lengths. We may further track the manner in which the solution is built to get the overall path. The Bellman Ford algorithm that we write uses the same notion as Algorithm 2.

Algorithm 2. Bellman Ford algorithm

```
Bellman Ford(G, source)
f(S) ← 0
S->parent ← NULL
for k = 1 to N-1
   for every edge (x, v) ∈ E
          if f(v) > f(x) + w_xv
                 f(v) ← f(x) + w_xv
                 v->parent ← x
          end
   end
end
```

The parent attribute stores the information about how the smaller solutions make the larger solution. In other words, it tells which vertex contributes to enable the vertex v have smallest value from the source. The path may be traced in the same manner as we traced the path in the graph search approaches. This is given in Algorithm 3.

For related works in robot motion planning, please refer to Barbehenn and Hutchinson (1993, 1995), Goto et al. (2003), Liu and Arimoto (1990), and Lozano-Perez (1987).

Algorithm 3. Path retrieval algorithm

```
Path(goal)
P ←φ
I ← goal
While I ≠ null
       P ← I∪P
       I ← I -> parent
end
return P
```

Results

We execute the Bellman Ford algorithm for a number of maps. All the maps were imported into JAVA programming platform using a bmp file. The algorithm was coded as a separate module. The simulation results using a number of scenarios are given in Figure 1. Following the discussions of chapter 2, one of the scenarios is further analyzed for different number of allowable moves. The results are shown in Figure 2.

It may be clearly seen that the presented results are like the ones obtained from the heuristic approaches in chapter 1. However, the algorithm in this case took a lot longer time to give the results. If paths to multiple goals were needed, this algorithm may have performed better, since this algorithm gives as result all the generated paths to all the vertices. Here we nicely present a tradeoff between guarantee of path optimality and execution time. Dijkstra's algorithm as well as Bellman Ford algorithm takes reasonably longer than most heuristic algorithms. However, these algorithms do guarantee that the resultant path to be optimal in any type of situation. This is not the case with the heuristic set of algorithms. The heuristics does ensure a faster computation of path for most simple scenarios; however, these paths may not always be optimal. Our notion of heuristic here is of a general heuristic and not an admissible heuristic which is the case with path planning problem as discussed in chapter 2.

If some heuristic can be formulated which is admissible, the path can be guaranteed to be optimal. However, this demands that heuristic values should not overestimate. This means that

Figure 1. Experimental results for Bellman Ford algorithm

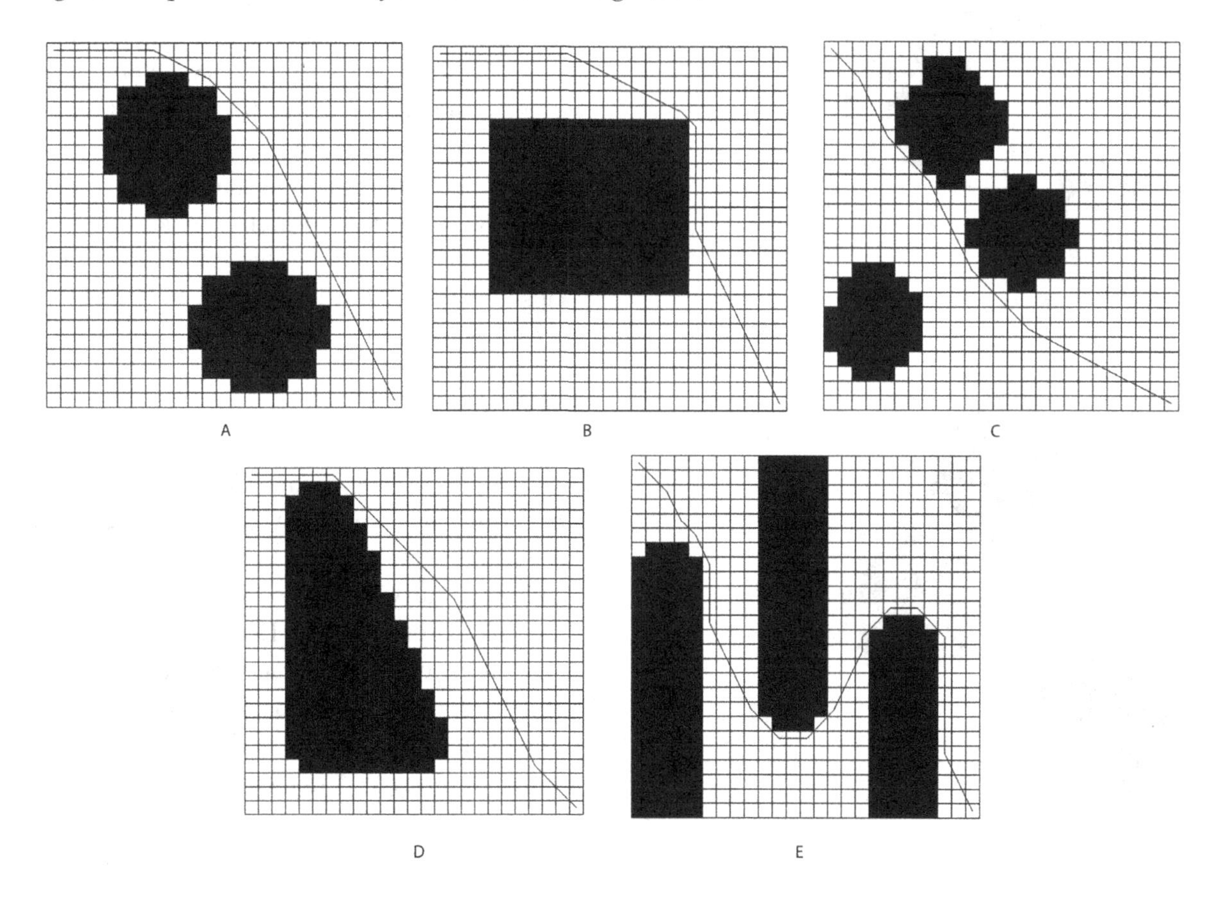

Figure 2. Path traced by robot: (a) 4 allowable moves, (b) 8 allowable moves, (c) 16 allowable moves

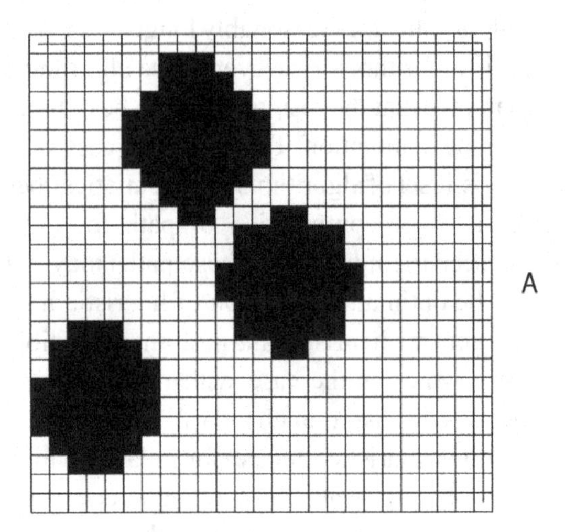

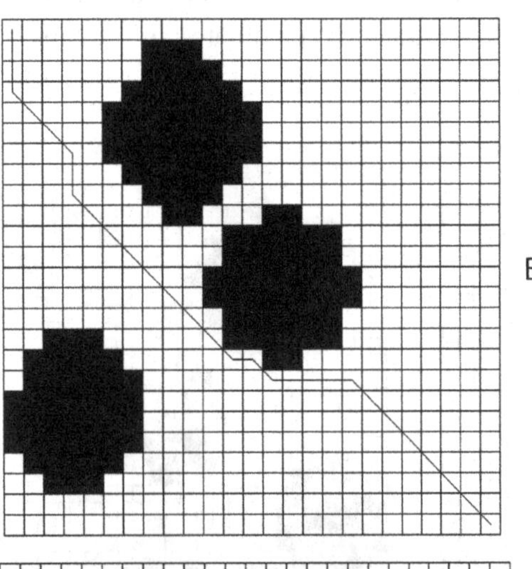

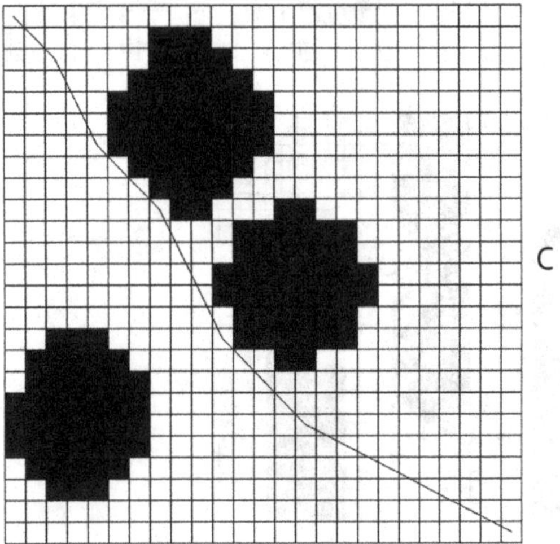

for every node, the heuristic cost or the estimate of distance to goal should be lesser than or equal to the actual smallest distance to goal. The lesser the magnitude of heuristic values, the more close is the behavior of algorithm to Dijkstra's and Bellman Ford algorithm and larger is the execution time. Using a heuristic function that generates larger magnitude values for nodes in the optimal path may not ensure that algorithm does not overestimate. If such an admissible heuristic function can be determined, which we saw in chapter 2 for robot motion planning problem, the path would be optimal. In all other cases, optimality is not guaranteed.

RAPIDLY EXPLORING RANDOM TREES

In this section, we study another exciting manner of dealing with the problem of motion planning. This is by using Rapidly Exploring Random Trees (RRTs) (Kuffner & LaValle, 1999). Here, we carry an iterative search for the goal starting from the source node, where the exploration is stored in the form of tree. A number of variants of the RRTs have been formulated, each differing to some degree regarding the exploration of the maps. The RRT algorithm along with some of its variants is given in the following sub-sections.

RRT

The RRT is another manner in which we may search for a goal from a given source. The RRTs try to explore the entire map in hunt of the goal. While they carry down the exploration process, they make a tree-structure. Every node of the tree is laid in the part of the map where the exploration was done by the algorithm. Every node is further connected to the nearest possible node in the tree by an edge. A random RRT may look as shown in Figure 3.

Figure 3. A random RRT

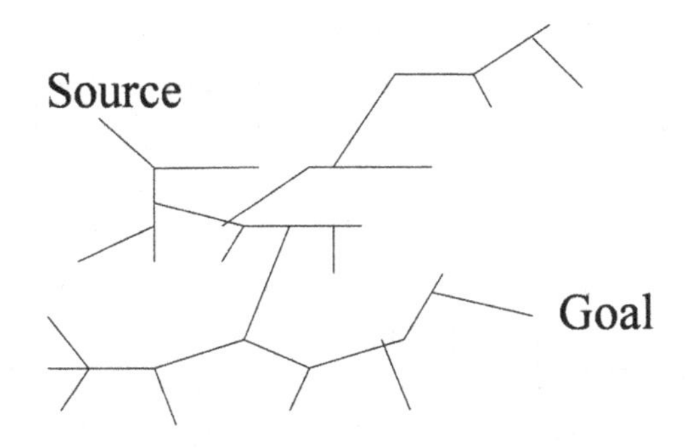

It may be clearly seen that basically the algorithm tries to generate the tree structure or the RRT. Every node of this tree has a location that determines its position in the map as well as a parent pointing towards the previous node or vertex. The root of the tree is taken as the source. Since this is the root of the tree, its parent is null. Initially the root or the source is the only vertex available in the map. As the algorithm proceeds with the exploration process, it finds newer and newer areas in the map. All of these are added as children nodes.

The complete algorithm that carries forward the task of exploration is iterative in nature. At every iteration of the algorithm, a new node is generated and added to the tree structure RRT. The procedure to generate a new node and add it to the RRT data structure is simple. We generate a random point in the entire map space. We then query the RRT and find out the node already existing in the RRT that lies closest to the newly generated random point. This nearest node in the RRT is then used and is extended to generate the new node. The extension takes place by a magnitude of step size or δ. The new node formed by the extension of the nearest node becomes a new node of the RRT. This node is added to the RRT with the nearest selected node as the parent. This addition of node into RRT however only takes place if this node was not earlier lays into the RRT. Nodes very close to each other may sometimes be interpreted as same nodes. Further path from parent to the newly generated node must be collision-free.

In this manner, the algorithm keeps adding nodes by extending its present nodes. The result is that the tree keeps growing and soon covers the entire search space. During any stage if the goal is discovered, the algorithm may terminate. Additionally the algorithm may sometimes be made to terminate upon meeting any other stopping criterion like number of iterations, computational time, etc. This is important for problems where a path does not exist or is very difficult to find

Algorithm 4. RRT algorithm

```
GenerateRRT(Source, Goal)
Root ← Source
Root -> parent ← null
for i = 1 to no_of_iterations
        v ← random position
        y ← node in RRT nearest to v
        x ← extend y in direction of v by amount δ
if y→x is feasible and x not exists in RRT
                x -> parent ← y
                add x to RRT
                if x = goal return x
        end
end
```

out. The complete algorithm to generate the RRT in this iterative manner is given by Algorithm 4.

Here, we explicitly check that x should not already exist in the RRT already. This check not only checks for the exact location of v to be equal with any of the existing nodes of the tree, but rather the search checks that x should not lie close enough to any of the nodes within a distance of δ.

Here, δ is the minimal separation that a node must have in order to be entered into the RRT. This puts a threshold and stops the possible generation of two nodes in the RRT very close to each other. This ensures that every node would contribute towards better exploration of the map without two nodes conveying the same information, which may mean a waste of time. δ must hence be large enough. At the same time, large values of δ may mean that nodes have to lie far apart. In such a case, it may be possible that the algorithm is unable to generate flexible paths surpassing all big and small obstacles making the complete path optimal.

The other part of the problem is computation of the path from the goal to the source. Since the tree structure is already known, with every node pointing to its parent, we may simply backtrack the tree structure. The goal would always be the leaf node (assuming the search stops when the goal is found) and the source would be the root. We trace back the path in similar way as we did for the graph-based approaches. This is given by Algorithm 5.

Algorithm 5. Path retrieval algorithm

```
Path(goal)
P ← φ
I ← goal
While I ≠ null
        P ← I∪P
        I ← I -> parent
end
return P
```

The generation of new nodes is an important step in the algorithm. Generating completely random points in case of the problem of motion planning may lead to the entire map being explored. In such a random exploration of the entire map, whenever the goal happens to be met, we may terminate the algorithm with success. However, this would naturally mean a lot of wastage of computational time. We may in turn be biased towards construction of the tree towards the goal node. In such a case, the algorithm would try to explore more towards the goal, and thus probably reach the goal earlier. At the same time we must keep a provision of random exploration, because going straight towards the goal based on heuristics may meet some obstacle and it would need a lot of time to backtrack and come towards the earlier branches from the source nodes for a possibility of the way. Further, it may make the algorithm reach goal by a sub-optimal path.

A popular method of generation of the new states is that we first select a random node from the entire RRT and then extend it towards the goal with some step size δ. This action may take place with a probability p. With probability 1-p, we generate any random point, find the nearest node already existing in the RRT and extend the same for the generation of the new node. This factor gives the algorithm some exploration powers for possible generation of paths, where no almost direct path connects the source and the goal and the map is complex. At the same time, the algorithm is also biased towards the goal to some extent. The resultant algorithm is shown in Algorithm 6.

A typical RRT generated by this process is given in Figure 4.

In another similar scheme, we generate n points randomly into the entire workspace instead of only generating a single point v as done in the previous approaches. The attempt is to be biased towards moving towards goal, but at the same time attempting to explore new areas. If points generated near to the goal are selected, the exten-

Algorithm 6. RRT algorithm biased towards goal

```
GenerateRRT(Source, Goal)
Root ← Source
Root -> parent ← null
for i = 1 to no_of_iterations
        with probability p
                y ← node in RRT
                x ← y extended towards goal by step size δ
        with probability 1-p
v ← random position
                y ← node in RRT nearest to v
                x ← extend y in direction of v by amount δ
        if y→x is feasible and v not exists in RRT
                x -> parent ← y
                add x to RRT
                if x = goal return x
        end
end
```

sion done to the nodes would be such that the algorithm quickly attempts to reach the goal. If the path from current location to goal is straight or almost straight, this strategy would be useful. However, there is an obstacle somewhere ahead, after some extensions the algorithm would have to find an alternative route. For this reason, the exploration is also important. Points generated not so close to the goal enable the nodes to explore new areas. These areas may be directly or almost directly connected to the goal, and hence the

Figure 4. Extension of the RRT

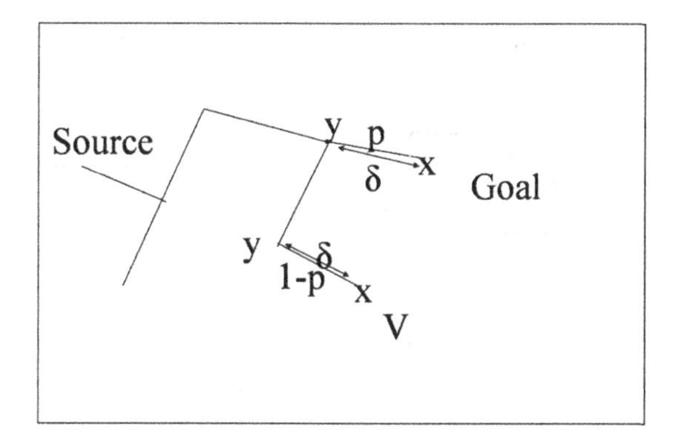

exploration may be successful. We need to balance the two strategies and accordingly select one out of the n generated random points for further selection.

To carry out the selection, we use the stochastic uniform selection scheme. This selection scheme asks assignment of weights or expectation values to all the generated points. A higher weight or expectation value means more probability of being selected for further processing. In this problem, weight or probability of selection is inversely proportional to the distance of the node from the goal. The weight of any point v may be given by Equation 2.

$$w(v) = \frac{1}{1 + \|v - g\|} \tag{2}$$

Here, g is the goal.

The details of stochastic uniform selection are being omitted here, and would be dealt in chapter 4.

Execution Extended RRT

Another variant of the RRT is known as the Execution Extended RRT (ERRT) (Bruce & Veloso, 2003). In this algorithm, the RRT tree is made in an iterative manner in quest of the smallest path from the source to the goal. However, in this approach, we additionally try to benefit from the previous iterations of the algorithm where it was successful in finding the solution. Hence, it is assumed that the algorithm has a memory structure where it stores the successful paths and all the nodes visited to make the successful path called as the waypoint.

At any moment of time, in this case, we have three possible moves to make, unlike the two possible moves in the previous case. The first possibility is that we try to go straight towards the goal. This move is executed with a probability p. Here, we select a random node from the RRT and extend it towards the goal. The other possibility is that we try to go towards the goal using a previously successful path already stored in the memory. This move is executed with a probability of q. In this approach, we select a random node from the waypoints stored in the memory. We then try to add this point into the RRT by finding the nearest node. The other possibility is that we do a random exploration. This move is followed with a probability of 1-p-q. Here, we select a random point in the entire map and join it to the nearest node in the RRT. A typical scenario with the implementation of the algorithm is given in Figure 5.

Figure 5. Execution extension RRT

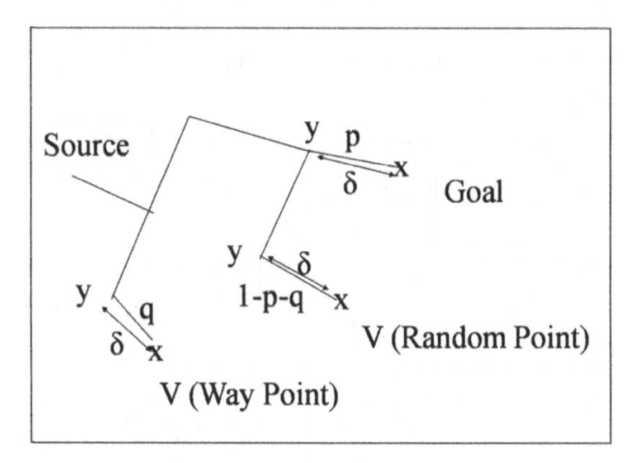

RRT Connect

The other variant of the RRT is known as RRT-Connect (Kuffner & LaValle, 2000). Here, the purpose is to accelerate the exploration of the algorithm. The conventional algorithm may take reasonably long to reach the goal, as it is driven by both heuristic and non-heuristic or random means. The step size puts a restriction to the maximum that the algorithm may traverse. Hence, the speed may be low. In many applications with high-resolution maps, the algorithm may not work. RRT-Connect tries to accelerate this exploration by allowing the algorithm to extend in the direction until the goal is found or an obstacle is met. Hence, once the decision has been taken to expand a node in any direction, the expansion would be carried out for multiple steps, which increases the exploration speed. However, the algorithm may as well get trapped by an obstacle, in which case it would waste some time.

Use of a single tree being generated from source and moving to the goal may sometimes also be problematic. Another major concept of RRT is hence to give it a bi-directional nature. This means that the search not only starts from the source to the goal, but also from the goal to the source. The algorithm in this manner tries to connect the source and the goal by a small path, which is the resultant output of the planner.

A bidirectional RRT algorithm maintains two trees for the two searches in both directions. One of the trees is rooted at the source and the other tree is rooted at the goal. At every iteration, the RRT algorithm tries to extend one of the two trees. In the first iteration, it tries to connect the tree rooted at source, in the second iteration it tries to extend the tree rooted at goal, and so on. The extension to the tree may be applied using any of the methods discussed earlier. These may

Algorithm 7. RRT-connect algorithm

```
RRT-Connect(Source, Goal)
Root1 ← Source
Root2 ← Goal
Source -> parent ← null
Goal -> parent ← null
for i= 1 to no_of_iterations
        v ← random position
        y ← node in RRT nearest to v
        x ← y extended towards goal/source by step size δ
        if y→x is feasible and v not exists in RRT
                x -> parent ← y
                add x to RRT
                if connect(Root1, Root2, x) return Root1, Root2
        end
        swap(Root1, Root2)
end
Connect(Root1, Root2, v)
Do
        Extend x in root1
While source/goal not reached and no collision with obstacles
if source/goal reached
return true
else
        return false
end
```

be heuristic biased to enable the goal reach the source and vice versa. The algorithm then tries to check whether the extension results in the two tress being connected to each other. In case a connection is possible, the algorithm terminates with a success. The corresponding path is returned. In case the connection is not possible, the algorithm continues.

The complete algorithm of using a bidirectional RRT-Connect is given in Algorithm 7.

Here, the extension is similar to the addition of new nodes in the RRT. We carry this process indefinitely until the goal is reached or an obstacle is found. A typical scenario of planning with RRT-Connect is given in Figure 6. The bi-directional RRT is given in Figure 7.

For related work in the domain of motion planning please refer to Brandt (2006), Kim and Ostrowski (2003), Martin et al. (2007), and Tokuse et al. (2008).

Results

The approach was experimented via simulations. The complete system was developed using MATLAB as the programming platform. The map was read as a bmp image and was further used for computation. The path was displayed using simple graphics functions of the platform. The experimen-

Figure 6. RRT connect

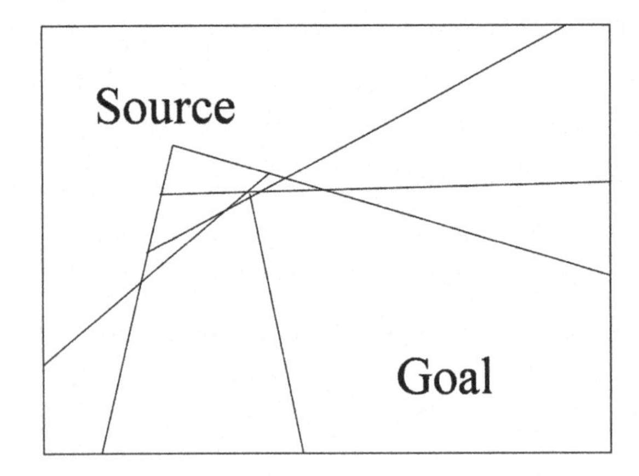

Figure 7. Bidirectional RRT

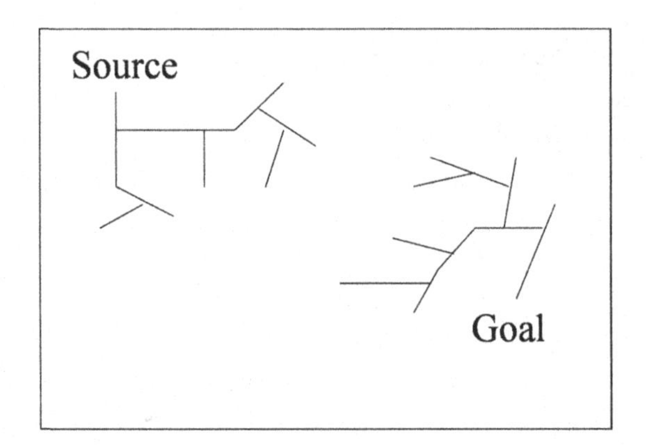

Figure 8. Experimental results for RRT

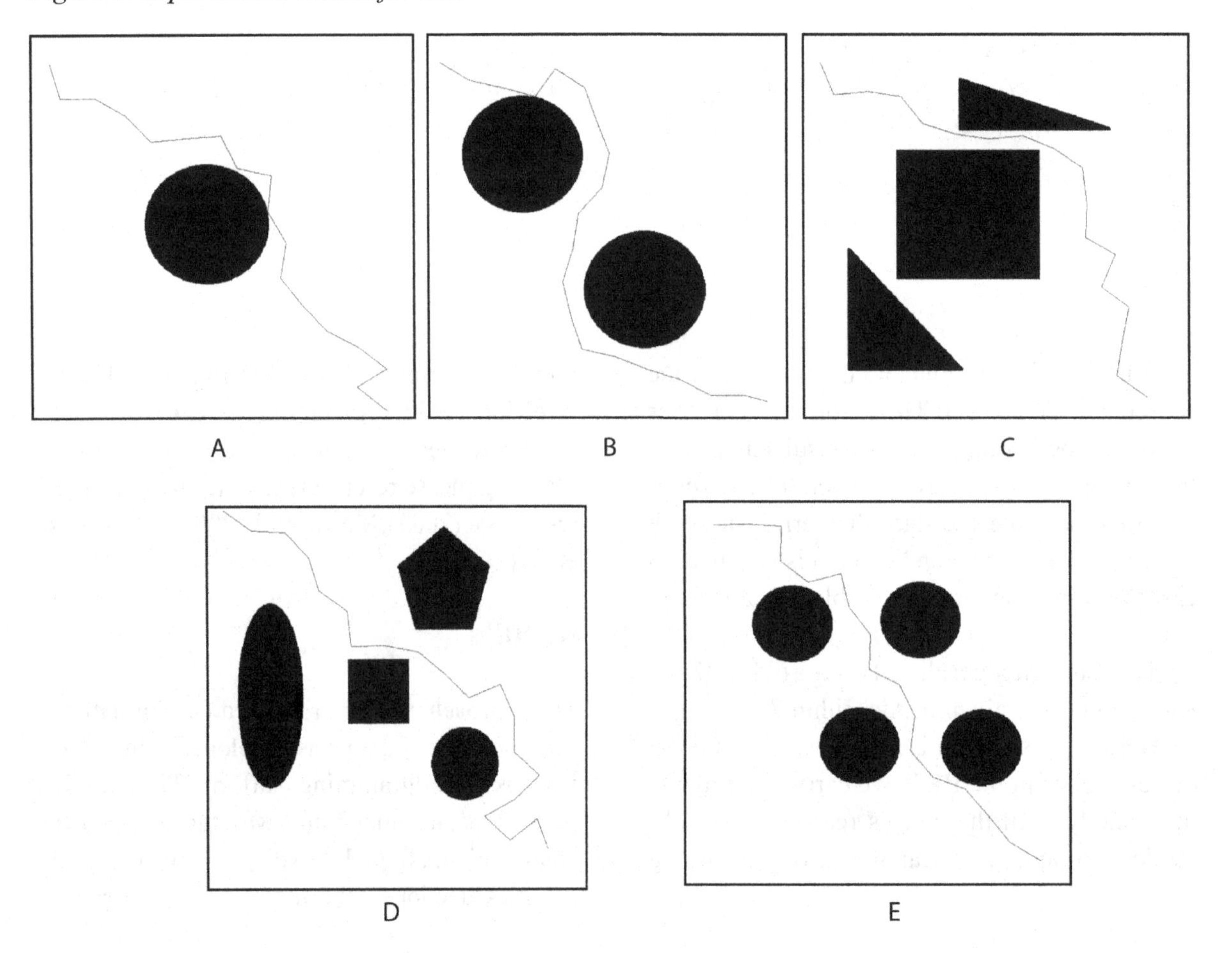

tation was first done on a single direction RRT algorithm. Heuristics were used by generating a number of points the final selection being done by the Stochastic Uniform Selection. The results to some of the maps are shown in Figure 8. The RRT generated are shown in Figure 9.

The other task was to make the algorithm bidirectional to study the advantages of the bidirectional approach as compared to the uni-directional approach. The uni-directional algorithmic framework was extended to the bi-directional framework and the resultant module was simulated. The results in the use of the bi-directional algorithm are given in Figure 10. The RRT generated is shown in Figure 11. Clearly, it may be seen that in both algorithms results are not optimal in terms of path length. However, the solutions are returned fast even for high-resolution maps. Further bidirectional search results in lesser number of nodes being expanded. This should mean a lesser computational time. However if the path joining source to goal is complex it may result in too much memory utilization and too many cross lookups between the two trees wasting a lot of time.

EMBEDDED SENSOR PLANNING

In the approaches discussed so far, we assumed that there is a map made available by some map-building algorithm, and the problem is to find the path of a robot from a given source to a given goal in this map. In this approach, the robot carries forward the task of motion planning and all the related computations. Since the computation is

Figure 9. RRT generated

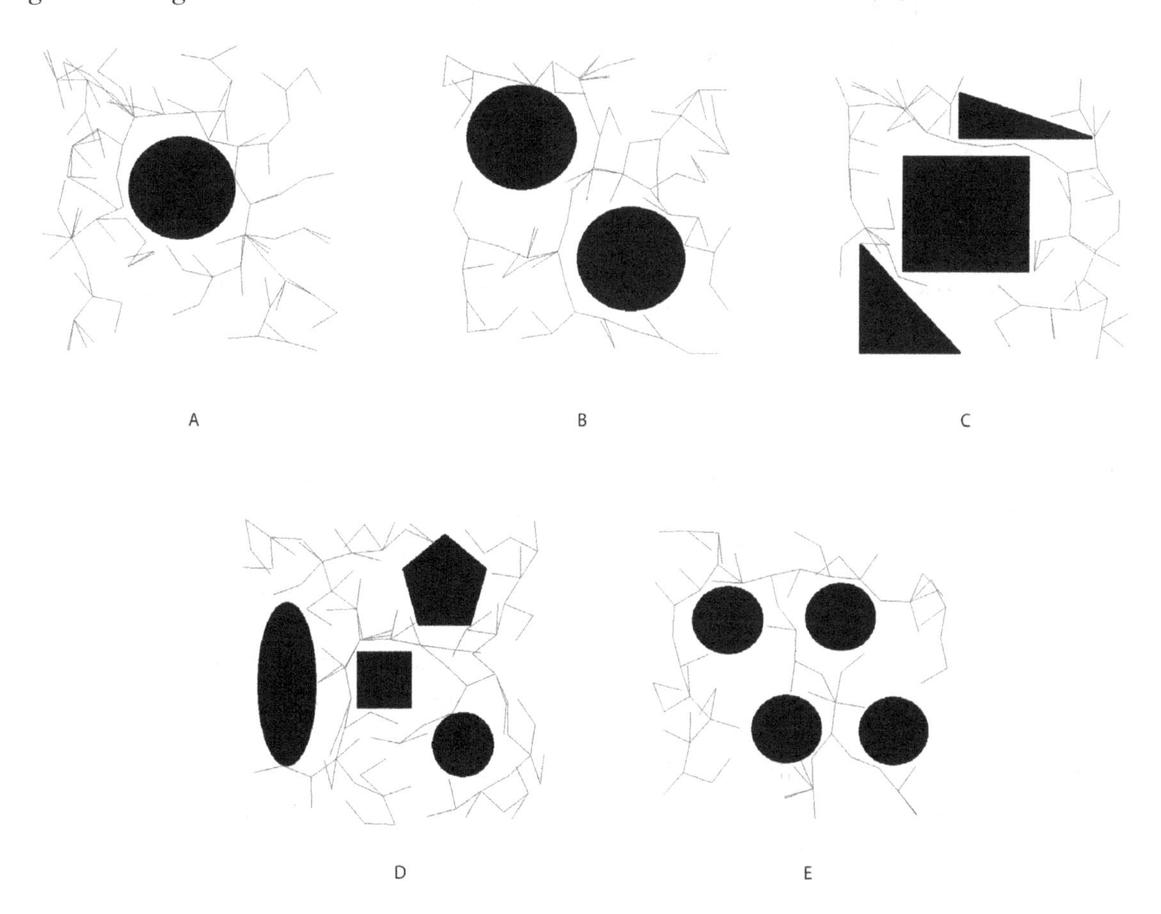

Figure 10. Experimental results for bidirectional RRT

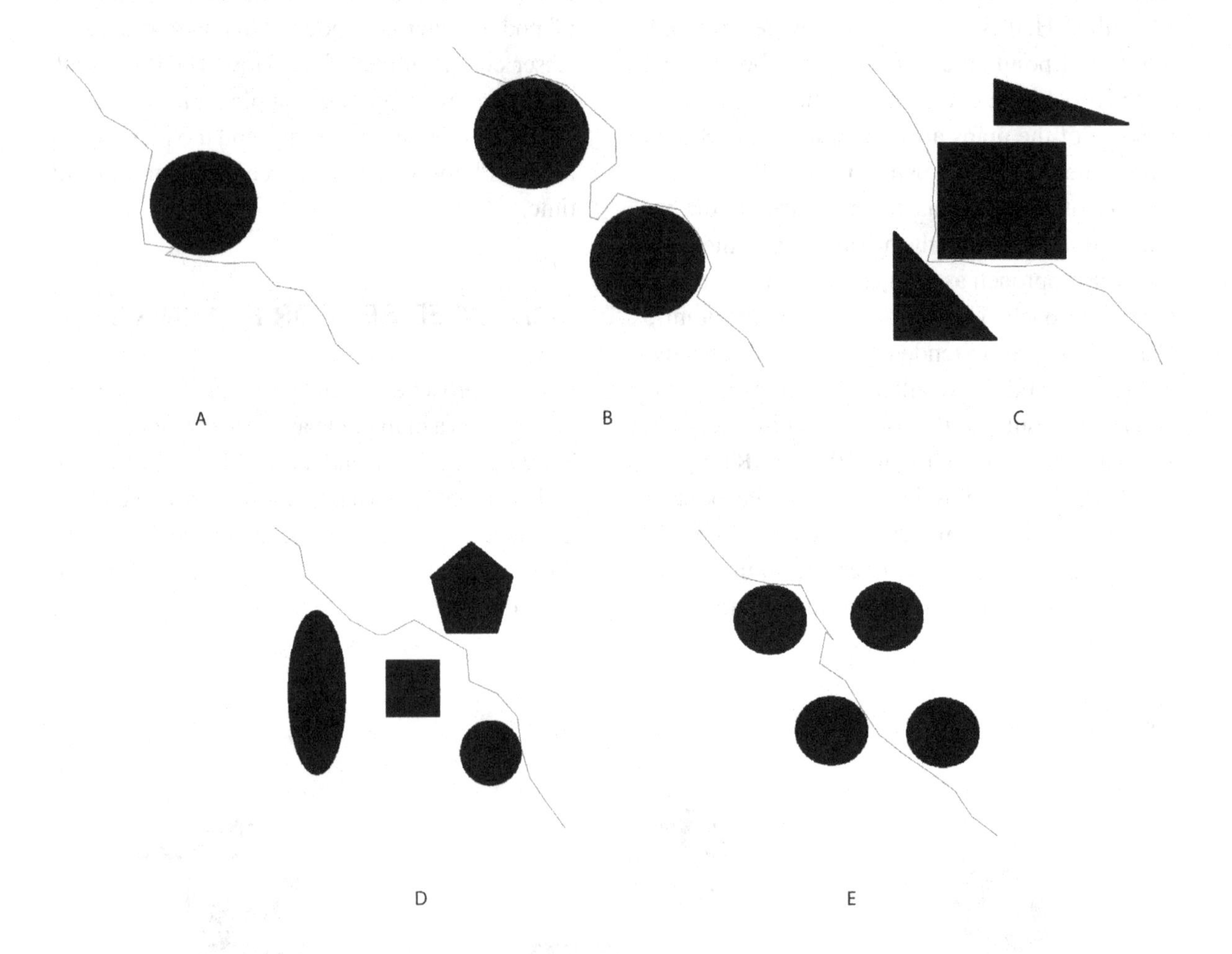

done by a single processer (or a limited number of processers in case of a multi-processor robot), the computational time required is reasonably high. The embedded sensor planning technique (O'Hara, et al., 2008) tries to eliminate the same problem in case of robot motion planning to enable the robot plan fast. Further, it enables the robot to re-plan very fast in case there is some sudden occurrence of obstacles.

Concept

The basic concept behind embedded sensor planning is that the planning should not be done centrally at the robot. Instead it must be done by a large number of sensors distributed throughout the map. Since the entire computation is distributed, the workload over a single unit or sensor is much less. Fundamentally, this concept makes use of two entities, sensors and robots. Sensors are sensing agents that are spread all across the map. These are sufficient in number to allow a good and flexible planning for the robot. All the sensors have transmission and receiving facilities. Hence, they can sense each other. The transmission and reception is highly limited and each sensor only senses the nearby few sensors. This converts the entire map embedded with sensors into a small network with the sensors as the nodes. Two sensors that sense each other are said to be connected and can transmit and receive information from each other. It is assumed that if any obstacle comes

Figure 11. Bidirectional RRT generated

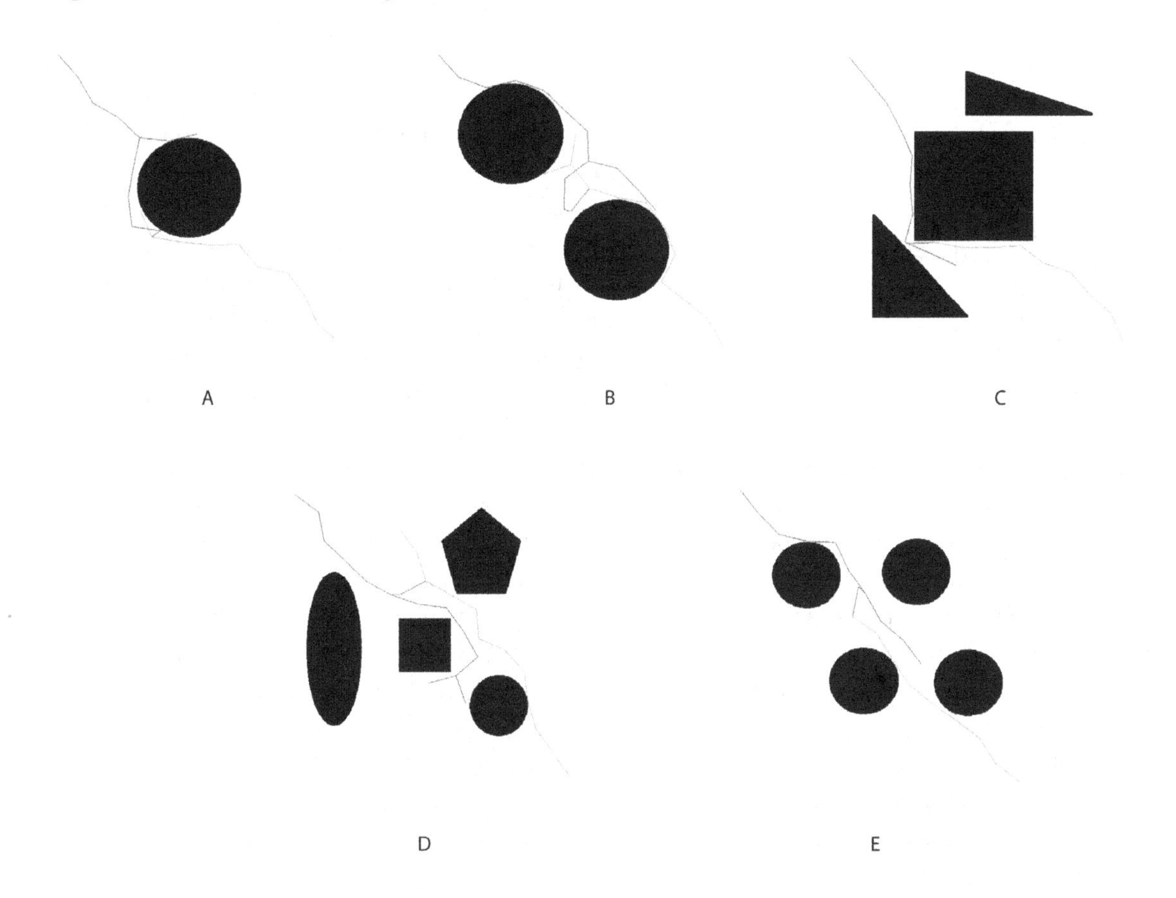

between any two sensors, they would no longer be able to sense each other and hence the link between them would fail. Hence, the obstacles considered in this scenario are the ones, which can break links between the sensors.

Working

The entire planning is this system is primarily carried by the sensors. Each sensor receives the distance from the goal transmitted by the other sensors as per their computations, and uses this measure to compute its own distance from the goal. The distance computed in this manner is transmitted by this robot, which would be useful for the planning of the other sensors for their distances from the goal. In this manner, each sensor becomes aware of the distance between it and the goal. This distance is the mechanism by which the entire planning takes place. The robot in this planning is just an agent. The robot simply needs to traverse to the position of the sensors with the least distance from the goal. The robot moves to one sensor. The robot then looks at the surrounding sensors to find the sensors with the least distance in the vicinity of the current sensor. It then traverses to this sensor and the process goes on. Finally, the last sensor directs the robot to reach the goal.

Take a small example with a map embedded with some sensors as shown in Figure 12. The connections shown in figure are the sensors that can sense and communicate with each other. It may further be seen that no connections are drawn

Figure 12. Planning with embedded sensors

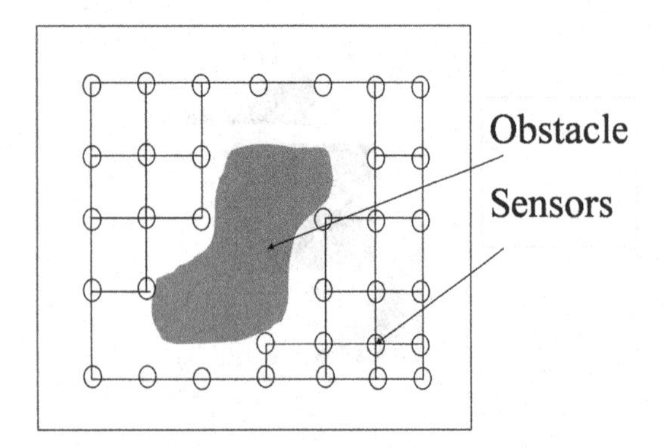

between two sensors if they have an obstacle in between. The various sensors start transmitting the distances to each other, once the connections have been made. Initially the sensors may have to wait for the distances to be propagated from the goal. First, the sensors nearby to the goal may compute their distances. The distances transmitted by these would be useful for the other sensors a little farther off to compute their distances. In this manner, the distance propagation takes place until the sensor at the largest distance from the goal.

The entire computation may sound expensive. This was the case when we dealt with the problem using graph algorithms. The same concept of searching was regarded as expensive. In this case, however, the computation is spread across multiple sensors which are all processing agents unlike the case of graph where everything was being performed centrally. Hence, even though the method might look similar to a graph problem with sensors playing the role of vertices and the sensing and network connections playing the role of edges, this approach is very computationally efficient. It may be further seen that every sensor has a very basic computation to perform. Hence, the entire algorithm runs upon little computations being widely distributed in the entire environment.

The computation at a single node or sensor follows a simple Bellman Ford equation. This is given by Equation 3.

$$f(V) = \begin{cases} \min\{f(X) + w_{vx}\} & V \neq G \\ 0 & V = G \end{cases} \quad (3)$$

Here, V is any sensor, X is any other sensor connected to V. The distance of the goal is always taken to be 0. w_{vx} is the physical distance between the sensor v and the sensor x. In this manner, the entire algorithm may even be taken to be similar in working to the Bellman Ford algorithm. The major difference is again the fact that the sensors are all processing agents spread across the map.

The other major point with this planning is that it is real time or online. Suppose that there is a sudden emergence of an obstacle between any two sensors that led the path from source to goal. Now, ideally, as per the working guidelines of any planning algorithm, the robot is supposed to re-plan its journey. In this, the sudden emergence of the obstacle is marked by cutting of the connections. The distance propagation equations continue to run, but now since the un-connected nodes cannot communicate to each other, these run for the updated map. All the nodes soon update and compute their new distances from the goal. There is no change required in the planning by the robot, which is simply an agent that goes towards sensors with minimum distance from the goal. In this manner, using this concept, the robot is capable of moving from the source to the goal in a path of optimal length.

However, it needs to be realized that the planning may require multiple sensors to be placed at multiple places of the map. For better results, the map needs to be packed with these sensors. Having lower number of sensors may have problems leading to sub-optimal paths. It would be similar to planning with low resolution maps where it is not perfectly known how to surpass an obstacle in the optimal manner, as the visibility is limited. Lesser sensors may further have problems of sensing of the obstacles. Further, the planning cannot be carried out in any general environment where the sensors would not be pre-embedded. This is unlike the other algorithms that use video cameras, sonar sensors and other such devises for navigation in any general environment.

MORE PLANNING TECHNIQUES

The entire planning domain does not restrict itself to the discussed algorithms or the ones that would be discussed later. There are a wide variety of approaches. Besides there is always a possibility to design algorithms based on current algorithms that suite some scenarios more than the existing approaches. In this section, we briefly describe some more planning approaches that are extensively used.

Artificial Potential Field

Artificial Potential approaches make use of a potential field based modeling of the map to enable a robot reach goal from source (Hui & Pratihar, 2009; Pozna, et al., 2009; Tsai, et al., 2001; Yagnik, et al., 2010). Potential is a kind of force that every obstacle and goal apply. The goal always has a strong attractive force or a positive potential. It always tries to attract all particles towards itself. The obstacles however have a strong repulsive force or a negative potential. They always try to repel the particles from coming close to them. The force or the potential decreases as the distance of the particle increases from the source of the potential. As per the laws, the decrease is inversely proportional to the square of the distance of separation. The robot in this case of planning is simply a particle that behaves as these potentials guide it.

At any time in the map, the robot is affected by all the potentials that are applied to it. The goal tries to attract it and the obstacles try to repel it. In such a context, the resultant direction of motion of the robot is one as computed by the resultant potential vector by adding all the individual potentials. As the robot moves, it gets affected by various potentials, and always the resultant potential is computed. It is natural that it is very difficult for the robot to collide with any obstacle. This is because as soon as the robot comes close to an obstacle, a repulsive potential is applied. This increases as the robot comes even more near. Soon the potential becomes large enough to surpass all other potentials being applied by the other obstacles or goal. In such a case, the robot is strongly repelled. The different obstacles may apply potential of different magnitudes that may in turn depend upon their size.

It is important to realize that the algorithm in this modeling just has a map as an input. The entire concept of potential is artificially generated only for the purpose of trajectory planning of the robot. While implementation of the approach or generation of the robotic trajectory, the application of potential and all the other computations are performed by the planning algorithm. Accordingly, the next motion of the robot is computed.

Once the obstacles and goal along with the potential they apply is known, we may compute the potential at every point in the map. The points of same potential may be joined which gives the potential contour. This is a two dimensional representation that gives an idea of the three dimensional graph of potential with the two image axis. Drawing the potential curve for any map gives an understanding of the motion of the robot when placed at various points in the map. The curve for any general map is given in Figure 13.

The planning with artificial potential takes into account all static and dynamic obstacles and the planning online in nature. Whenever any obstacle comes, we may consider it in the computation of the potential at next time step. This enables the robot to navigate in uncertain and dynamic environments. The entire computation of the move of the robot by considering the various potentials is also not a time costly process. The path traced by the robot is non-holonomic. Hence, the controller can smoothly trace the path without the requirement of any smoothening of path operators. The

Figure 13. Planning with potential fields

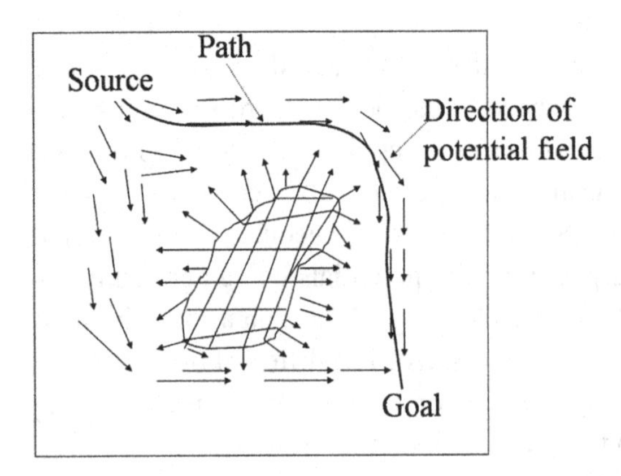

non-holonomicity of the path is due to the smoothness of the increase and decrease of the potentials by the agents as we move along the path. As we move forward, the repulsive forces by some agents decrease and by others increase giving the entire path a smooth trajectory.

The major problem with this planning technique is that it is localized in nature. The planning is entirely done on the basis of the current potential functions. It is possible that the robot follows the potential values to get struck at some location. In such a case, it may never reach the goal. These locations are equi-potential where resultant of all forces adds up to zero. In many other scenarios, it may reach the goal by some non-optimal path. Hence, these approaches may only work for simple maps, which do not have a complex path from the source to the goal.

Reinforcement Learning

The next mechanism of solving the problem of motion planning that we study is Reinforcement Learning (RL). As the name suggests, reinforcement learning is a learning algorithm where the learning takes place by reinforcements. The basic approach of reinforcement states that we have an agent that performs an action out of all the possible actions into the system. The action of the agent is judged by a reinforcement agent that rewards or punishes the agent in magnitude of the goodness of the action. The agent is supposed to learn the selection of the actions such that the net reward given to it is maximized.

We use the same concept of reinforcement learning in our problem of motion planning of mobile robots. In this case, we have the agent as the robot whose plan we are trying to figure out. The entire system has a number of discrete states. The concept of states is same that we discussed in the graph-based approach for the motion planning. There it was stated that every state is a vertex and the states or vertices accessible from each other are connected by an edge. Similarly, the discussion of the discreteness of the states is similar as in the case of graph based search approaches. The discreteness applies to the map, the robot state, as well as the time.

The agent or a robot in case of reinforcement learning has a number of actions. The actions in case of the motion planning of robots correspond to the moves allowed for the robot per unit time. In this case, again, the notion to some extend resembles the notion of edges in graphs for our graph based planning. The purpose of reinforcement learning is to designate an action or move to every state or a robotic position. The designation of action is done on the basis of the rewards.

A popular learning mechanism used in RL is the Q-learning algorithm. The algorithm computes the expected reward from every state and action, and uses it for the learning. A table is maintained in terms of every state and action over which the learning takes place. The system learns in an iterative manner, where the learning keeps improving with time. After some iteration, the learning converges to a point and further learning does not change the state of the system. This is stated as the optimal policy of the learning. Towards the end, we know the actions that the robot needs to make at every state in its journey. The robot may hence travel from one state to the other as per the mapped actions to reach the goal position. For more details regarding RL please read more general texts (Singh, 2002; Sutton & Barto, 1998). For applications of RL to robotic planning refer to Gullapalli et al. (1994), Li et al. (2002), and Mahadevan and Connell (1992).

Voronoi Map-Based Planning

The graph approaches discussed had a major problem of dimensionality. The problem can be simplified if we convert a map to a Voronoi map and use it for the purpose of motion planning (Choset & Burdick, 1995; Garrido, et al., 2006; Tomono, 2003). The Voronoi maps segment out regions, which do not contain any obstacles and can be used by the robot for motion planning from the entire map. Hence, we have a clear idea of the regions where the robot may travel. This reduces a very heavy map of high dimension into a very small map that may be easily planned. There are a large variety of techniques that convert a map into a Voronoi map. After conversion, we get the locations of the Voronoi centers. These may then be converted into a graph with each center as a vertex and the connections between the centers that have a passage between them.

Any graph search technique may be used over the graph so formed. One of the very common search techniques is the use of Dijkstra's algorithm. This algorithm could not be used over a map of large resolution as the algorithm is very sensitive to the number of vertices. It is only able to handle few vertices. The advantage of the algorithm is however that it always returns the shortest path from the source to the goal. Since we have a limited number of vertices, it may be used over the converted graph. The output of the search algorithm is the path to be traced by the robot.

The problems with the Voronoi-based planning is that one primarily gets a broader path planned by the graph search algorithm. It may not be able to narrate the exact vertices at finer level or the exact locations to visit to give the minimal path length for the robot. In such a case, it may be possible that the robot goes through regions that are far from the optimal path, keeping the region of travel the same as the optimal path. These planning techniques may often be used with other techniques for more effective planning.

CONCLUSION

A large number of algorithms exist for the task of motion planning of mobile robots. All these vary in their problem modeling, working methodology, and concept of working. In this chapter, we extended our discussions of the previous chapter on motion planning using graph-based approaches. A number of related methods of motion planning were presented. These include the dynamic programming and Bellman Ford algorithm, planning with Rapidly growing Random Trees (RRTs), planning with embedded sensors, artificial potential fields, and some other planning techniques. The dynamic programming used principles of dividing the problem into smaller problems and solving them in an iterative bottom up manner from lower complexities to the higher complexities. The basic approach is divide and conquer. Bellman Form algorithm is a graph search algorithm that uses the principles of dynamic programming to find the shortest distance from a source to all other vertices of the graph.

The other algorithm of study was the RRT. Here, we looked at the fundamental data structure of the RRT and the mechanism in which they are initialized and grown. A number of variants of the RRT were discussed that better suit the problem of motion planning of mobile robots. All these tried to better use the available heuristics to enable generation of optimal paths. The next algorithm was the embedded sensor network. Here, we used the concept of using a large number of sensors distributed within the robotic map to form a network. In this manner, distributed computing all around the map led to figuring out an optimal robot path in a real-time mode fairly early. The robot in this case was just an agent to find the sensor closest to the goal and move towards it.

The next algorithm of study was artificial potential fields. Here, the goal and the various obstacles applied a potential to affect the robot move. The robot moved as per the resultant potential. This was a localized planning technique but worked in the real-time scenarios. The other discussed approaches included the reinforcement learning where every robot state was learnt to be mapped to a corresponding action. The learning may be performed by the Q-Learning algorithm. The last approach was the Voronoi map-based motion planning where a robotic map is first converted into a Voronoi map and then planning is used over the map to get the shortest path from source to goal.

REFERENCES

Barbehenn, M., & Hutchinson, S. (1993). Efficient search and hierarchical motion planning using dynamic single-source shortest paths trees. In *Proceedings of the IEEE Conference on Robotics and Automation,* (Vol. 1), (pp. 566-571). IEEE Press.

Barbehenn, M., & Hutchinson, S. (1995). Efficient search and hierarchical motion planning by dynamically maintaining single-source shortest paths trees. *IEEE Transactions on Robotics and Automation*, *11*, 198–214. doi:10.1109/70.370502

Brandt, D. (2006). Comparison of A and RRT-connect motion planning techniques for self-reconfiguration planning. In *Proceedings of the IEEE Conference on Intelligent Robots and Systems,* (pp. 892-897). IEEE Press.

Bruce, J., & Veloso, M. M. (2003). Real-time randomized path planning for robot navigation. *Lecture Notes in Computer Science*, 288–295. doi:10.1007/978-3-540-45135-8_23

Choset, H., & Burdick, J. (1995). Sensor based planning: The generalized Voronoi graph. In *Proceedings of the IEEE International Conference on Robotics and Automation,* (vol. 2), (pp. 1649-1655). IEEE Press.

Cormen, T. H., Leiserson, C. E., Rivest, R. L., & Stein, C. (2001). *Introduction to algorithms* (2nd ed.). Cambridge, MA: MIT Press.

Garrido, S., Moreno, L., & Blanco, D. (2006). Voronoi diagram and fast marching applied to path planning. In *Proceedings 2006 IEEE International Conference on Robotics and Automation*, (pp. 3049-3054). IEEE Press.

Goto, T., Kosaka, T., & Noborio, H. (2003). On the heuristics of A* or A algorithm in ITS and robot path-planning. In *Proceedings of the IEEE Conference on Intelligent Robots and Systems,* (vol. 2), (pp. 1159-1166). IEEE Press.

Gullapalli, V., Franklin, J. A., & Benbrahim, H. (1994). Acquiring robot skills via reinforcement learning. *IEEE Control Systems*, *14*(1), 13–24. doi:10.1109/37.257890

Horowitz, E., & Sahni, S. (1978). *Fundamentals of computer algorithms*. Baltimore, MD: Computer Science Press.

Hui, N. B., & Pratihar, D. K. (2009). A comparative study on some navigation schemes of a real robot tackling moving obstacles. *Robotics and Computer-integrated Manufacturing*, *25*(4–5), 810–828. doi:10.1016/j.rcim.2008.12.003

Jihong, L. (1995). A dynamic programming approach to near minimum-time trajectory planning for two robots. *IEEE Transactions on Robotics and Automation*, *11*(1), 160–164. doi:10.1109/70.345949

Kim, J., & Ostrowski, J. P. (2003). Motion planning a aerial robot using rapidly-exploring random trees with dynamic constraints. In *Proceedings of the IEEE Conference on Robotics and Automation,* (vol. 2), (pp. 2200-2205). IEEE Press.

Konar, A. (2000). *Artificial intelligence and soft computing: Behavioral and cognitive modeling of the human*. Boca Raton, FL: CRC Press.

Kuffner, J. J., & LaValle, S. M. (1999). Randomized kinodynamic planning. In *Proceedings of the IEEE International Conference on Robotics and Automation*, (pp. 473–479). IEEE Press.

Kuffner, J. J., & LaValle, S. M. (2000). RRT-connect: An efficient approach to single-query path planning. In *Proceedings of the IEEE International Conference on Robotics and Automation,* (vol. 2), (pp. 995-1001). IEEE Press.

Kuwata, Y., Karaman, S., Teo, J., Frazzoli, E., How, J. P., & Fiore, G. (2009). Real-time motion planning with applications to autonomous urban driving. *IEEE Transactions on Control Systems Technology*, *17*(5), 1105–1118. doi:10.1109/TCST.2008.2012116

LaValle, S. M., Gonzalez-Banos, H. H., Becker, C., & Latombe, J. C. (1997). Motion strategies for maintaining visibility of a moving target. In *Proceedings of the IEEE International Conference on Robotics and Automation,* (vol. 1), (pp. 731-736). IEEE Press.

Li, Y., Zonghai, C., & Feng, C. (2002). A case-based reinforcement learning for probe robot path planning. In *Proceedings of the IEEE Conference on Intelligent Control and Automation,* (vol. 2), (pp. 1161-1165). IEEE Press.

Liu, Y., & Arimoto, S. (1990). A flexible algorithm for planning local shortest path of mobile robots based on reachability graph. In *Proceedings of the IEEE Conference on Intelligent Robots and Systems*, (vol. 2), (pp. 749-756). IEEE Press.

Lozano-Perez, T. (1987). A simple motion-planning algorithm for general robot manipulators. *IEEE Journal on Robotics and Automation*, *3*(3), 224–238. doi:10.1109/JRA.1987.1087095

Mahadevan, S., & Connell, J. (1992). Automatic programming of behavior-based robots using reinforcement learning. *Artificial Intelligence*, *55*(2-3), 311–365. doi:10.1016/0004-3702(92)90058-6

Martin, S. R., Wright, S. E., & Sheppard, J. W. (2007). Offline and online evolutionary bi-directional RRT algorithms for efficient re-planning in dynamic environments. In *Proceedings of the IEEE Conference on Automation Science and Engineering,* (pp. 1131-1136). IEEE Press.

O'Hara, K. J., Walker, D. B., & Balch, T. R. (2008). Physical path planning using a pervasive embedded network. *IEEE Transactions on Robotics*, *24*(3), 741–746. doi:10.1109/TRO.2008.919303

Pozna, C., Troester, F., Precup, R. E., Tar, J. K., & Preitl, S. (2009). On the design of an obstacle avoiding trajectory: Method and simulation. *Mathematics and Computers in Simulation*, *79*(7), 2211–2226. doi:10.1016/j.matcom.2008.12.015

Shin, K., & McKay, N. (1986). A dynamic programming approach to trajectory planning of robotic manipulators. *IEEE Transactions on Automatic Control*, *31*(6), 491–500. doi:10.1109/TAC.1986.1104317

Singh, S. (2002). *Reinforcement learning*. Reading, MA: Kluwer Academic Publishers.

Sniedovich, M. (2011). *Dynamic programming: Foundation and principles*. Boca Raton, FL: CRC Press.

Sutton, R. S., & Barto, A. G. (1998). *Reinforcement learning: An introduction*. Reading, MA: Kluwer Academic Publishers.

Thornton, C., & du Boulay, B. (2005). *Artificial intelligence: Strategies, applications, and models through search.* New Delhi, India: New Age Publishers.

Tokuse, N., Sakahara, H., & Miyazaki, F. (2008). Motion planning for producing a give-way behavior using satiotemporal RRT. In *Proceedings of the IEEE Conference on System Integration,* (pp. 12-17). IEEE Press.

Tomono, M. (2003). Planning a path for finding targets under spatial uncertainties using a weighted Voronoi graph and visibility measure. In *Proceedings of the IEEE Conference on Intelligent Robots and Systems,* (vol. 1), (pp. 124-129). IEEE Press.

Tsai, C. H., Lee, J. S., & Chuang, J. H. (2001). Path planning of 3-D objects using a new workspace model. *IEEE Transactions on Systems, Man and Cybernetics. Part C, Applications and Reviews, 31*(3), 405–410. doi:10.1109/5326.971669

Yagnik, D., Ren, J., & Liscano, R. (2010). Motion planning for multi-link robots using artificial potential fields and modified simulated annealing. In *Proceedings of the IEEE Conference on Mechatronics and Embedded Systems and Applications,* (pp. 421-427). IEEE Press.

KEY TERMS AND DEFINITIONS

Artificial Potential Fields: Planning technique involving the use of a strong attracting potential or force by the goal node and a strong repulsive potential or force by all obstacles. Robot moves as per the resultant potential vector.

Bellman Ford Algorithm: A graph-search algorithm that returns the shortest path from source to any other node.

Embedded Sensor Planning: A motion planning technique where multiple sensor nodes are distributed throughout the map and communicate with each other to compute a path to the goal. Any change in the map is rapidly reflected.

Offline and Online Planning: Offline planners are ones that may have high computational time, work for stationary environments, and are the opposite for online planners. High processing time may enable offline planners to be complete and optimal.

Rapidly Exploring Random Trees: Motion planning algorithm for robots, where planning proceeds by construction of a tree-like data structure whose every node is a point in the map. At every iteration, a node in the tree is expanded towards some location in the map.

Chapter 4
Evolutionary Robotics 1

ABSTRACT

Evolution is a widely used paradigm of computing that is finding an application in all domains. Evolutionary algorithms use an analogy from natural evolution to model the problem solving approach. They consist of a set of individuals that are the solutions to the problem and make a population pool. The various operators are used to modify the population pool of individuals. This consists of scaling, selection, mutation, crossover, and elite. Evolutionary algorithms are used for the problem of motion planning of the robots. Non-holonomicity is a major issue associated with mobile robots. The paths returned by the planners need to be smooth to ensure easy tracking by the robot control algorithms. The authors make use of BSplines and Bezzier curves to solve the problem. These are smooth curves that are controlled by the control points on the maps. Adaptation is another major aspect studied in evolutionary algorithms. The authors use evolutionary strategies to solve the problem of motion planning. These are self-adaptive in nature and are able to adapt the evolutionary parameters during optimization. Covariance Matrix Adaptation Evolutionary Strategy is the studied method of implementation.

INTRODUCTION

Evolution is a major computing paradigm. Most of the problems that we see all around are fairly complex in nature. They have a lot many constraints that must be met. This makes it impossible for the conventional problem solving techniques to solve the problem and return the output within finite time. Hence, better techniques for solving these problems are required. The essence is that it may not be desirous for the solution generated to be perfect, but it is desirable for the solution to be returned within finite time. In this manner, we may convert an unsolvable problem to a solv-

DOI: 10.4018/978-1-4666-2074-2.ch004

able problem with some loss to the accuracy of the solution.

The evolutionary algorithms are inspired from the natural evolution process. All the individuals have different capabilities of acting and surviving in the world. The set of all the individuals is known as population. Different individuals in the population have different characteristics. All these individuals, however, compete with each other for survival. The basic philosophy comes from Darwin's theory of survival of the fittest, which states that the individuals in a population that are most adaptive to change survive. Hence, the fittest individuals from a generation go to the next generation. They interact with the other fitter individuals and generate their offspring. The offspring are believed to be better than the parents as per the changing scenario or the environment. In this manner, evolution goes on and on. This evolution has resulted in the generation of different kinds of species that suit their environment and are able to survive well.

The evolutionary algorithms make the same analogy. Here the algorithm generates a number of random solutions to the problem. These solutions may solve the problem in different degrees that determines their fitness. Better solutions have better fitness values. The attempt is to carry forward changes in this set of individuals or population to make a higher generation population. The changes are carried out with the help of a number of genetic operators. These operators attempt to generate better solutions as possible for the next generation.

Optimization is a major problem of study for most of the problems. Optimization problems deal with maximization or minimization of a functional value, where the functional value changes as per some parameters. The optimization algorithm is supposed to find the correct set of parameters as well as the optimal (minimal or maximal) value of the function for this combination of parameters. Many problems that we see all around can be converted into optimization problems, and hence this class of problem is under heavy study. Evolutionary algorithms are extensively used for optimization problems. These become very valuable tools when the problem has too many parameters that need to be tuned, over which the performance of the function depends. Further, many times the function may not be differentiable and all derivative approaches that try to find a slope of the function to figure out the maxima or minima fail.

Genetic Algorithm is a member of the general class of evolutionary algorithms and is widely used for solving problems like optimization. While the conventional optimization problems only deal with simple parameter tuning, the genetic algorithm is able to perform a wider variety of tasks, one being the generation of the robotic path, which we shall be studying in this chapter. We look in detail regarding the working methodology, terms, and concepts of this algorithm and the manner in which it is used for solving the problem of robot motion planning in the following sections.

The problem of robot path planning can be easily stated to be a problem where the path of a robot is to be produced, given the source point, goal point, and the map. It may be seen that for every map a variety of paths may be possible. Some of these may be feasible while the others are infeasible. Infeasible paths mean that these do not reach from source to the goal without collision. Hence, these can never be used for the motion of the robot. The feasible paths may again be of variable lengths, depending upon the landmarks they use to reach the goal. Some paths may be using obvious routes to reach the goal, while others may be at some bigger routes. Further, the paths going through obvious routs may have different landmarks where they make a turn. These may go to different extents near or farther from the obstacle in their entire journey. Hence, there are always wide possibilities of paths from the source to the goal for any map. The optimal path that leads to the goal from the source may have any kind of shape. So far, the only intention is to have the path of the least possible length and the validity of the non-holonomic constraints as far as possible.

It may naturally not be possible to generate all types of paths that may exist between any source to the goal. This becomes especially difficult since any point in the path can lie at any position of the map. The planning algorithm need not only ensure that the path is optimal when seen at a broader level, but also that the path must be optimal at a finer level. Many times, two paths may seem to be of similar lengths at a broader level. In such scenarios, the optimization becomes very difficult, and the finer level optimization reveals the optimal path. Since all the possible paths cannot be figured out, the use of a genetic algorithm is a natural consequence.

The use of evolution in robotics is a very commonly used feature. The evolutionary algorithms aid the designers of the robotic algorithms or the entire robot at multiple levels. At each level, these algorithms try to automate some process, which could not be done automatically by the trivially known algorithms and humans had limitations to perform the same. The evolutionary algorithms are used in robot shape optimization, design of control mechanism, for motion planning, designing coordination strategies and protocols, etc. This extensive use of evolutionary algorithms in robotics gives rise to the field of evolutionary robotics. This puts us forward towards the time when machines would be sophisticated enough to design the complete robot for whatever task the robot is intended for. It would then not involve any human intervention or expertise at the design level. The machines run by an evolutionary technique would be able to figure out the ideal design for the task requirements.

Evolutionary computing is increasingly being used for solving more and more complex tasks. To understand the potential of this computing paradigm, we look into the domain of evolutionary music (Miranda & Biles, 2007). Music forms a very complicated sound wave pattern. Even the simplest of music pieces represents a very complicated signal. These days evolutionary computing is helping in automatic generation of music. While these techniques may not be able to generate a complete song because of its complicated nature, the evolutionary techniques help in generation of music pieces, beats, and nodes, which fit into songs well and are liked by the composer. Evolutionary computing paradigm is able to come up with tunes, out of an alarming number of possibilities, which sound good and can be used in songs.

In this chapter, we make use of evolutionary techniques for the purpose of motion planning of the robot. Two major techniques are used. These include the genetic algorithm and evolutionary strategies. We first model the problem of motion planning without considering any non-holonomic constraints. A genetic algorithm solution in such a scenario is designed and presented. We further consider the non-holonomic constrains and try to solve the problem with these constraints. This is done by the application of the Bezier and the Bspline *curves*. The path figured out by these approaches is near optimal as well as smooth. The solution is further extended to the use of evolutionary strategies for the problem of motion planning. We present the motivation behind the use of this technique of evolutionary algorithm and explain the associated concepts and working. All this follows in order in this chapter.

PRINCIPLES OF GENETIC ALGOIRHTM

The Genetic Algorithm (GA) is an evolutionary approach to problem solving. Here, the main attempt is to keep a set of solutions or individuals and keep improving them so that they get better and better as the algorithm proceeds. The entire algorithm is motivated from the natural evolutionary process, and hence, the entire terminology has been taken from the same. We first discuss the various terms associated with the algorithm, and then we build the genetic algorithm based on the lines of the natural evolution. For more details, readers may refer to original works and books by Baker (1985), Chakraborty and Dastidar (1993), Davis (1987), Fogel (1995), Forrest and Mitchell (1993), Goldberg (1989), Holland (1975), Michalewicz (1992), Mitchell (1999), and Shukla et al. (2010a, 2010b).

Individual

The first term is the individual. In case of a natural world, the individual corresponds to a unit entity of existence. In GA, we use the term individual to denote a unit solution to the problem, which we are trying to optimize. The solution may be good or bad, feasible or infeasible. For every problem, it is important to decide how an individual is represented. The solution in a native state that is understandable to the human expert is generally referred as the phenotype. The phenotype may be directly used as a solution to the problem. Many times, however, the GA may not be able to work over the phenotype representation, which may take any arbitrary form depending upon the problem.

The phenotype solutions are, hence, cast into a standard form of individual representation called as the genotype. The genotype contains all the specifications of the solution. This is usually in the form of a string of real numbers, string of binary numbers, or sometimes a tree-based representation. The GA works over the genotype and optimizes it. The final genotype solution may be converted into the phenotype solution to be used as the solution to the problem.

Population

Population is a collection of individuals at any point of time in a natural world. In the context of GA as well, we use the term population to denote a set of solutions to the problem, or a set of individuals. The population set is worked upon as the algorithm proceeds. The intent is to iteratively improve the individuals in the population pool along with time. The population may further have properties like the best individual, worst individual, etc. This enables us to understand the type of individuals inside the population pool as well as the manner in which the population pool is changing along with time.

Fitness

Every individual in a population has some fitness, which measures its possibility of survival. Individuals with higher fitness are more likely and more dominant than individuals with lower fitness. The fitness in the case of the GA refers to the goodness of the solution or individual to solve the problem. Every problem being considered may have its own mechanism of assigning the goodness or the fitness to a solution. The fitness if assigned to the individuals by a fitness function in the GA. Further, the fitness is measured over the phenotypic representation of the problem, and hence, the conversion of the genotype to phenotype may be required in case both of them have different formats.

Generations

In a natural world, one generation of population is responsible for the construction of the next generation, which further constructs a still higher generation. In this case, the entire population pool keeps being generated with a higher generation population being generated from a lower generation population. Similarly, in GA we use generation as the number of iterations of the algorithm. At each iteration, the current set of solutions or population pool is used to generate the next set of solutions or the higher generation population pool. In this manner, the population is expected to improve as the iterations or the generations go on. The improvement is measured in terms of the fitness values of the individuals.

Evolutionary Operators

At every generation or iteration, the algorithm needs to use the previous population of previous generation and generate the next generation population. This is done by application of a variety of operators called the evolutionary operators. Each evolutionary operator takes some individuals and returns the individuals of the next generation. A number of evolutionary operators are applied at every generation that gives the entire population pool of the next generation. We discuss these operators in the next section.

Stopping Criterion

The continuous loop of evolution may continue on and on. In fact this is what happens in a natural scenario where one generation leads to the other and the process goes on and on in this mechanism. The evolution results in continuous improvement of the individuals as per the changing conditions. However, in the artificial systems that we develop, the evolution must stop. This is to enable the algorithm return the best individual which can be actually used for the purpose of problem solving.

The stopping of the process may be carried out in a variety of ways. One of the most widely used mechanisms is to fix the total number of generations or the iterations of the evolutionary process. The other mechanism used is to fix the total time of execution of the algorithm. Fixing the total time of execution would make the algorithm stop after the allocated time has expired. This is a very important criterion when the fitness function evaluation takes a lot of time and we have a limited amount of time. The large computation time of the fitness function increases the time of the algorithm execution, which in turn forces a just fixing of the stopping criterion. The other option is that the stopping criterion be triggered externally by a human or by other events.

The other set of stopping criterions as slack generations and slack time. The slack generation stopping criterion causes the algorithm to stop when no improvement can been seen for some specified number of generations. At this time, it is observed that the individuals have converged and no improvement may be observed on executing the algorithm further. Similar is the criterion of slack time. Here, the specification is of the time within which no improvement was observed. Accordingly, we may also specify the minimum improvement in terms of fitness value of the best individual in the population for the stopping criterion.

While designing the genetic algorithm, we may specify one of more of the stopping criterions. The algorithm or the optimization loop is stopped when either of these stopping criterions is met. The algorithm searches for the best individual in the population pool and returns the same. This is the output of the optimization algorithm, which may be used as per the problem. In case of the problem of motion planning of mobile robots, this would be the path that may be traversed by the mobile robot.

SIMPLE GENETIC ALGORITHM

The simplest form of evolutionary algorithm is the Simple Genetic Algorithm (SGA). The algorithm uses the principles discussed in the previous section for the task of optimization. The complete optimization process is designed so as to enable the algorithm to optimize along with generations. Metrically it is expected that the individuals keep being optimized in terms of their fitness values. The individual here is a reference to the solution of the problem in hand. We deal with the different aspects of the algorithm in the coming sub-sections.

Broadly, the entire framework of the genetic algorithm is given by Figure 1. Initially, we start with a randomly generated population, each of which is a potential solution to the problem. These individuals may differ in their ability or effective-

Figure 1. Simple genetic algorithm

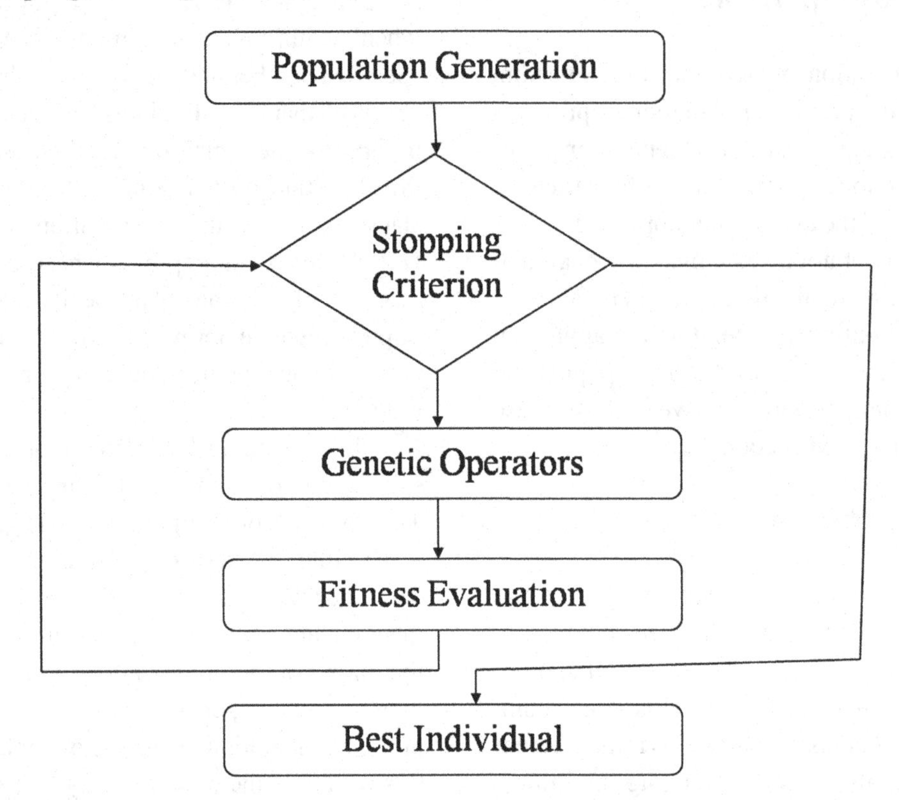

ness to solve the problem. This is measured by passing them through a fitness function, which returns their fitness values. The individuals are then placed inside the loop of the evolution process. At every iteration or generation of the loop, the individuals of higher generation are generated using the individuals of the lower generation. In other words, a higher generation population is obtained from a lower generation population.

The complete process of generation of higher generation population is done using genetic operators. These operators are applied in combination for producing each of the individual of the higher generation. The operators attempt to make fitter individuals for the next generations, fitter than the individuals of the previous generation. This is an attempt to improve both the average fitness of the population pool as well as the best fitness. The major evolutionary operators used are selection, which selects the individuals that participate in the evolution. The selected individuals are passed through the evolutionary operators of crossover and mutation. Each of these operators is discussed in the subsequent sub-sections.

Individual Representation

The first and foremost task in solving any problem is an individual representation. The individual representation specifies the manner in which an individual of the population pool may be represented. The individual representation depends upon the problem being handled; hence, this step is problem specific. This step of the algorithm defines the fundamental logic or the bridge between the problem being handled and the genetic algorithm.

The individual is a working solution to a problem, which is intended to be optimized by the genetic algorithm. The problem specific individual

Figure 2. Sample genotype

1.3	4.8	-2.3	8.7	2.9	2.8

or the solution to the problem is known as a phenotype. Examples of this would be a robotic path consisting of points in the map, in case of problem of robot motion planning, a neural network in case of problem of evolution of a neural network, and likewise. The phenotype may be directly used for solving the problem. However, the genetic algorithm may not be able to work on optimization of the phenotype directly. For these reasons, it is sometimes better to convert the phenotype into a form that the genetic algorithm can optimize. This form is known as the genotype. The genotype in most of the cases of genetic algorithm is a vector of real numbers. This is known as a real coded genetic algorithm. Sometimes a vector of binary numbers us also used, which is known as a binary genetic algorithm. Some other implementations enable a tree-based representation of the problem.

The conversion of the problem in hand into either of these types may be done by the problem logic. For most of the problems simple conversion logic may be enough to carry forward this conversion. This would involve identification of the parameters of the problem and to embed them together in a linear array. Every parameter becomes a gene of the complete individual. All the embedded values are later optimized by the evolutionary process. Each gene has some specific meaning as per the representation scheme. For the problem of evolution of the neural network, the scheme would involve jotting down all weights and biases in a linear manner. Each gene corresponds to a specific weight or a specific bias. The process of optimization involves finding the set of values for each of these genes, such that the resultant solution in its phenotype representation is the most ideal to solve the problem.

Figure 2 shows a sample genotype for a hypothetical problem. It may be seen that a real number representation scheme is followed here. The complete evolutionary process returns the optimized set of values of these genes, which may be easily converted back into the phenotype for the task of problem solving.

Initial Population

One of the initial steps is to generate an initial population. In most of the cases, the initial population may be generated randomly. This means we make a pool of individuals, each of the type as discussed in previous section, with random values to all the genes. In some cases, we may use heuristics to generate the population. An easy application of heuristic would be to bias the individuals towards some range of values for every gene. This means the probability of generation of higher values may be large for some cases and likewise. The heuristics may again depend upon the genes and their meaning in the phenotypic representation. It is assumed that we already have an idea of the upper and the lower bounds of each of the genes. This may be obtained by considering the individual representation technique, the meaning of the gene, and its role in the phenotype. The total number of individuals to be generated is known as the population size. This factor may be constant, or in some implementations it may vary along with time as per the situation of the evolution. This creates the pupation at the 0th iteration or the initial population.

Scaling

One of the first tasks associated with the generation of a higher generation population from a lower generation population in the evolutionary pool is scaling. Scaling assigns an expectation value to each of the individuals in the evolutionary process. The expectation value is an indication of the chance of the individual being selected for participating in the evolutionary operations. A higher expectation value means that the individual is more likely to be selected for the evolutionary process. Once the individual gets selected, it or its produced offspring go to the next generation of population.

In the selection process, as we shall see later, an individual may be selected multiple times. It is even possible that an individual does not get selected at all. This depends upon the expectation values and their distribution amongst the individuals. Two popularly known schemes are fitness-based scaling and rank-based scaling. Fitness-based scaling assigns expectation values to individuals proportional to their fitness values. The expectation of any individual may be given by Equation 1.

$$Exp(I) = ExpMin + \frac{ExpMax - ExpMin}{FitMax - FitMin}(Fit(I) - FitMin) \quad (1)$$

Here ExpMin is the minimum expectation value, ExpMax is the maximum expectation value, Fit(I) is the fitness of the individual, and Exp(I) is its expectation value. FitMin is the minimum fitness and FitMax is the maximum fitness value. A higher fitness value and a higher expectation value are considered better.

It may be easily seen that the complete fitness spectrum containing the fitness values of the various individuals is re-distributed to the expectation scale. This scheme results in the selection of fitter individuals, which is a prime requirement of the process. However, the problem with this scheme is that individuals with reasonably high fitness values result in being selected too many times, and the other solutions with lower fitness values become extinct from the population pool very soon. Hence, the next generation of the population is dominated by few individuals that possessed exceptionally large fitness values. The other good characteristics that existed with the other individuals are all lost, as these are extinct now. The only few characteristics that exist with the few dominating individuals are left. This, as we shall see later, results in the problem of premature convergence.

The other scheme used for the problem is a rank-based scaling. In this scheme, the expectation value of any individual is proportional to its rank in place of the fitness value. This avoids the domination of the population by a few fit individuals. The expectation value in this case may be given by Equation 2.

$$Exp(I) = ExpMin + \frac{ExpMax - ExpMin}{N - 1}(N - Rank(I) - 1) \quad (2)$$

Here ExpMin is the minimum expectation value, ExpMax is the maximum expectation value, Rank(I) is the rank of the individual and Exp(I) is its expectation value. N is the population size. A lower rank and a higher expectation value are considered better.

One of the other schemes to be used is the top scheme. Here, we select the top few individuals and give them the same expectation values. Hence, these are equally likely to be selected.

Selection

The first genetic operator used in the evolutionary loop is selection. This operator selects the individuals that participate in the evolution process, which involves the application of the other genetic operators. The selection operator is inspired from the Darwin's principle of survival of the fittest. The purpose of this operator is to probabilistically select individuals based on their expectation values. This means that the probability of individuals with higher fitness value is higher, and vice versa. However, it is possible for highly fit individuals not to be selected and a few very low fitness individuals to be selected. However, generally the fitter individuals are selected multiple times and most of the lower fitness individuals are not selected. This plays a major role in enabling the algorithm converge, as we shall see later. This further means that the fitter individuals would be more dominant in the population pool (and likely not being over-dominated). Their characteristics would be spread around many individuals, which is good as it is expected that the fitter individuals had better characteristics that resulted in higher fitness.

The selection may be done by a variety of schemes. One of the schemes is tournament selection. In this scheme, selection is done by a set of tournaments. Each tournament is by a set of matches being played between two or more individuals, as per the scheme used. For every match, the set of individuals with high expectation values have more probability to win. The individuals winning the tournament are selected. These participate in making the population of the next generation by the other operators.

The other commonly used technique is roulette wheel selection. Here all the individuals are represented on a roulette wheel. The circumference of the individuals is made proportional to the expectation value of the individual. The roulette wheel has a pointer used to decide the selected individual. The wheel is rotated, and when it comes to a stop, the individual appointed by the selector is selected. In this manner, to select n individuals, we rotate the wheel n times. At each rotation, we get a winning individual. It is natural that the winning probability of an individual is proportional to its circumference, which again is proportional to its expectation value.

Figure 3. Stochastic uniform selection

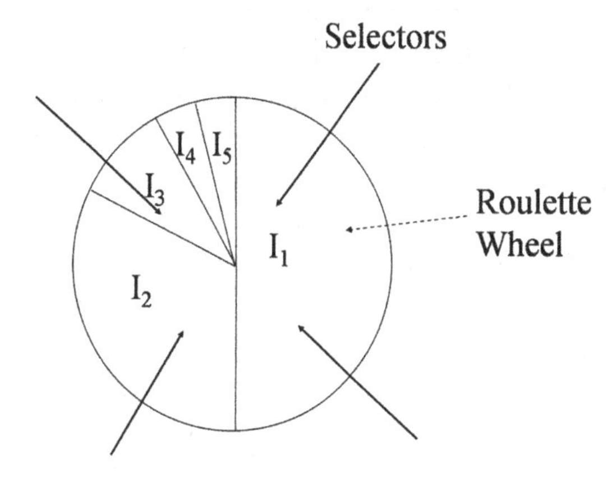

The other scheme used for selection is a stochastic uniform selection. The problem with roulette wheel selection is that fitter individuals are likely to be selected very large number of times, thus over-dominating the solution. In this scheme, we use the same concept. However, we take n number of pointers in place of one pointer. Due to the same, all the individuals are selected at a single run of the wheel. This scheme is summarized in Figure 3.

Crossover

The next operator we discuss is crossover. The crossover operator tries to fuse parents, which belong to the previous generation, to generate children for the next generation. The crossover operator carries out an exchange of characteristics between the parents to produce the children. This operation may be carried out between any number of parents to generate any number of children. However, the most commonly used is the two-parent crossover. This crossover technique uses two parents to generate two children. This is inspired by the natural evolution process in which a sexual reproduction between two parents results in a child.

There are a number of schemes for carrying out crossover. One-point crossover selects a point over which the crossover is to be carried out. The genes of the parents are copied from the parents to the two children across this point. The genes lying to the left of the point of crossover are copied from the first parent and the others from the second parent for the first child and vice versa for the other child. This scheme is shown in Figure 4a. The other scheme used is the two-point crossover. Here, two points are taken around which crossover is carried out in place of one. The genes from the extremes are taken from one parent, and the middle genes are taken from the other parent. This scheme is shown in Figure 4b.

The discussed schemes have positional dependencies, which is a hazard for the optimization processes. Two genes together in the parent are likely to be in the same order in the children, unless the point of crossover lies in between these genes. The optimization process is affected by the ordering of the genes. The next scheme for crossover attempts to remove these limitations. The scheme is called the scattered crossover scheme. In this scheme, any gene from any of the parents may go to any of the children. The decision is taken randomly for every gene. The corresponding gene from the other parent goes to the other child. Hence, there is no point of crossover here, or in other words, a random number of points of crossover are scattered all around the individual. This method of crossover is given in Figure 4c.

Heuristic crossover is another scheme that may be applied to the individuals, especially those with a real number genotype representation. Here the crossover takes place gene by gene, and the two individual genes are considered to give the gene value of the child chromosome. Suppose two corresponding genes of the parents are x_1 and x_2, the child may get a value of $w_1x_2+w_2 x_2$ where w_1 and w_2 are normalized weights.

At any time the number of individuals produced by this operator is given by cr x N, where cr is the crossover rate and N is the number of individuals in the population pool.

Mutation

The other major operator is mutation. While the crossover focuses upon the exchange of characteristics in the individual, this operator attempts to generate new characteristics. A mere exchange of characteristics amongst the individuals may not always result in good optimization. Many times many characteristics may be missing from the population pool. Many other times some characteristics may require some kind of tuning for optimal performance. All these needs are catered by the use of a mutation operator.

The mutation operator may be different depending upon whether the problem representation is in the form of a sequence of bits or real numbers. For a bit sequence representation, some of the bits of the individual undergoing mutation are flipped. For a real number representation, the values of the real numbers are altered by some degree. The number of bit flips or the magnitude of alteration depends upon a rate known as the mutation rate. This operator is shown in Figure 5.

One of the commonly used techniques of mutation is Gaussian mutation. This technique generates random Gaussian numbers. The probability of generation of low numbers is very high and high numbers is very low. These numbers are used as mutation rate in the algorithm execution. This avoids the algorithm having ill effects of high mutation rate and a low mutation rate. This we shall study in our discussion on convergence.

Other Genetic Operators

A number of other genetic operators are also used for the purpose of optimization. Many times a number of genetic operators may be customized, or entirely new operators may be developed, as per the problem understanding. This may result

Figure 4. Crossover schemes: (a) one point crossover, (b) two point crossover, (c) scattered crossover

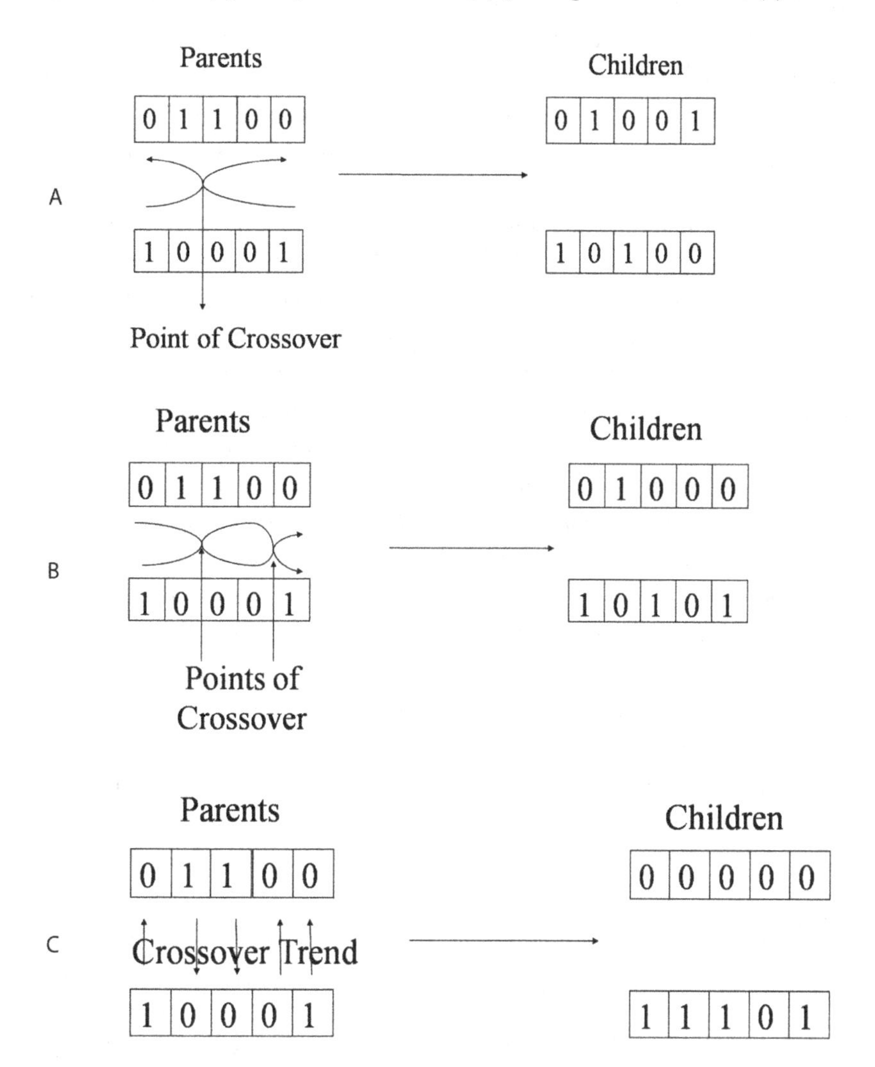

in getting better individuals for the next generation. The entire evolution is greatly benefited if we can fit in some problem logic that guides the evolution towards a generation of better and fitter individuals. This is usually in the form of some customized genetic operators.

Elite is another commonly used operator. Many times the best individual of a population might get damaged by the evolutionary operators or may not be selected at all. In such a case, there is a drop in the best fitness of the population pool. This may be dangerous as the ultimate aim of the evolutionary

Figure 5. Mutation in binary individual representation

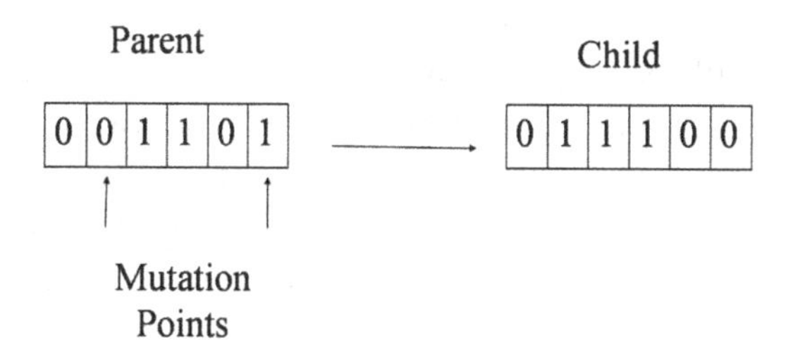

process is to return the fittest individual. Hence, using elite operator, we directly transfer the best individuals of a generation to the next generation without any change. The number of individuals is called as the elite count.

Sometimes a repair operator is also used. The evolution process or the random initial population generation process may sometimes result in individuals that are not feasible as per the constraints mentioned in the problem. These individuals may hence be repaired, where their parameter values are altered such that the resultant individual is feasible.

Other operators include insert, where new individuals are added to the population pool; delete where some individuals may be deleted from the population pool; etc. Many times mutation operators may be split into soft mutation and hard mutation with low and high mutation rates, respectively. The soft mutation may be frequently applied, while the hard mutation is applied with very low frequency. This avoids the problem of small and high values of the mutation rate.

CONVERGENCE IN GENETIC ALGORITHM

In the previous section, we discussed the genetic algorithm and its working in order to get the optimal solution of the problem. We further analyze the working of the algorithm in this section. Imagine a high dimensional space with m+1 dimension where m is the number of genes in the individual representation. Every axis in this space corresponds to one gene. As every gene has an upper and a lower bound, the entire space is bounded. Let the vertical axis of this space be reserved for the fitness value. For every combination of genes, we may find a fitness value and mark it as a point in the space constructed. This space is known as the fitness landscape. The purpose of the genetic algorithm is to compute the highest point (or lowest point for minimization problems) in this landscape. It is natural that since the number of points, even for a fair resolution, are very high, this landscape cannot be physically constructed. The fitness landscape for a sample problem is given in Figure 6.

The initial population is randomly generated and randomly spread across the landscape. At any generation individuals not selected (the ones with lesser fitness value) ultimately die. The ones selected are usually those that lie at better areas in the fitness landscape. At any generation, crossover operation between two individuals results in the generation of an individual in between these two individuals in the fitness landscape. Since the two individuals possess good fitness, they lie at better areas in the fitness landscape. Hence, the resulting solution is also likely to lie at better locations and possess better fitness. Since the individuals are produced at intermediate positions, the entire population set at any generation is found to lie within a smaller region of the fitness landscape as compared to the previous generations. In this manner, crossover operator results in convergence along with generations, with the different individuals ultimately converging to have their positions within the same point in the fitness landscape.

The other major operator discussed was mutation. This operator makes the selected individuals move around in the fitness landscape. The amount of movement depends upon the mutation rate.

Figure 6. Fitness landscape

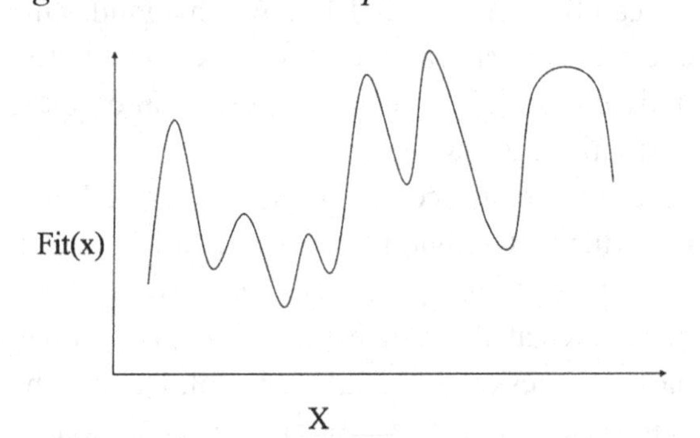

The selected individual usually has a good fitness value and, hence, motion of the individual in this landscape by small amounts results in small changes to this fitness value. If the global optimum is far off, many mutations would be required to discover it. Very large movements may make the individual move at any random point, which may have good or bad fitness value. Mutation may result in expansion of the entire region where the population is found. This avoids convergence. In the initial stages of the algorithm run, since the individuals are spread across a wide area of the fitness landscape, crossover dominates mutation, and the various individuals quickly attempt to focus at a point. However, this might lead to a pre-mature convergence. This is balanced by the mutation operator, which plays a mild role in initial generations. Towards the end, the various individuals get converged within a small area of the fitness landscape with similar fitness values. In such a scenario, the mutation operation is responsible for exploration for search of the global optima.

Another manner of analysis of the algorithm is by the schemata theory. Imagine the individual representation is in the form of a bit string. In this theory, we represent the complete population by a template or schema. The number of genes in this template or schema is equal to the number of genes in the individual representation. A schema may contain 0, 1, or *. Presence of a 0 means all individuals of the population contain a 0 at the gene. Similarly, presence of a 1 means all individuals of the population contain a 1 at the gene. Presence of * means some individuals may contain 0 while the others contain 1. In other words, all the individuals of the population do not contain the same bit.

If the schema of the initial population contains all *, no convergence has happened. If presence of a 0 as a characteristic at some gene is highly unlikely, all the individuals would have very low fitness and would soon die off. This is when 1 would appear in the schema at the specific gene representing some characteristic. Hence, as the optimization proceeds, the schema starts getting loaded with 0s or 1s. Towards the end the schema has very little *s and mostly 0s and 1s. This is when convergence has happened in the population.

PATH PLANNING BY GENETIC ALGORITHM

The studied concept of optimization by use of Simple Genetic Algorithm is applied to the problem of path planning. The problem description remains the same as discussed in the preceding chapters. Since the algorithm used is a genetic algorithm, we do not assume that the dimensionality of the problem is reasonably small. However, at the same time, some restrictions do exist in the dimensionality of the problem, as path needs to be worked over, and very large dimensionality would require a very long time in going through the path and computing its metrics. Further restrictions exist in terms of the complexity of the path. We assume the map is not like a maze where considerable effort may be required to compute a path from the source to the goal. The number of turns from the source to the goal needs to be limited in nature.

The complete problem of motion planning of the mobile robot may be taken as a standard genetic algorithm problem for the purpose. Hence, taking this problem in mind, we design the complete genetic algorithm that computes a robotic path, given the map. For related works, readers are directed to Hu et al. (2004), Mahjoubi et al. (2006), Manikas et al. (2007), Naderan-Tahan (2009), Petrinec and Kovacic (2005), Xiao et al. (1997), Xing et al. (2007), and Yan et al. (2007).

Individual Representation

The first task in problem solving with genetic algorithm is of individual representation. Here the phenotype is a robotic path, which the robot controller can use for the physical navigation of the robot. We first attempt to convert this robotic path into a linear array of real numbers that can be used for the purpose of optimization. Any robotic path may be represented by a set of points on the robotic map. These points when joined to each other in order represent the robotic path. The source is always the first point in this set of points and goal is the last point. Hence, consider the robotic path as <$S \rightarrow P_1 \rightarrow P_2 \rightarrow P_3 \rightarrow P_4 \rightarrow$ ....$P_N \rightarrow G$>. Here, S is the source, G is the goal, and P_1, P_2, ... P_N are any points in the robotic map. N is the number of points, besides the source and the goal. Consider a random path shown in Figure 7.

Suppose the map that we take is of 2-dimensions of size m grids x n grids. The source $S(s_x, s_y)$ and the goal $G(g_x, g_y)$ are already known. Any point in this map may be represented as $P_i(P_{ix}, P_{iy})$. Since the point lies inside the map, $1 \leq P_{ix} \leq m$ and $1 \leq P_{iy} \leq n$. We further suppose that the robot may take a maximum of N turns in its path from the source to the goal. This means that the algorithm would not be able to generate paths greater than this complexity of N turns. This is one of the major limitations of the algorithm that this complexity needs to be specified and finite. It may be observed that every point serves as a point of turn. Hence, the complete path is the specification of these points of turn.

The purpose of the algorithm is hence to optimize the location of the points P_1, P_2, ... P_N. Each point may be represented by its x and y coordinate $P_1(P_{1x}, P_{1y})$, $P_2(P_{2x}, P_{2y})$, (P_{3x}, P_{3y})... $P_N(P_{Nx}, P_{Ny})$. The complete individual to be optimized may be represented by ($P_{1x}P_{1y}P_{2x}P_{2y}P_{3x}P_{3y...}P_{Nx}P_{Ny}$). This forms the individual for the optimization purpose. The individual contains a total of 2N genes.

The complexity of the optimization process depends directly on the number of variables or parameters being optimized. In this case, the problem has a total of 2N variables for optimization. Having too much complexity results in the algorithm capable enough of generating very complex paths, but the time of convergence is very large, and the algorithm is likely to get struck at some local optima. Further exploratory potential is limited and the algorithm may not converge well into the optima. Accordingly, the variable may be set as a tradeoff between the algorithmic capability to generate complex paths, and the problems of high dimensionality in case of evolutionary algorithms.

Figure 7. Individual representation for the problem

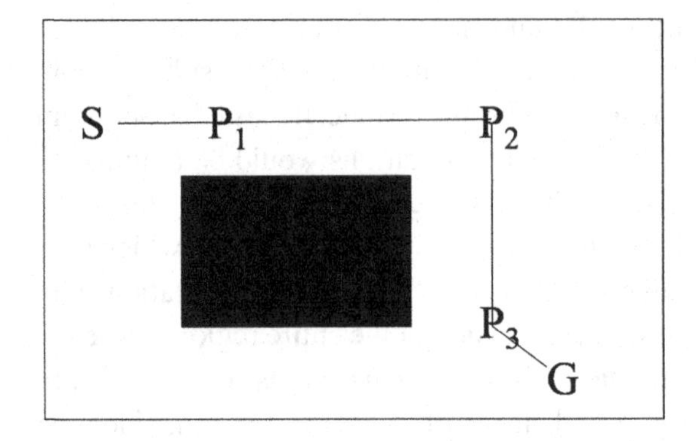

Genetic Operators

The other major factor in the implementation of the genetic algorithm is the formulation of genetic operators. Here, since we have already formulated the problem as a vector of real numbers, we may easily use all the conventional genetic operators for the task of optimization. The used genetic operators consist of a rank-based scaling, stochastic uniform selection, scattered crossover, Gaussian mutation, and small elite count. This carries forward the evolution process as per the discussed concepts of the simple genetic algorithm.

However, we make another operator for the problem. This operator is known as shorten, and it is a problem specific operator. It is natural that the solution to any problem should have as few turns as possible. This is an incentive to the robot for traveling in a straight line without useless turns. With this operator, we attempt to reduce the length of the path by doing away with some turns. This concept is given in Figure 8.

Figure 8: Shorten operator

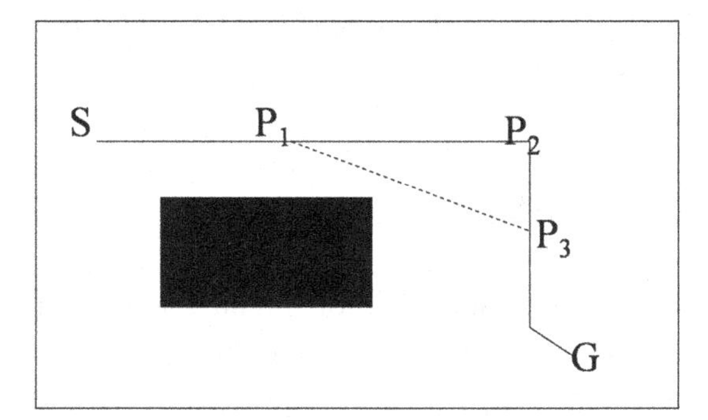

Path S-P_1-P_2-P_3-G
Shortened to S-P_1P_3-G

The process of application of this operator is simple. One by one we try to eliminate all points in the path. If the elimination results in a better path, the newer path is accepted, and the process is repeated with this path. On the other hand, if the elimination does not result in a better path, the path is not changed and the attempt of shortening is repeated. The algorithm stops when it is no longer possible to eliminate any other point from the path. The best possible shortened path is then used. This operator is applied to a path just before its fitness evaluation. This gives the path as few turns as possible. This may then be converted into its phenotype physical path for the fitness computation.

Fitness Evaluation

The final step towards the use of the genetic algorithm for the problem is to make a fitness function. The GA works over the genotype individual representation, which is a linear vector of real numbers. The genotype first needs to be converted into a phenotype or a robotic path for the fitness computation. We first obtain a set of points that the individual represents. This set of points is then used to figure out the robotic path over the map under consideration. Now, we first need to find out whether the path is feasible or not. The path is known as feasible if moving on this path does not result in any collision with any of the known obstacles. The path is infeasible if some of the points in the path lie within obstacles. To compute the path feasibility, we need to travel the entire path of the robot grid by grid. Hence, a very large dimensionality of the map may mean a very large computation time as the number of grids would be very large.

The fitness of the path may be given by Equation 3:

$$Fit = \left(\sum_{i=0}^{i=N} || P_i - P_{i+1} || \right) + \alpha X \quad (3)$$

This assumes P_0 is the source and P_{N+1} is the goal. $||x||$ is any norm, which may usually be taken as the Euclidian norm. α is the penalty constant. X is the total number of grids in the path that lie inside obstacles. This is obtained by physically traversing upon the path.

This is an interesting mechanism of handling infeasible solutions by penalty. The penalty deteriorates the fitness value by the extra term. In this case, it is a minimization problem and smaller fitness is better. The magnitude is kept proportional to the amount by which the solution disobeys the problem constraints. This means that the feasible solutions may amount to be better than the infeasible solutions as the penalty constant is usually kept high, but when there are a large number of infeasible solutions, the algorithm encourages a generation of solutions that obey the problem constraints to as large degree as possible, if not completely. In this case, the encouragement is to reduce the part of the solution lying in the obstacle region, by minimizing the number of points within obstacle. Hence, one may observe that if there are a very large number of infeasible paths in the population process, the infeasibility is reduced, and, ultimately, it may completely vanish off.

Results

The approach as discussed was implemented in MATLAB. The Genetic Algorithm toolbox was used as a pre-available module for Genetic Algorithm. The map was fed into the algorithm as a bmp image file. Unlike the previous graph search approaches, in this algorithm we fed maps of higher resolution. Each map used was of size 500 x 500 pixels. The maximum number of turns was fixed to 5, which meant that the genetic algorithm had a genomic length of 10 real numbers. Each of these could lie anywhere in the map. In all the cases, the source was the top left point of the map and goal was the bottom right corner of the map. The optimization was carried out for 20 generations with a population pool of20 individuals. Crossover rate was 0.7 and Gaussian mutation was used. The results to some of the scenarios are shown in Figure 9.

CURVE SMOOTHENING

The discussed approach for solving the problem of motion planning works well under the stated conditions. However, one of the major limitations of the discussed approach is that the paths are very steep at the points of turn. Any robot would not be able to follow these paths as these do not obey the non-holonomic constraints. For the same reasons we need to smoothen these paths, before the controller being used for the purpose of making a robot move. However, as all these points in which

Figure 9. Experimental results with genetic algorithm

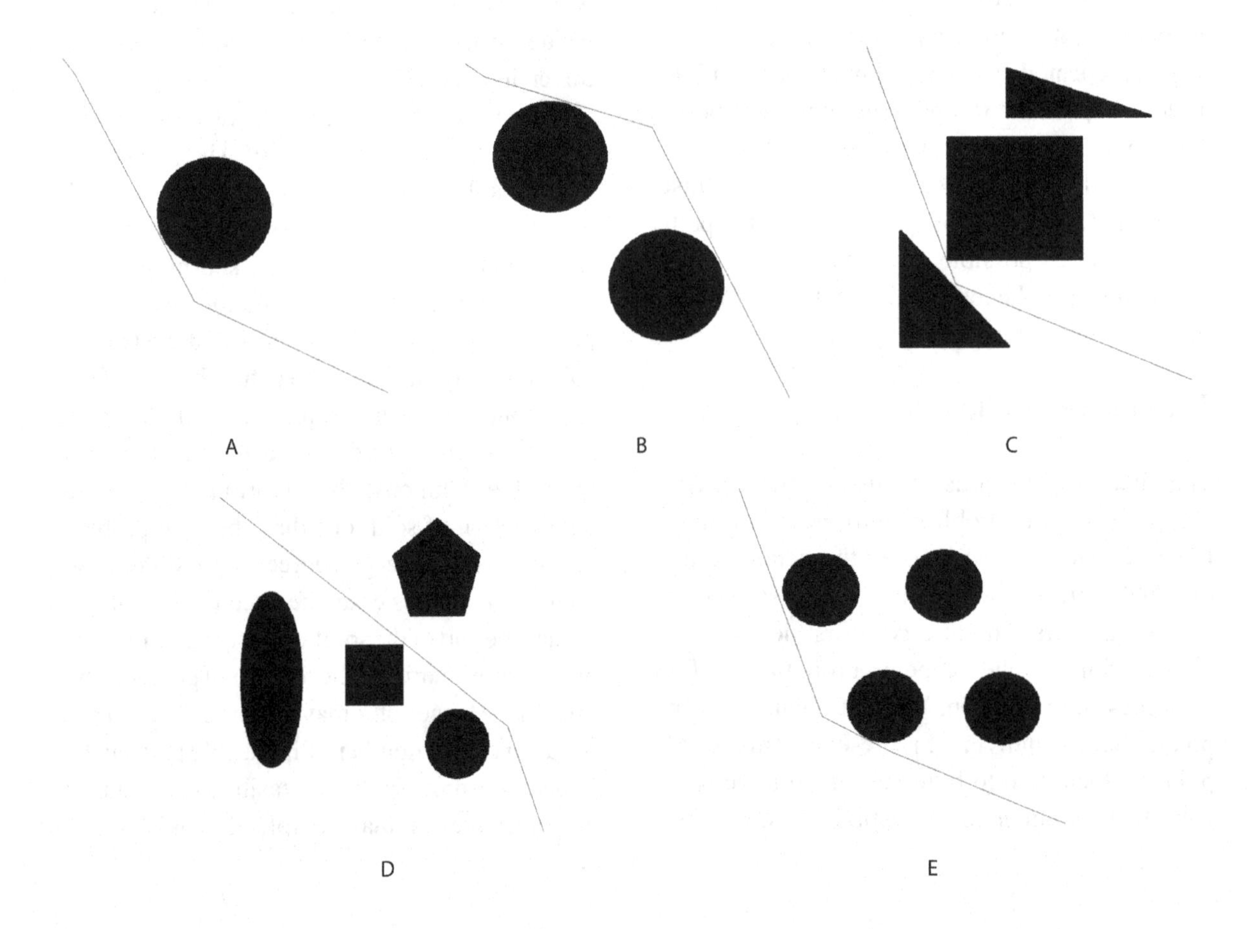

turns are made are very close to the obstacles, smoothening of the paths may result in possible collisions. Smoothening may make some part of the path to lie over an obstacle. This would make the planning ineffective.

Hence, the part of smoothening of the curve needs to be integrated into the planning algorithm by the genetic algorithm. This would ensure that the paths generated are smooth and may be easily traversed physically by the robot using the robotic control. The smoothening of the path of the robot is done by application of Bezier curve or by BSpline curves. Each of these curves takes as an input a set of control points. The output of these curves is a smooth curve that is generated using the input control points. In this section, we further improve the algorithm and use the Bezier and BSpline curves for the motion planning. Again, the entire problem would be taken as a standard genetic algorithm problem. In fact, all modules from problem representation to genetic operators and fitness evaluation are used as it is from the previously built solution. The only difference we introduce is an additional step applied before the fitness evaluation. This uses the set of points as the control points. This converts a straight path into a smooth curve. All fitness computations are then carried out on this smooth curve.

Individual Representation

The first problem is the mechanism of individual representation. Again, we represent the path of the robot as a set of points. The only difference in the individual representation is the meaning of these points. In the previous approach, these points represented the physical position where the robot makes a turn. However, in this case the points represent control points over which the curve is constructed. Hence, these points may themselves lie on an obstacle with the resultant path still feasible, if all the points on the curve do not lie over obstacles. However, the restriction is that still all the points must lie inside the map. Hence, the bounding conditions remain the same. The individual may be given by (P_{1x} P_{1y} P_{2x} P_{2y} P_{3x} $P_{3y...}$ P_{Nx} P_{Ny}).

Genetic Operators

The same genetic operators along with the shortening operator may be used in this case as well. This is attributed to the fact that the basic approach remains the same. The operator set consists of rank-based scaling, stochastic uniform selection, scattered crossover, Gaussian mutation, small elite count, and shorten.

Bezier Curve

Bezier curves (Bartels, 1998; Weisstein, 2010) are the first smoothing technique to be used. These curves take a set of control points and return a smooth curve more or less bounded by the control points. The construction of the curves depends upon the degree of curve used. As the degree of curve increases, the complexity of generation and the curve behavior increases. Liner curves with a degree of 2 can just connect two control points by line. The next are the quadratic curves. Consider three consecutive points A, B, and C. The aim is to draw a smooth curve from A to C. Every point in the smooth curve is constructed using two intermediate points, one laying in straight line from A to B and the other in straight line from B to C. Using these points, every point in the Bezier curve is sought. Similarly, for higher order Bezier curves, more consecutive points are considered and lower order Bezier curves are used for generation of intermediate points. By piecewise generating the curve, a general curve can be computed across the entire set of control points.

BSpline Curve

The BSpline curves (Bartels, 1998; de Boor, 1978) are also used for curve smoothing and take a set of control points as the input. These curves attempt to connect one point to the other using piecewise polynomial functions. The manner of connecting is like interpolation of the supplied control points, which gives a smooth curve joining the supplied control points.

Fitness Evaluation

From the previous two sub-sections, we get an idea of the concept and working of Bezier curves and BSplines curves for the smoothening of the curves around the control points. In this sub-section, we use the same properties of these curves for the task of computation of fitness given these points. Here, for the computation of the fitness, we first pass the set of control points P_1, P_2, ... P_N to these curve generating functions to get the entire curve, as a set of continuous points B_1, B_2, B_3,... B_b. We physically traverse the path by these points, find out the number of infeasible points that lie within the obstacles, and compute the total path length. The fitness of the path may be given by Equation 4.

$$Fit = \left(\sum_{i=1}^{i=b-1} || B_i - B_{i+1} || \right) + \alpha X \quad (4)$$

This assumes B_1 is the source and B_b is the goal. || x || is any norm, which may usually be taken as the Euclidian norm. α is the penalty constant. X is the total number of grids in the path that lie inside obstacles.

Results

The discussed approach is further experimented. The entire experimental structure is similar to the implementation of the genetic algorithm as discussed earlier. The major difference is the smoothening carried out by the use of these two curves. The same parameters and experimental procedure were used as with the Genetic Algorithm. First experimentation was done by the use of Bezier curves. The results for various maps are shown in Figure 10. The same maps are further used for experimentation. Here, the smoothening is done by the use of BSpline curves. The results for this smoothening technique are shown in Figure 11. It may be easily seen that the paths are better than those given by the genetic algorithm. These paths obey the non-holonomic constraints. Further, these are smooth enough to allow the robot to move at high speeds.

EVOLUTIONARY STRATEGIES

The next topic of discussion is evolutionary strategies. The entire paradigm of evolutionary algorithm is big and consists of four basic pillars of genetic algorithm, genetic programming, evolutionary programming, and evolutionary strategies. While we cannot cover the complete domain of evolutionary computing in a couple of chapters, the attempt is to cover as much detail as possible, due to its widespread use in evolutionary robotics.

Evolutionary strategies are the attempt to make evolutionary computing adaptive in nature. One of the major problems associated with genetic algorithm is the need to specify the values of mutation rate, crossover rate, and other similar parameters. These parameter values are specified by a human expert, and hence it is natural that the values may not be optimally set. Too large or too small value for any of these creates a problem in the optimization process. Further, the ideal values of each of these parameters depend upon the values of the other parameters. This further makes it impossible to compute or set these values. The likely values change with every problem, which is because of the change in the dimensionality and characteristics of the fitness landscape.

Another major problem associated with the parameters is the fixed nature. The ideal values of the parameters may change along with time. Keeping these values constant throughout the

Figure 10. Experimental results of Bezier curve with genetic algorithm

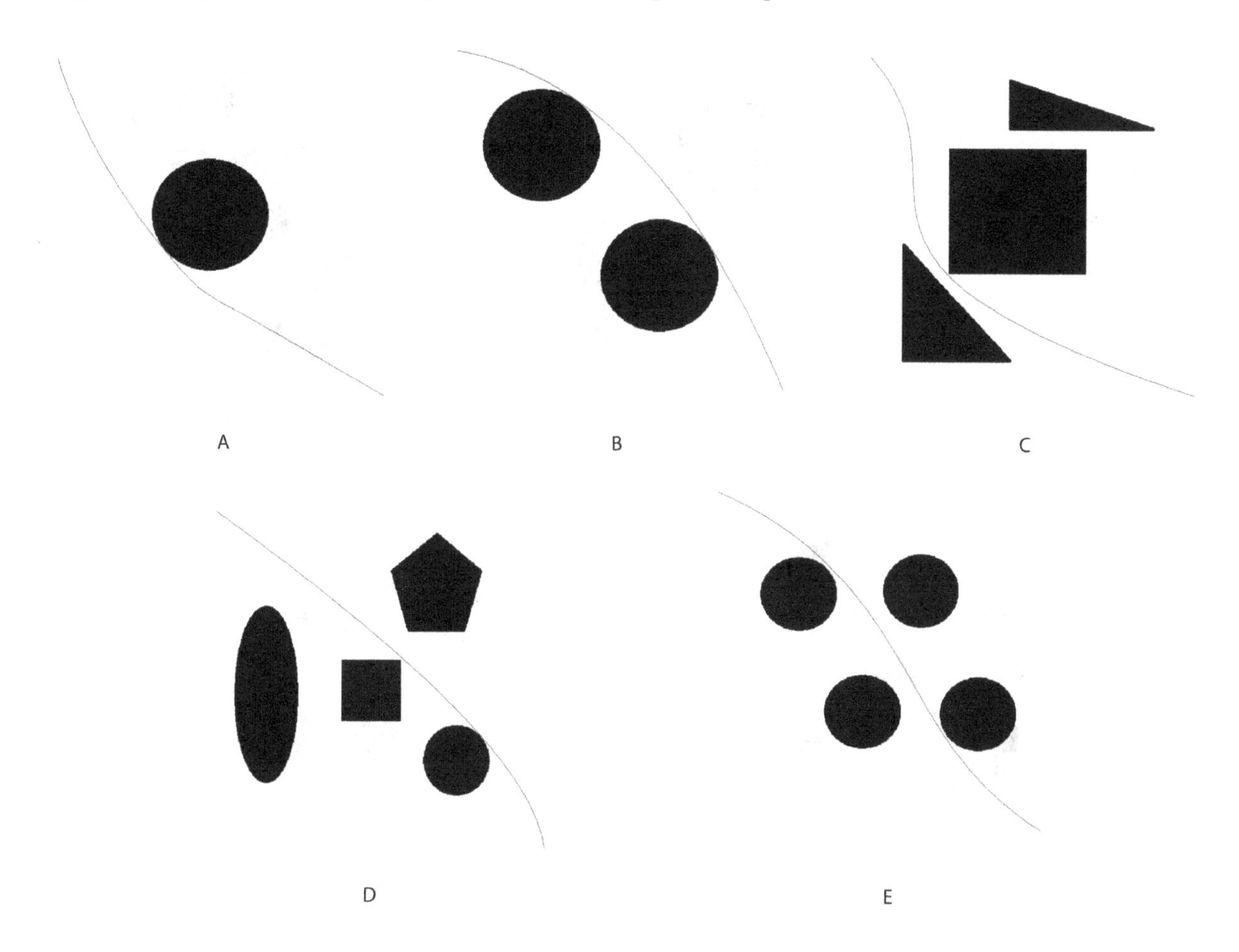

algorithm results in making the evolution sub-optimal. Hence, there must be some mechanism by which the parameter values may be automatically computed at every instance of time. In other words, the attempt is to make the values of these parameters adaptive in nature. Adaptation may be of various kinds and may be implemented at various levels. The adaptive systems are fundamentally classified into deterministic adaptive systems, adaptive systems, and self-adaptive systems. In deterministic adaptive systems, the values of the parameters changes as per some set heuristic defined at the start. In adaptive systems, the change in values of the parameters is as per the current situation indicated by the various metrics of the algorithm. Hence, situation of the evolution has a direct impact in deciding the parameter values. The self-adaptive systems have the parameter values embedded right inside the genetic individual. These values are also optimized as the algorithm proceeds.

Evolutionary strategies are an improvement over the genetic algorithm, as they are self-adaptive in nature. Mutation rate is a major parameter identified in the genetic optimization that plays a major role in deciding the behavior of the evolutionary algorithm. In evolutionary strategies, we intend to make this parameter adaptive in nature. Hence, apart from the normal genes or variables, there are some extra variables that are stuffed into the individual representation. These variables are known as strategy variables. For every gene, we

Figure 11. Experimental results of SPline curve with genetic algorithm

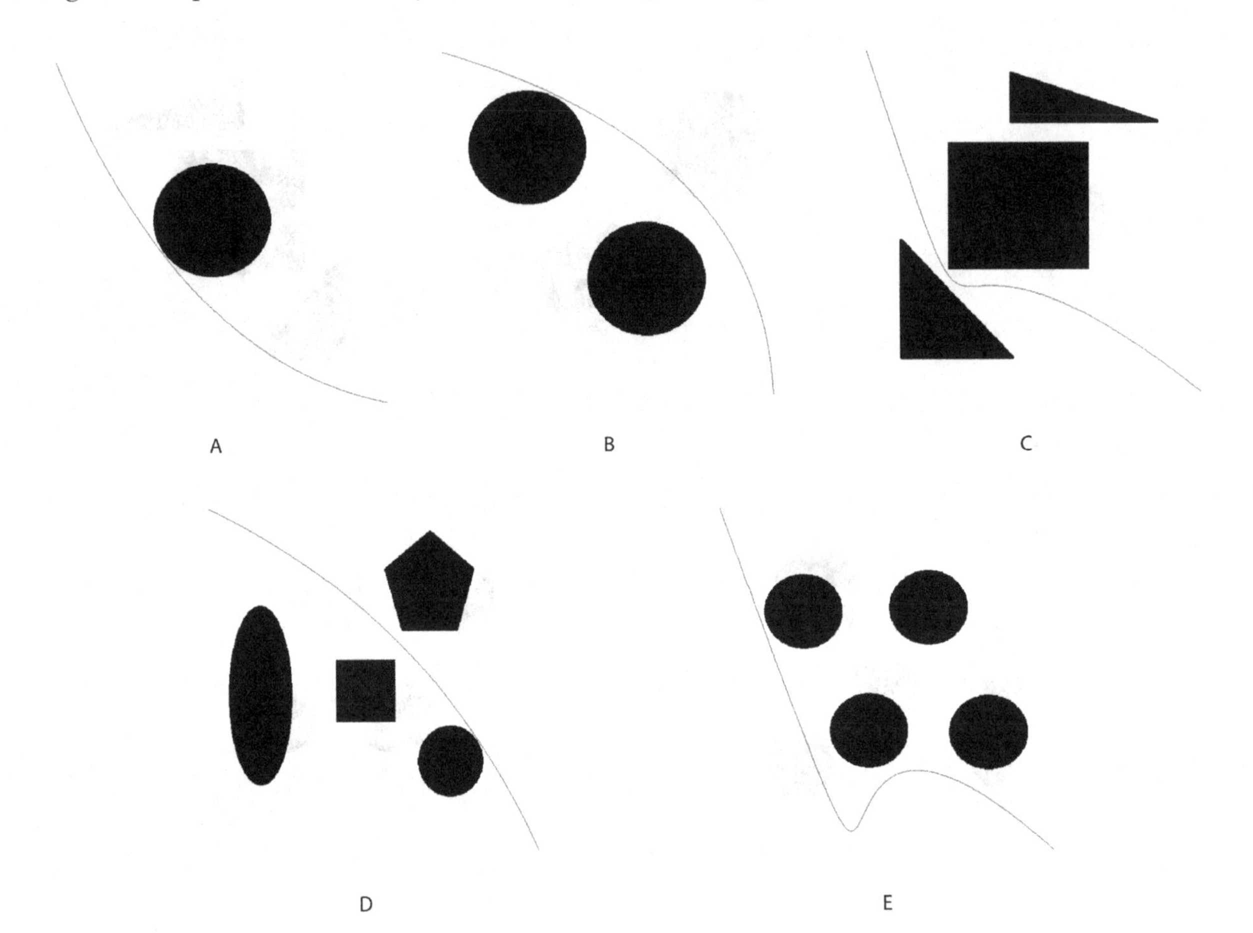

stuff two strategy variables that collectively decide the mutation rate of the variable.

The major genetic operators used in evolutional strategies consist of recombination and mutation. The recombination is equivalent to the crossover operator of the genetic algorithm. This has many parents that together generate a number of children. In this algorithm, all the individuals are generated by the recombination process. The produced individuals are then subjected to mutation. Mutation in case of this algorithm is done in steps. The first step involves the computation of the mutation rate from the strategy variables. After the mutation rate has been computed, this rate is used for the specific gene. Every gene has its own separate strategy variables and hence mutation rate.

The evolutionary strategy may be of various types. One of the simplest forms of evolutionary strategy is represented by (1+1) evolutionary strategy. This consists of only one individual in the population pool, which goes through the operation by the mutation operator. This generates a new individual. The old individual and the new individual are combined and the fittest individual in this lot goes to the next generation. The general representation of the evolutionary strategy is (λ μ/ρ). This means that the population pool consists of λ individuals. These individuals undergo recombination and mutation to generate μ ($\geq \lambda$) children. Here recombination takes place in groups of ρ parents. The best λ individuals of the resultant children are selected that go to the next generation.

The representation may sometimes be (λ+μ/ρ). Here also the population pool consists of λ individuals. Recombination in pairs of ρ parents is carried out along with mutation to generate μ children. A pool of individuals is made consisting of the λ parents and μ children. The best λ individuals from this population pool are selected for the next generation. We do not deal with the complete procedure of working with evolutionary strategies and their various variants in this text. For more details, please refer to the original works and texts (Arnold, 2001; Back, 1996, Back, et al., 1991, 1997; Hansen & Ostermeier, 1997; Schwefel, 1995; Schwefel & Rudolph, 1995).

The problem of path planning of mobile robots may be likewise solved by evolutionary strategies, the same mechanism of individual representation works. The extra strategy parameters, however, need to be stuffed, which get optimized as well as the algorithm proceeds. This algorithm also operates in iterations or generations. The fitness evaluation is done in the same manner. The working and results obtained by the use of evolutionary strategies is similar to the use of genetic algorithms. Hence, we do not discuss the same again.

CONCLUSION

This introduced the domain of evolutionary robotics. The applications of evolutionary computing into the disciple of robotics are immense. Together these domains result in effectively solving the various problems in robotics. The chapter first introduced the principles, concepts, and working of genetic algorithms. A deep insight into the natural inspiration from evolution was taken and developed as a complete computing paradigm. Analysis of the algorithm in terms of working of various operators and convergence enabled the understanding and conceptualization of the algorithm.

The algorithm was later used for the motion planning of the robots. The major step involved the individual representation, where a set of points in the path of the robot were used to effectively represent the robot path. These points were optimized by the genetic algorithm. A customized genetic operator called the shorten operator was designed. This operator eliminated points in the path of the robot, without which the solution could have been shorter. The problem identified in the approach was the sharp turns, which meant not obeying the non-holonomic constraints. In such a case, the path needs to be smoothened. The next approach was to use genetic algorithms over a path represented by a Bezier curve or a BSpline curve. This resulted in a generation of smoother paths, which can be easily traversed physically by the robot. Towards the end, a discussion over evolutionary strategies was done. Here, self-adaptation of the mutation rate was the basic concept used. This was done by embedding extra variables in the individual representation, known as strategy variables. These variables were used for the computation of the mutation rate. A recombination operator was used for the exchange of characteristics amongst the individuals.

REFERENCES

Arnold, D. V., & Beyer, H.-G. (2001). Local performance of the (μ/μI,λ)-ES in a noisy environment. In Martin, W., & Spears, W. (Eds.), *Foundations of Genetic Algorithms* (*Vol. 6*, pp. 127–141). San Francisco, CA: Morgan Kaufmann. doi:10.1016/B978-155860734-7/50090-1

Back, T. (1996). *Evolutionary algorithms in theory and practice*. Oxford, UK: Oxford University Press.

Back, T., Hammel, U., & Schwefel, H. P. (1997). Evolutionary computation: Comments on the history and current state. *IEEE Transactions on Evolutionary Computation, 1*(1), 3–17. doi:10.1109/4235.585888

Back, T., Hoffmeister, F., & Schwefel, H. P. (1991). A survey of evolution strategies. In *Proceedings of the Fourth International Conference on Genetic Algorithms*, (pp. 2–9). IEEE.

Baker, J. E. (1985). Adaptive selection methods for genetic algorithms. In *Proceedings of the First International Conference on Genetic Algorithms and their Applications*. Mahwah, NJ: Lawrence Erlbaum.

Bartels, R. H., Beatty, J. C., & Barsky, B. A. (1998). Bézier curves. In *An Introduction to Splines for Use in Computer Graphics and Geometric Modelling* (pp. 211–245). San Francisco, CA: Morgan Kaufmann.

Chakraborty, U. K., & Dastidar, D. G. (1993). Using reliability analysis to estimate the number of generations to convergence in genetic algorithm. *Information Processing Letters, 46*, 199–209. doi:10.1016/0020-0190(93)90027-7

Davis, L. (1987). *Handbook of genetic algorithms*. Reinhold, NY: Van Nostrand.

de Boor, C. (1978). *A practical guide to splines*. Heidelberg, Germany: Springer Verlag.

Fogel, D. B. (1995). *Evolutionary computation: Toward a new philosophy of machine intelligence*. Los Alamitos, CA: IEEE Press.

Forrest, S., & Mitchell, M. (1993). What makes a problem hard for a genetic algorithm? Some anomalous results and their explanation. *Machine Learning, 13*, 285–319. doi:10.1023/A:1022626114466

Goldberg, D. E. (1989). *Genetic algorithms in search, optimization, and machine learning*. Reading, MA: Addison-Wesley.

Hansen, N., & Ostermeier, A. (1997). Convergence properties of evolution strategies with the derandomized covariance matrix adaptation: The (μ/μI,λ)-CMA-ES. In *Proceedings of the 5th European Congress on Intelligent Techniques and Soft Computing*, (pp. 650–654). Aachen, Germany: Verlag Mainz. Holland, J. H. (1975). *Adaptation in natural and artificial systems*. Ann Arbor, MI: University of Michigan Press.

Hu, Y., Yang, S. X., Xu, L., & Meng, Q. H. (2004). A knowledge based genetic algorithm for path planning in unstructured mobile robot environments. In *Proceedings of the IEEE Conference on Robotics and Biomimetic,* (pp. 767-772). IEEE Press.

Mahjoubi, H., Bahrami, F., & Lucas, C. (2006). Path planning in an environment with static and dynamic obstacles using genetic algorithm: A simplified search space approach. *IEEE Congress on Evolutionary Computation,* (pp. 2483-2489). IEEE Press.

Manikas, T. W., Ashenayi, K., & Wainwright, R. L. (2007). Genetic algorithms for autonomous robot navigation. *IEEE Instrumentation & Measurement Magazine, 10*(6), 26–31. doi:10.1109/MIM.2007.4428579

Michalewicz, Z. (1992). *Genetic algorithms + data structures = evolution programs*. Berlin, Germany: Springer.

Miranda, E. R., & Biles, J. A. (2007). *Evolutionary computer music*. Heidelberg, Germany: Springer. doi:10.1007/978-1-84628-600-1

Mitchell, M. (1999). *An introduction to genetic algorithms*. Cambridge, MA: MIT Press.

Naderan-Tahan, M., & Manzuri-Shalmani, M. T. (2009). Efficient and safe path planning for a mobile robot using genetic algorithm. In *Proceedings of the IEEE Congress on Evolutionary Computation,* (pp. 2091-2097). IEEE Press.

Petrinec, K., & Kovacic, Z. (2005). The application of spline functions and Bezier curves to AGV path planning. In *Proceedings of the IEEE Conference on* Industrial *Electronics,* (vol. 4), (pp. 1453-1458). IEEE Press.

Schwefel, H. P. (1995). *Evolution and optimum seeking*. New York, NY: Wiley.

Schwefel, H. P., & Rudolph, G. (1995). Contemporary evolution strategies. *Lecture Notes in Computer Science, 929*, 893–907. doi:10.1007/3-540-59496-5_351

Shukla, A., Tiwari, R., & Kala, R. (2010a). *Real life applications of soft computing*. Boca Raton, FL: CRC Press. doi:10.1201/EBK1439822876

Shukla, A., Tiwari, R., & Kala, R. (2010b). *Towards hybrid and adaptive computing: A perspective*. Berlin, Germany: Springer-Verlag.

Weisstein, E. W. (2010). *Bézier curve*. Retrieved from http://mathworld.wolfram.com/BezierCurve.html

Xiao, J., Michalewicz, Z., Zhang, L., & Trojanowski, K. (1997). Adaptive evolutionary planner/navigator for mobile robots. *IEEE Transactions on Evolutionary Computation, 1*(1), 18–28. doi:10.1109/4235.585889

Xing, X., Jia, Q., Ling, L., & Yuan, D. (2007). A novel genetic algorithm based on individual and gene diversity maintaining and its simulation. In *Proceedings of the IEEE Conference on Automation and Logistics,* (pp. 2754-2758). IEEE Press.

Yan, X., Wu, Q., Yan, J., & Kang, L. (2007). A fast evolutionary algorithm for robot path planning. In *Proceedings of the IEEE Conference on Control and Automation,* (pp. 84-87). IEEE Press.

KEY TERMS AND DEFINITIONS

Curve Smoothing: Process of conversion of a path in the form of distant points, which denote sharp turnings into a smooth curve, which can be traversed by a non-holonomic robot.

Evolutionary Robotics: The application of evolutionary algorithms to the domain of robotics. Evolutionary algorithms are used in most of the modules of robotics, including mechanical design, control, planning, etc.

Evolutionary Strategies: A self-adaptive form of evolutionary algorithm where extra strategy parameters are added to the genetic individual for every gene. These determine the mutation rate for the gene.

Fitness Evaluation: The task of assigning fitness values to a solution or individual of a problem, where fitness determines how well the solution solves the problem. In the case of a problem of path planning, a genetic individual in the form of an array of distant points is first converted into a robotic path, and fitness corresponds to the path length along with magnitudes of regions of path in some obstacle.

Genetic Algorithm: An evolutionary algorithm used extensively for the task of optimization, where the task involves tuning values of some parameters, or tuning the complete design of a solution to a problem.

Chapter 5
Evolutionary Robotics 2

ABSTRACT

Evolution is a major computing tool. Its effectiveness can be seen in a large variety of problems. A number of evolutionary algorithms exist that differ in the mechanism in which they represent the problem and carry out the evolutionary process. In this chapter, the authors extend the ideas of the previous chapter to look for more evolutionary concepts and algorithms. The first major class includes swarm intelligence algorithms, where they primarily study particle swarm optimization, ant colony optimization, artificial bee colonies, and probability-based incremental learning. In all these algorithms, the basic motive is to use a number of particles that form a complete population, and optimization is carried out by their mutual interaction. The other major algorithm that the authors study in the chapter is Genetic Programming, which forms a major pillar of evolutionary computation. Here, the individual is a program whose execution results in the estimation of fitness denoting the goodness of problem solving. The various operators are modified to enable the representation of the individual to work. A tree-based representation is one of the most commonly used representations. They further study a linear representation of the program, and the underlying technique is known as grammatical evolution. All the algorithms are applied to the problem of motion planning of mobile robots.

INTRODUCTION

Evolutionary computing is a very versatile computing paradigm that finds applications in a very large variety of problems. It is an easy endeavor to model the problem and use an evolutionary computing technique for problem solving. Evolutionary techniques are widely used, especially for the purpose of optimization. This computing technique is able to evolve the complete system, with fewer or no difficulties. The evolution may be of a machine design, pattern design, or a ro-

DOI: 10.4018/978-1-4666-2074-2.ch005

botic path. These algorithms may, however, have large computational times in some occasions. This is especially true when the fitness function that measures the goodness of a solution is time consuming. In such a case, the maximum number of evaluations we may do of the fitness function is limited. This puts a heavy restriction on the maximum number of individuals and generations that the entire system may have. Hence, the task is the generation of effective individuals or solutions to the problem with as few fitness function calls as possible.

The entire paradigm of evolutionary computing may be bifurcated into four major pillars. These are genetic algorithms, genetic programming, evolutionary programming, and evolutionary strategies. We studied the genetic algorithm and the evolutionary strategies in the previous chapter. The Genetic Algorithms had a real representation of the problem. Here, the problem parameters in their phenotypic form were stuffed into a genetic individual, which made its genotypic form. This mechanism of individual representation was used in the entire cycle of evolution. In evolutionary strategies, the basic motive was self-adaptation. Here, the complete individual was stuffed with some extra parameters known as the strategy parameters. These parameters enabled the optimization of the parameters used for the purpose of evolution. These parameters let to the computation of the desired mutation rate. The problem parameters were later mutated using this mutation rate computed by the strategy parameters.

The basic focus of this chapter is to introduce other exciting evolutionary paradigms. Here, we would discuss swarm intelligence and genetic programming. The basic motivation behind the swarm intelligence is to understand the mechanism in which a collective effort of multiple particles or problem solutions can lead to optimization. Here, we take a real representation of the problem. While most of the evolutionary concepts remain similar to the discussions with genetic algorithms in the previous chapter, we take a greater look at the problem solving methodology and the evolutionary operators used in this case. Under this head, a variety of algorithms exist, each one of them inspired from some biological phenomenon.

The other major topic of discussion in the chapter is genetic programming that forms one of the major pillars of evolutionary algorithms. Here, the discussion involves the mechanism of representing an evolutionary individual as a program and to use the same for optimization. We discuss the tree-based representation and later advance to the linear representation, which is used in the grammatical evolution technique.

The basic problem being dealt with is the problem of robot path planning. Here, we are given a robotic map, and need to find a path from the source to the destination for the robot. This path needs to be without collision. Further, the path so evolved needs to be preferably as smooth as possible. The smoothness enables a robotic controller to physically move the robot at high speeds. There is always a limit to the time the planning algorithm may take to generate the trajectory. Taking a large amount of time may change the map, and the output of the planner may not be valid. Ideally, the planning time needs to be as small as possible. However, the plans generated in a short amount of time may not be optimal or complete.

This chapter first presents the basic computing paradigms of particle swarm optimization, genetic programming, and grammatical evolution. A number of swarm techniques have been duly explained, including the particle swarm optimization, ant colony optimization, etc. While the swarm intelligence algorithms have the similar mechanism of problem representation and working as the genetic algorithms, we briefly discuss their mechanism of solving the problem of robot motion planning. The solution with the use of grammatical evolution for the problem is taken explicitly. All these discussions follow in the chapter.

SWARM INTELLIGENCE

Swarm intelligence is taken as an inspiration from natural swarms. The natural swarms have a number of organisms that all live collectively and carry out the needed tasks together. In the process of looking into food or any other collective tasks, they all share their findings, information, and experiences. This directs the other members of the swarm who adjust themselves as per the information received. Hence, the swarms are able to solve the task efficiently. A wiser look at the process reveals the existence of a lot of intelligence inside the swarm that enables the swarm to collectively solve the task. This collective intelligence is the result of information found by each member and communicated to the other members.

Flocking birds, fish schools, ant colonies are all natural inspirations to the researchers. The ease with which all these collectively solve problems of finding food or motion is simply astonishing. Hence, there is a considerable effort in understanding these natural phenomena and to use the same understanding to build artificial systems. The attempt is to model the intelligence possessed by these swarms and use the same for problem solving. We carry forward a large number of concepts from the discussion of Genetic Algorithms from the previous chapter. In fact, the Genetic Algorithm is also an example of swarm intelligence with the natural evolution as the inspiration. The carried concepts include the individual, individual representation, generations, fitness landscape, etc.

In all these algorithms, however, there would be a different method of using these concepts and carrying forward the optimization. In other words, entirely new genetic operators would be designed. Further, each of the discussed algorithms is inspired from swarm or another. In the coming sub-sections, we separately study the different algorithms under this class of swarm algorithms. The discussed algorithms include particle swarm optimization, ant colony optimization, artificial bee colonies, and probability-based incremental learning.

Particle Swarm Optimization

The first algorithm of study is Particle Swarm Optimization (PSO). This algorithm is inspired by flocking birds. The birds always carry out the process of flocking in groups. As they fly about in the environment, they get to know places, which may be good or bad for flocking. Each bird records the kind of places it has visited in the past. Hence, each bird has memory storing its experiences during flocking. However, the flocking of a single bird may not be efficient. The presence of different birds has a possibility in which the different birds can interact and hence coordinate themselves. This coordination is provided by the communication between the birds. The communication enables the birds to broadcast the type of places they are in, which enables the other birds to plan their moves.

In our artificial systems, we attempt to use the same inspiration for better problem solving. As per the evolutionary computing principles, each bird or particle is a solution to the problem. This is analogous to the individual of the genetic algorithm. Each of these particles has a velocity and position. The position denotes the current state of the particle or its position in the fitness landscape. This is the same as the genotype used in the genetic algorithm. The velocity denotes the speed by which this individual or particle is moving in the fitness landscape. Hence, the PSO models the problem of optimization or searching the global minima in the fitness landscape as a population of particles moving with certain velocities.

The concept of generations is replaced by iterations or time. As the particles move, they change their positions. Suppose the position of any particle at time or iteration t is given by $X_i(t) <x_i^1(t), x_i^2(t), x_i^3(t), \ldots x_i^n(t)>$. Here, n is the number of variables to be optimized or the dimensionality of the problem. Further, let the velocity of the particle at any time t be given by $V_i(t) <v_i^1(t), v_i^2(t), v_i^3(t), \dots v_i^n(t)>$. The position of the particle at time t+1 is given by Equation 1.

$$P_i(t+1) = P_i(t) + V_i(t) \quad (1)$$

However, the particles do not continuously move in the same direction with the same velocities. As these particles move, their velocities are changed, which further changes their direction of motion. At any instance of time, the particle expects itself to be oriented towards the global minima. The basic problem is that the location of global minima in the entire fitness landscape is unknown, which is the problem being solved. Hence, we attempt to modify the velocity of the particle at any time to enable it reach the global minima.

We assume that each of these particles has memory and it knows the fitness of places it searched before, or the places that came on its journey so far. It stores the location of the best fitness place it has visited in its entire path. Let this position be $P_{il}<P_{il}^1, P_{il}^2, \dots P_{il}^n>$. As the particle visits new places in its journey, it keeps updating this value. This position P_{il} is the local best position that the particular particle has encountered, and is an indicative of the local optima as per the particle searching in the nearby areas. This may be the global optima or somewhere very near to the global optima. However, this cannot be ascertained as the single particle has a limited vision comprising only the limited places it visited in its journey.

Further, we know that in an evolutionary approach, the individuals or particles must converge to the global optima at the last stages of the optimization process. This results in better search around the global optima and good convergence. For all these reasons, we use some coordination mechanism. The various particles further keep a copy of the place corresponding to the best fitness found by any particle in its journey P_g. This value is common for all particles. If a particle succeeds in exploring a location in the fitness landscape better than this value, P_g is updated. P_g is an indicative of the global optima as per the current exploration of all the particles combined.

A particle at any time may hence have three intentions for better exploration. Accordingly, its velocity is modified. The first attempt is not to change the velocity at all. This results in better exploration by the particle in unknown lands. This has its resemblance to the use of mutation operator in conventional Genetic Algorithm, where the intent was to search at a place, even though it may be on the other side of recorded optima. The second attempt is to modify the velocity in such a manner that the particle gets close to the position where it recorded the best fitness P_{il}. It is possible that as a result of this behavior the particle gets a possibility to explore the area near its local optima P_{il}. As a result, the algorithm may converge inside this optima, which may at a later stage of time turn to be better than the global optima P_g explored so far. Both P_{il} and P_g, especially during the initial iterations or times, may only be vaguely recorded with a possibility of better convergence in each of them, before a concrete decision may be made regarding which of the optima is global optima. This is implemented by making the particle move towards P_{il}.

The last factor that influences in updating the particle velocity is the ability of the particle to move towards global optima recorded so far P_g. It is preferable that all the particles, towards later stages of the algorithm, concentrate on finding the optima near the best place recorded so far. This enables better convergence into the global optima. Hence, every particle further attempts to go inside the optima. The resultant formula of updating the particle velocity hence becomes as given in Equation 2.

$$V_i(t+1)= c_1 V_i(t) + c_2 (P_i(t) - P_{il}) + c_3 (P_i(t) - P_g) \quad (2)$$

Here c_1, c_2, and c_3 are constants denoting weights of the respective behaviors.

The speed of particle in any direction v_i^j has a great implication in deciding the algorithm performance. A higher speed means that the particle would make high jumps and check the fitness of a place after traveling large distances. It is possible that it completely jumps over some minima, which is not recorded. It would be further possible that the particle lands at places near the minima, which are not exciting, and does not converge. Unchecked speeds may attain very high values in the optimization process. To escape from all these problems, we limit the speed to some threshold v_{max}^j, giving Equation 3:

$$v_i^j < v_{max}^j \quad (3)$$

The complete algorithm is summarized in Figure 1. The analysis of the algorithm is quite similar to the case of genetic algorithm. Initially the particles are placed far apart. They all travel with the intensions, as stated before. They attempt to explore towards nearest minima. At later stages of time they convergence to global minima. At still higher stages there is little exploration for nearby places, which results in the optima getting improved, if a particle succeeds in obtaining a better place. For further details, readers are referred to Clerc (2002), Eberhar and Shi (2001), Kennedy and Eberhart (1995), and Shi (2004).

Figure 1. Particle swarm optimization

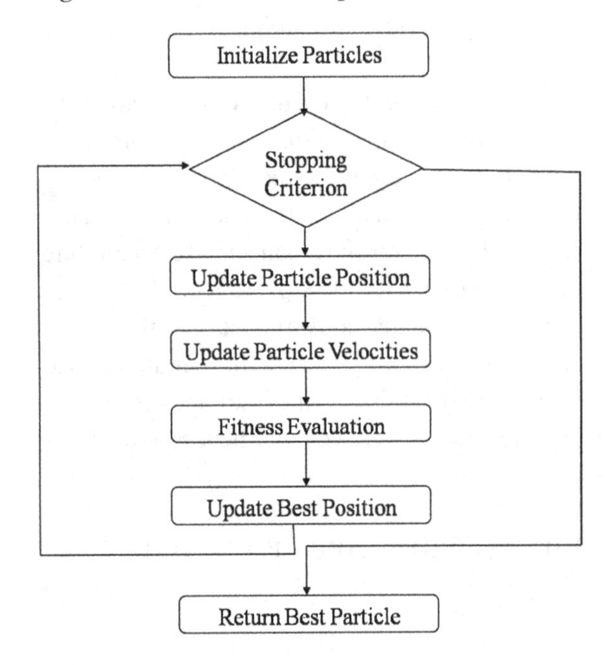

Ant Colony Optimization

The other algorithm that we study is the Ant Colony Optimization (ACO). Here, the basic inspiration is the ant colonies. Ants live in colonies, and the entire colony searches for food. It is common to see a large number of ants around any food item. This behavior of ants is a sure motivation, which can lead to good optimization, if modeled as an artificial system. The various ants in colonies start a random search for the food. These ants are known to deposit some pheromones in the path as they go around. This pheromone can be sensed by the various other ants, which acts as a means of coordinating them. The ants are able to decide which way to go for better possibility of getting the food.

The inspiration of ants is applied on the problem of graph search. Here, we are given a graph with a source and a goal. The problem is to compute the shortest path between the source and the goal. The graph has a number of vertices and edges. To solve this problem, we first take n number of ants. The ants in this case are analogous to the individuals in the genetic algorithm or particles in PSO. These ants are moved from the source to find goal in their own manner. Each ant attempts to move to the goal in the shortest possible manner. The ant is allowed a maximum number of steps within which it must reach the goal or else its journey is abandoned. This ends the indefinite exploration in case an ant is unable to reach the goal at all. Every ant needs to decide the edge that it must follow while located at any vertex. The choice of decision affects the next vertex that it gets and the complete path. Note that the problem here, unlike many graph search algorithms, is not to find a path between source and goal, but rather the shortest path. The choice of vertex is influenced by two factors. These factors are attractiveness and traits.

The attractiveness is like heuristics, which denote the goodness of a vertex. An ant is always attracted by edges that make it arrive at good vertices. These vertices are closer to the goal, and hence stand a chance to make the ant reach the goal in smallest possible path. This is similar to the arguments in heuristic search. However, heuristics may not always guide the ant in the correct direction to reach the goal in shortest possible way. This is again due to the uncertainties in heuristics that we studied in earlier chapters. Here, the second factor comes into play. This is called as traits. Traits are the amount of pheromone deposited by any earlier ant that happened to pay a visit in the earlier iterations of the algorithm. The edges of higher trait value or higher pheromone content means that the ants traveling through the edge earlier had a better path. This means potentially the edge lying on an optimal path or close to optimal path. Hence, the ant must follow this edge.

Let any ant be at vertex u. The probability that the ant uses the edge e to go from vertex u to a vertex v given by Equation 4:

$$p_{u,e} = \frac{\tau^{\alpha}_{u,e}\eta^{\beta}_{u,e}}{\sum \tau^{\alpha}_{u,e}\eta^{\beta}_{u,e}} \quad (4)$$

Here, α and β are parameters controlling the contributions of the two factors. $0 \leq \alpha, \beta \leq 1$. Denominator is for normalization of the expression. $\tau_{u,e}$ is the traits factor and $\eta_{u,e}$ is the attractiveness factor.

The ants carry out a probabilistic selection where every possible edge or possible subsequent vertex has a probability as described above. It may use a roulette wheel selection, tournament selection, or any other selection scheme. Once the selection is made, the ant selects the edge to carry forward its motion. The motion always starts from the source and ends once the goal is found. At every vertex that the ant lands in, this selection scheme is used to decide the next step.

If the ant reaches the goal, it needs to further deposit pheromones, which would help the ants of the next iteration. The ant first measures the fitness or the goodness of the path it traversed. Then the ant traces every edge it used to reach the goal from the source and deposits pheromones at every edge, where the quantity of the pheromones is proportional to the fitness of the path. Hence, if the path was highly good, there is an indication that the corresponding edges that form this path are good. This ant tries to make other ants come through this way by depositing large quantity of pheromone. However, if the path of any ant was not good, this means that the edges are not so good, but good enough that a path could be found. Hence, lesser amount of pheromone is deposited because of which those edges may not be the likely choice of other ants that come in next iterations.

Let the amount of pheromone deposited by an ant i at iteration t of the algorithm at the edge e from vertex u to vertex v be given by $\Delta\tau^{i}_{u,e}(t)$. This holds for all edges e in the path of ant i. The total pheromone deposited at the edge e at iteration t by all the ants may be given by Equation 5:

$$\Delta\tau_{u,e}(t) = \sum_{i} \Delta\tau^{i}_{u,e}(t) \quad (5)$$

The net value of pheromone at the edge may be given by Equation 6:

$$\tau_{u,e}(t+1) = \tau_{u,e}(t+1) + \Delta\tau_{u,e}(t) \quad (6)$$

In this manner, each of the iterations involves leaving of some ants from the source and enabling them to reach the goal. The ants take step by the factors of attractiveness and traits. The ants that reach the goal deposit pheromones at the edges they took for traversal. At the next iteration, again ants are left from the source. The pheromones deposited by the previous ants add to the traits factor. These reach the goal and add more pheromones to the edges. This process goes on.

Continuous deposition of pheromones at the edges may lead to excessive pheromone deposition. This is especially true for the edges that lie on the optimal path or the near optimal paths. This excessive pheromone value creates problems in computation, and excessive values need to be avoided. For the same, the fundamental of pheromone evaporation is brought up. Every edge continuously evaporates pheromones along with time or in our case along with iterations. This avoids any edge to have an excessively large value. Hence, the pheromones are added by traits and reduced by evaporation. Taking this factor into account the pheromone content at the edge may be given by Equation 7:

$$\tau_{u,e}(t+1) = \rho\tau_{u,e}(t+1) + \Delta\tau_{u,e}(t) \quad (7)$$

Here ρ is the evaporation constant.

In this algorithm, every ant does not know about the computations of the other ants. This creates a problem in computing statistics or taking measures where some kind of global picture is required that consists of knowledge about the states of all the ants. Examples include the computation of best or average fitness value. For the same, the concept of daemon actions is introduced into the ACO algorithm. Here, all the computations where the global information is required takes place explicitly.

The complete algorithm is summarized in Figure 2. The analysis of this algorithm follows somewhat different norms as compared to the other algorithms where the fitness landscape consisted of all the parameters drawn at the various axes. For the analysis of ACO, we consider the graph with the thickness of an edge denoting its trait value. Initially the traits value is same for all the edges, and hence their thickness. The first set of ants follow a more or less heuristic search and reach their goals. Since the choice of each vertex is probabilistic, the paths of different ants are different. When the ants deposit pheromones, the ones who reached by better paths have higher trait value and hence marked by more thickness. These may not have good heuristics, but since the paths are short, they are to be preferred.

Figure 2. Ant colony optimization

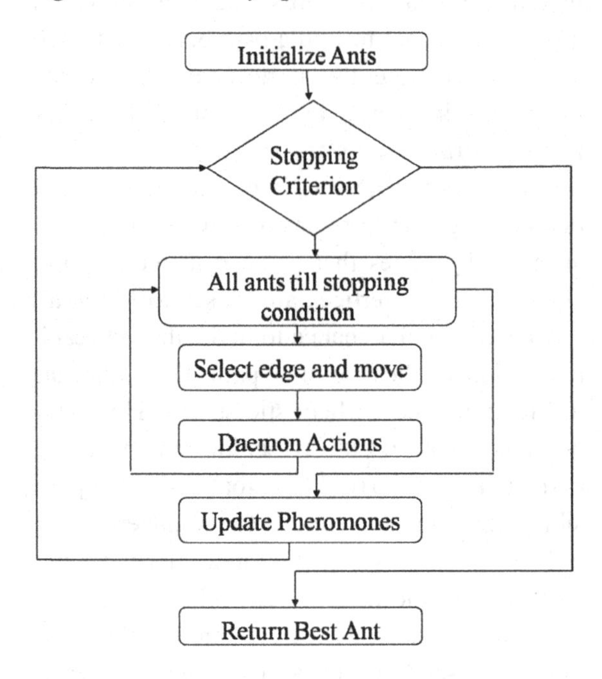

In this manner, we see that in some more iterations, the thickness of some edges increases, which lie at the optimal path, or close to the optimal path. As the genetic algorithm analysis showed the possibility of convergence at local minima, the same probability exists here that the local optimal path gets more pheromones. However, if an ant happens to explore a completely new path that is better, the traits of corresponding edges would increase, which would later attract more ants. In this manner, the traits keep increasing. Similar discussions were done in genetic algorithm analysis, with a fitter individual pulling the other individuals towards it. For a more detailed discussion, please refer to Dorigo and Caro (1999), Dorigo et al. (1996), Maniezzo et al. (2004), Parpinelli et al. (2002), Stutzle and Dorigo (1999), and Stutzle and Hoos (1997).

Artificial Bees Colonies

The other inspiration comes from another swarm leads us to the Artificial Bees Colonies (ABC) algorithm. This algorithm is inspired by honey bees. Honey bees are always found to be living in colonies from where they carry out their activities of searching for food. The different kinds of bees in the colony do various kinds of work, and the entire group or colony is able to work well. In ABC algorithm, we take the same inspiration of working of the honey bees and develop an algorithm for optimization. In the artificial algorithm, we take three kinds of bees. These are employed bees, onlookers, and scouts. Each of these bees has a different role to play in the search process. The problem is again the same that we are given a fitness landscape and we are asked to search for the global optima in this fitness landscape. The various bees of any kind are individuals that occupy different positions in this fitness landscape. Their movement inside this fitness landscape signifies the search strategy. The fitness is measured at every point they land upon in this landscape.

The different bees have different mechanisms of search. The employed bees are the ones that have discovered some kind of place of high fitness value in the fitness landscape. These bees attempt to search the nearby areas in quest of better places in the surroundings. Hence, these bees attempt to carry forward some kind of local search mechanism. The other type of bee is the onlookers. These bees attempt to move towards the employed bees. Every employed bee attracts the onlookers, where the attraction depends upon the fitness value where the employed bee is located. The onlookers travel towards the employed bees, and in their travel, they uncover information about the areas they visited in terms of their fitness values. There is a possibility of exploring better places, in which case an employed bee may be placed. The other type of bee is the scouts. These bees carry forward random searches. Similar to the scouts there is always a possibility of uncovering better places in this search process.

Any evolutionary technique always tries to match between exploration and exploitation. The exploration is an attempt to search more and more places in the fitness landscape, not necessarily restricted to the region of occupancy of the individuals of current population. This may result in the algorithm avoiding the local minima and getting close to the global optima; however, it may not be able to converge well into the global optima.

Exploitation attempts to converge the various individuals of the search or optimization algorithm into the minima. This may result in giving more accurate location of the minima, or reporting the best fitness value more precisely; however, the algorithm may get struck at some local minima. In conventional genetic algorithm, exploration is done by the use of mutation operator and exploitation is done by the use of crossover operator. In this algorithm, the different bees are modeled to provide exploration and exploitation features. The algorithm is shown in Figure 3. For further reading about the algorithm refer to Karaboga (2005), Karaboga and Basturk (2007), and Pham et al. (2006).

Figure 3: Artificial bee colony

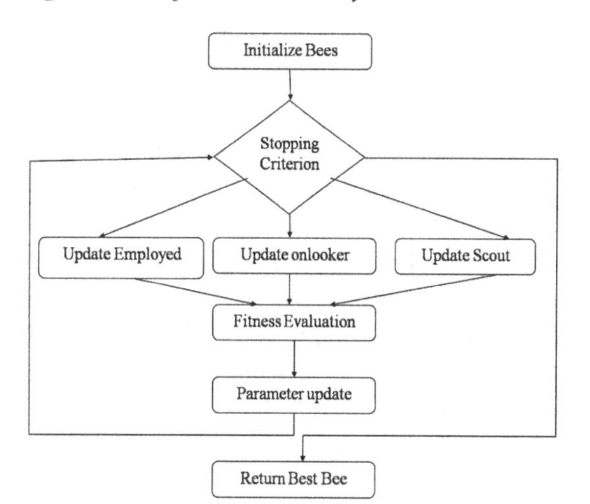

Probability-Based Incremental Learning

The other algorithm for discussion is the Probability-Based Incremental Learning (PBIL). So far, we have been maintaining a pool of solutions or individuals that migrate from one generation to the other. However, in this algorithm we do not do the same. Instead, we only maintain a vector called as the probability vector that migrates from one generation to the other. This algorithm is used for the optimization of binary representation of the individual. Hence, any phenotype must be first represented as a binary genotype for the application of this algorithm. Let the genotype have a total of N genes, where individual may have 0 or 1 as the allowed values for any gene. We take a probability vector of the form $<p_1, p_2, p_3, \ldots p_N>$. Here, any p_i denotes the probability of having a 1 at the gene i in the optimal solution. Initially the algorithm has no idea about the presence of 1 or 0 being preferable, and all the p_is in this probability vector are set to a value of 0.5.

As the name suggests, there is an incremental learning of this vector that is carried out. Hence, as the algorithm proceeds all the genes or probabilities in this vector keep aligning themselves towards the optimal solution. Initially, all genes have a value of 0.5. As the algorithm precedes the value increase towards 1 for the genes where presence of 1 is preferable, and towards 0 for the genes where the presence of 0 is preferable. Hence, this vector behaves similarly to the schema of the binary genetic algorithm.

At every iteration or generation, the first task is to compute the population. Assuming the population size is n, we generate n random binary vectors with 0 or 1 at all the genes. The allocation of 0 or 1 for the gene i of any individual is done probabilistically, where the probability of allocation of 1 is p_i. Once the set of individuals or the population pool is ready, the corresponding genotype individuals may be used for the fitness computation. This assigns a fitness value to every individual. The next step performed is the learning. In learning, we plan to modify the probability vector as per the recorded fitness values. We first select the best individual from the complete population pool. This individual has the best fitness value. Let this individual be form $<b_1, b_2, b_3, \ldots b_N>$. In learning, we attempt to incline the probability vector form $<p_1, p_2, p_3, \ldots p_N>$ towards this individual. Hence, the probability vector tries to take come characteristics from the best individual $<b_1, b_2, b_3, \ldots b_N>$. For every gene j this is given by Equation 8:

$$p_j(t+1) = (1-\alpha)\, p_j(t) + \alpha\, b_j(t) \qquad (8)$$

Here, α is the learning rate and t is the iteration of the algorithm. The algorithm is summarized in Figure 4. For a more detailed reading, refer to Baluja (1994), Hohfeld and Rudolph (1997), Southey and Karray (1999), Yang (2005), and Yang and Yao (2003, 2005, 2008).

Figure 4. Probability-based increment learning

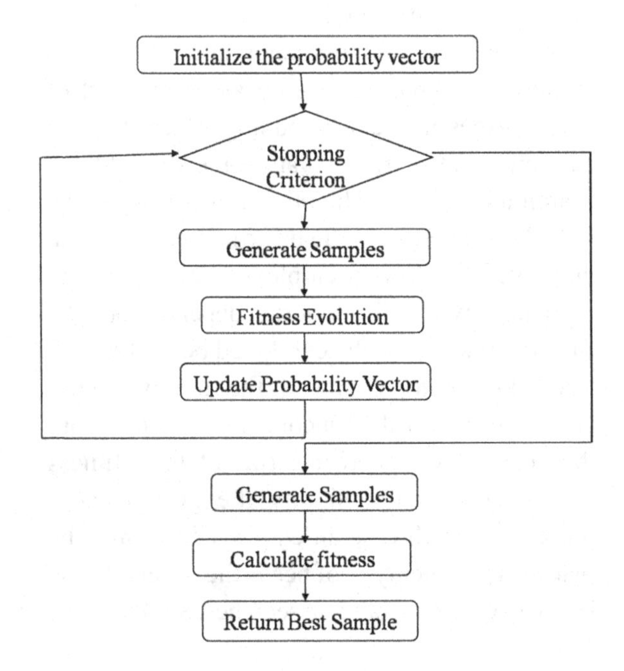

SWARM INTELLIGENCE IN PATH PLANNING

The entire notion of application of swarm intelligence in path planning follows the same concepts that we discussed in the previous chapters. We briefly summarize the experimental methodology for the sake of completeness of the chapter. Here, we discuss the application of particle swarm optimization and ant colony optimization only. The other algorithms may be used in the same form, or with minor modifications.

Particle Swarm Optimization

The first algorithm of implementation was Particle Swarm Optimization. This algorithm is quite similar in implementation to the genetic algorithm implementation discussed in the previous chapter. The same individual representation may be used. Further, the same fitness function may be used. The algorithm parameters of c_1, c_2, and c_3 are given equal values in the implementation. The results of some of the maps used are given in Figure 5.

Ant Colony Optimization

The next task is to experiment the ACO algorithm for its effectiveness for planning. Here, we first state a modification to the ACO algorithm, and then use it for the experimentation for motion planning. This section reports the work in Maurya and Shukla (2010). Excerpts from text, tables, and figures are taken from the same. © 2010 IEEE, reprinted with permission from Maurya and Shukla (2010).

The parameters α and β in the algorithm control the degree of contribution of factors of attractiveness and traits. Best results are usually found when both values are approximately in the

Figure 5. Experimental results of PSO

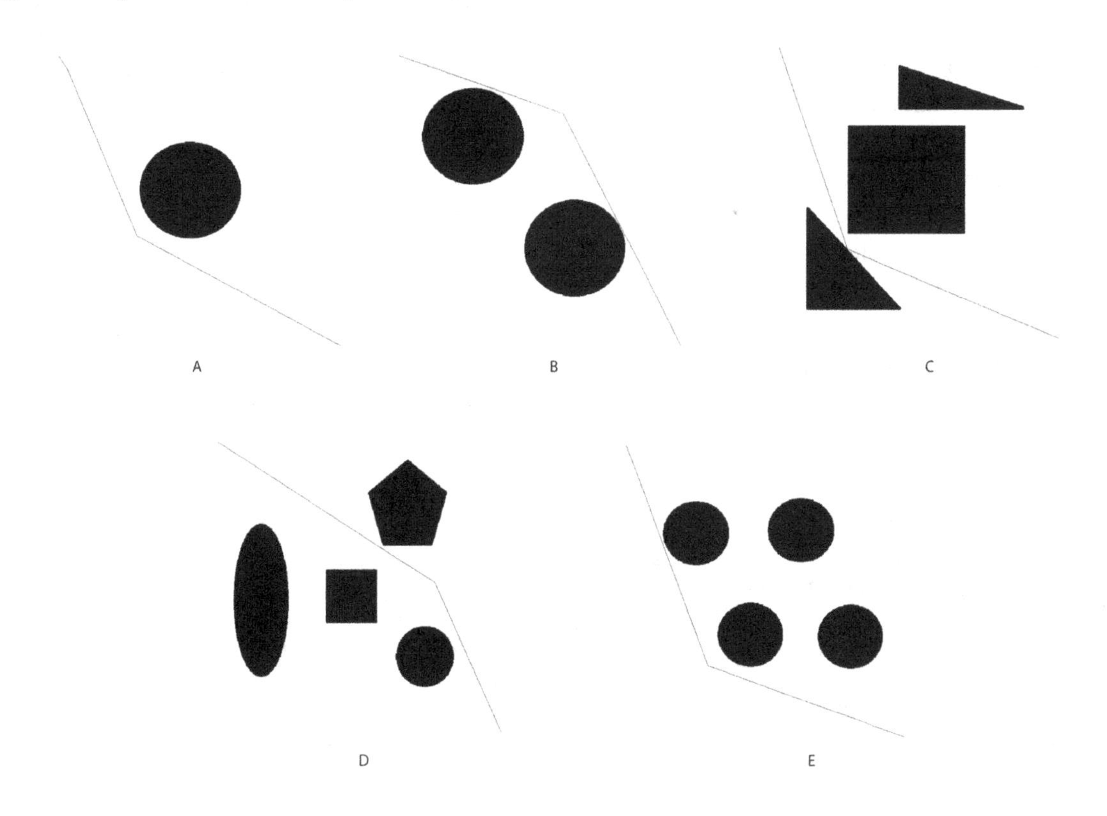

Table 1. Path lengths for different number of ants and number of allowed moves

Number of Allowed Moves	Path length for100 ants	Path length for200 ants	Path length for 300 ANTS	Path length for400 ANTS	Path length for 500 ANTS
8	33.799	31.3854	27.5563	27.3848	25.5657
16	32.247	30.1312	26.717	25.9596	23.2514
24	30.348	29.1869	21.8372	21.3225	22.4677
32	25.964	24.1935	21.5535	21.3782	21,1872

same range but there is usually a problem when the range of the attractiveness value and trait value are far from each other. In this case, the parameters α and β are selected in a way to near the two ranges so that both factors contribute in the algorithm. This approximation is added by inclusion of an exponential function in the probability computation. The modified probability is given by Equation 9:

$$p_{u,e} = \frac{\exp(\tau_{u,e})^{\alpha}.\exp(\eta_{u,e})^{\beta}}{\sum \exp(\tau_{u,e})^{\alpha}.\exp(\eta_{u,e})^{\beta}} \quad (9)$$

The exponential function is characterized by its large rate increase in values. That characteristic can be used to obtain the best result rapidly.

All the simulations were done in MATLAB. Here the concept of number of moves was further extended to 24 and 32 moves, unlike the earlier experimentations of chapter 2 where only 4, 8, and 16 moves were considered. The extended moves may be visualized extending the same notion. Path length for different number of allowable moves with different number of ants for the same scenario was noted, and is presented in Table 1. The corresponding computational time is shown in Table 2. The simulation results for map size of 15 x 15, for different number of allowable moves are shown in Figure 6. Each of the experiments used 400 ants. Note that since this is a graph-based approach, the resolution of map is kept significantly small. This is more because of the repetitive nature in which the entire map is explored, which is time consuming. We can conclude from the simulation that there is a considerable decrease in the path lengths with increase in the level of number of allowable moves, as shown in Table 1. However, the increase in number of moves also results in an increase in time, which puts a threshold on the magnitude by which the factor may be increased, as shown in Table 2.

GENETIC PROGRAMMING

The other major topic of study in the chapter is Genetic Programming (GP). This forms the other major pillar of evolutionary computing, besides Genetic Algorithm, Evolutionary Programming,

Table 2. Time taken for simple and modified ACO for different number of allowed moves

Number of allowed moves	Elapse time(100 ants)		Cpu time(200 ants)		Cpu time(300 ants)	
	simple	modified	simple	modified	simple	modified
8	270	143	340	246	450	345
16	334	165	362	258	490	367
24	456	193	485	395	521	432

Figure 6. Paths for different number of allowed moves: (a) 8, (b) 16, (c) 24, (d) 32

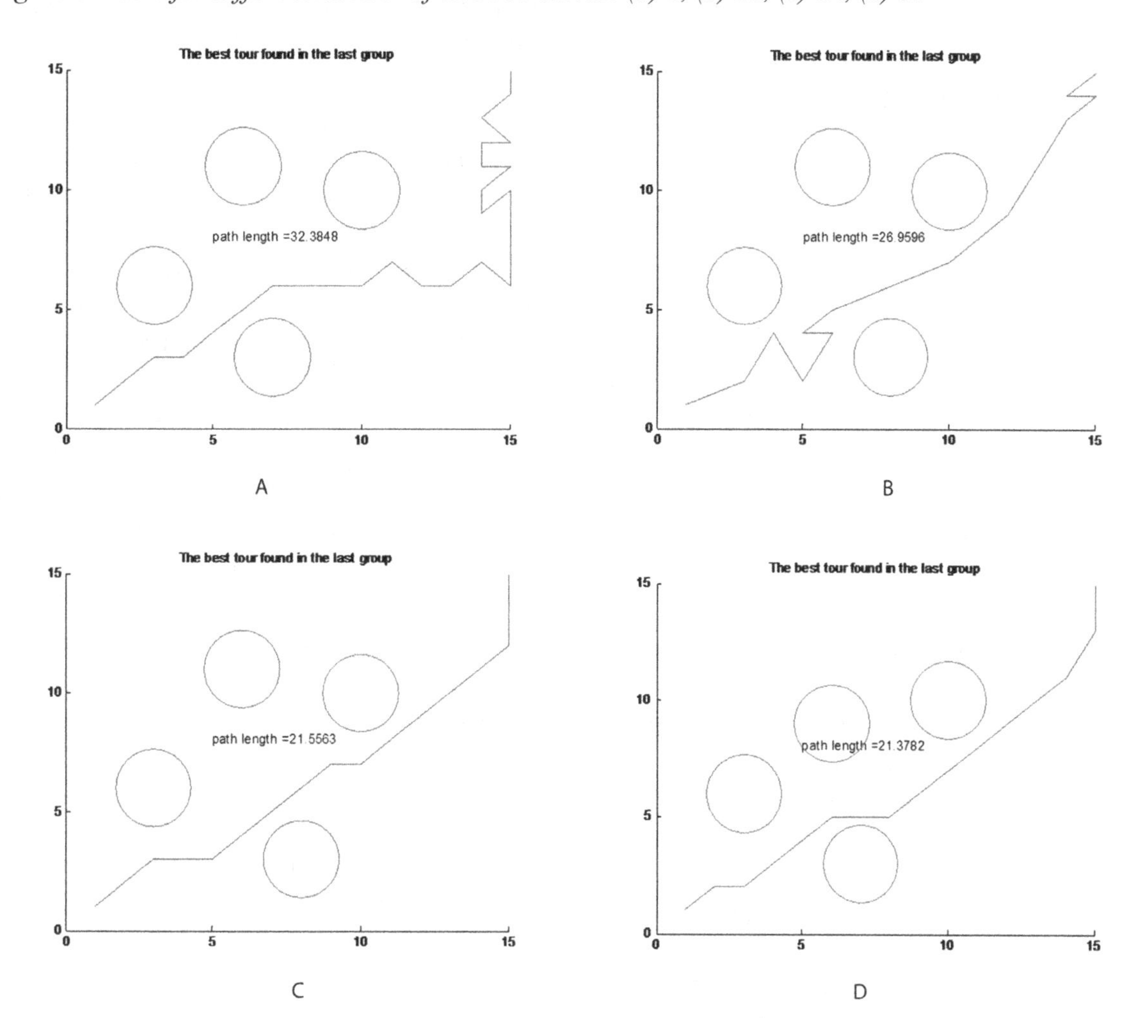

and Evolutionary Strategies. Being an evolutionary approach, most of the fundamentals of this technique remain the same as any evolutionary technique or Genetic Algorithm. The concepts of individual, population, fitness, and evolutionary operators are the same as the Genetic Algorithm. However, the major change in this algorithm is the mechanism in which the individual is represented. Unlike the real representation followed by the Genetic Algorithm, an individual of this algorithm is a program. That comes from the basic motivation behind genetic programming as a tool for automatic generation of programs. This notion of an individual representation as a program demands a change in the mechanism in which the various operators are defined as well as the mechanism in which the fitness value is computed.

We study about all these concepts in the upcoming sub-sections. The complete algorithm as a whole has similar architecture as the Genetic Algorithm. Though the notations and concepts are same, differences might come in the mechanism of application of the concepts. The algorithm is shown in Figure 7. For detailed discussion of the topic, refer to Altenberg (1994), Banzhaf et al. (1998a, 1998b, 1998c), Langdon (1998), Luke and Spector (1997), and Tackett (1993).

Figure 7. Genetic programming

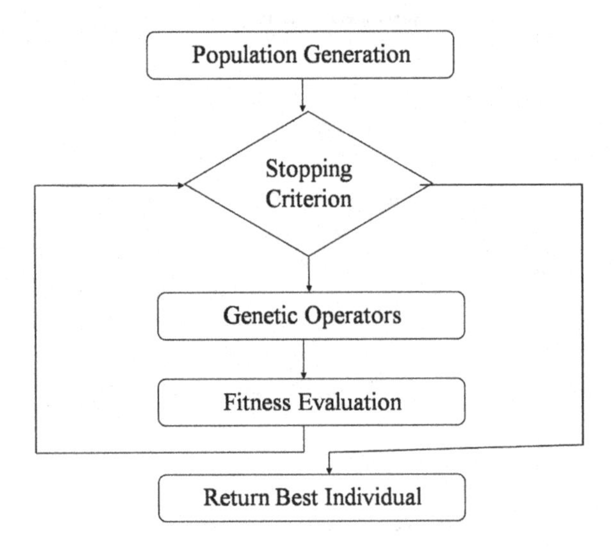

Individual Representation

The major change that comes in the use of genetic programming is the mechanism in which the individuals are represented. In this evolutionary technique, every individual is a program. Hence, the task in hand is to evolve the best program that solves the problem. The individual representation technique deals with the mechanism in which the problem phenotype may be represented as a program, which represents a potential solution to the problem and may be optimized in the evolutionary process.

One of the most commonly used techniques of individual representation in this technique is the tree-based individual representation. Here, the program or the genetic individual is given a tree-like structure. Unlike the genetic algorithm individuals that are normally of a constant length, the good part about these programs or trees is that they may have variable lengths. However, the maximum length or size is usually limited to some value. This restricts the individual from growing in size indefinitely.

The programs always have some syntax, which needs to be followed. The syntax is specific to the domain to which the program belongs, or in other words the problem being solved. Every individual representation technique may ensure that the syntax is being followed; else, it may be impossible to interpret the individual or use the individual as a solution to the problem. In most representation schemes, syntax is handled by specifying rules for the nodes of the trees, in terms of the types of children a node may have, the number of children, etc. In most other cases, the representation may allow some irregularities of syntax, which may be taken care of by the repair operator.

Consider a simple example of evolving a mathematical expression f(x,y) that returns its output based on two variables x and y as inputs. Further consider that the mathematical expression may only have +, -, unary +, unary -, * (multiply), and / as the six operations (and not functions like sin, cos, abs, etc.). In designing a representation scheme, we differentiate between the leaf nodes and the non-leaf nodes. In this scheme any non-leaf node may only be occupied by the allowed mathematical operators. Further, any leaf node may only be occupied by input variables (x and y) and constants that are real numbers. By this scheme, we may construct a large number of expressions as individuals of the genetic programming. Some of the expressions or individuals are given in Figure 8.

Seeing this scheme of representation, it is clear that the representation scheme is one-to-many. In other words, one phenotype or mathematical expression may be mapped to a large number of genotypes. Here, we assume that the phenotype is always in its shortest form or else the mapping scheme of phenotype to genotype would become many-to-many. An example in which various individuals map to the same phenotype is given in Figure 9. It may even be interpreted as the representation scheme allows redundancy. In other terms, many sub-trees of the tree may be replaced by a single node or a smaller sub-tree. As long as the individual tree is small, this redun-

Figure 8. Some individuals for sample genetic program problem

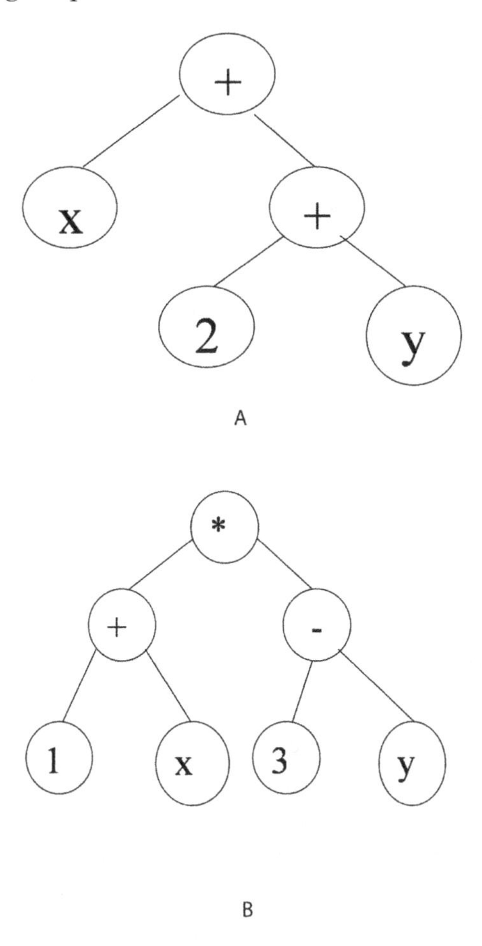

Figure 9. Individuals mapping to same phenotype in sample problem

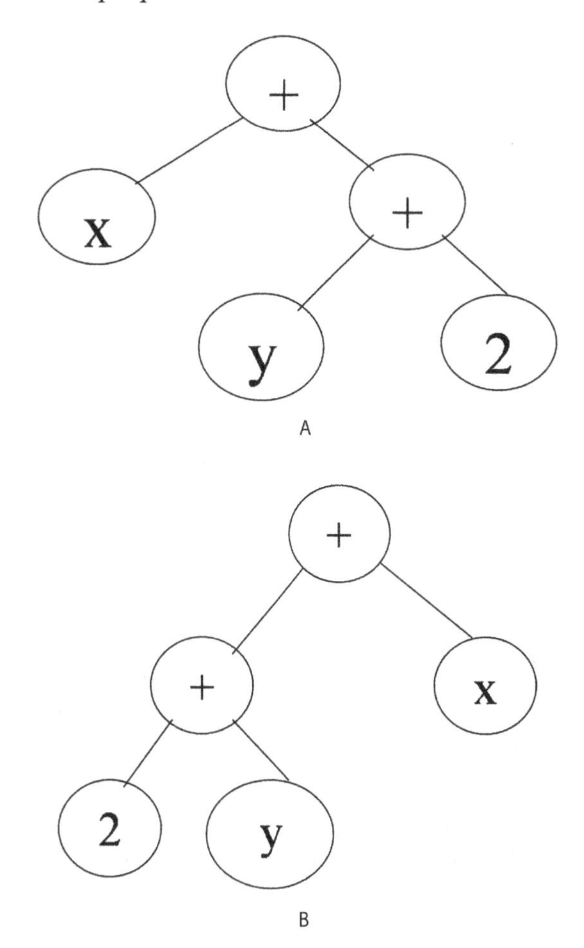

dancy may not have a problem. However, evolution in such a mechanism is likely to face problem that multiple individuals may possess a sub-tree or a node sequence that leads to high fitness (representing good characteristics) but is very long and can be greatly reduced.

Suppose that the sub-expression (or sub-tree) "1+5*5+3*x+4*x+10+12" is found to be very good for the individual. Hence, multiple individuals would copy this sub-expression in the evolutionary process. However, this sub-expression, as per this representation scheme, may take a fairly large amount of space in the entire tree-based representation. This leaves less scope for the rest of the tree to possess good sub-expressions. Hence, we may need to enlarge the maximum size of the tree, which has its own problems of making the dimensionality of the problem very large, which makes evolution very difficult and complex. The solution is to analyze the individual and attempt to reduce it to the smallest form, whenever possible. Designing such operators may naturally be complex task.

Crossover

The difference in individual representation across genetic algorithms and genetic programming leads to a difference in the use of genetic operators. As a result, we need to study the mechanism in

Figure 10. Crossover in genetic programming

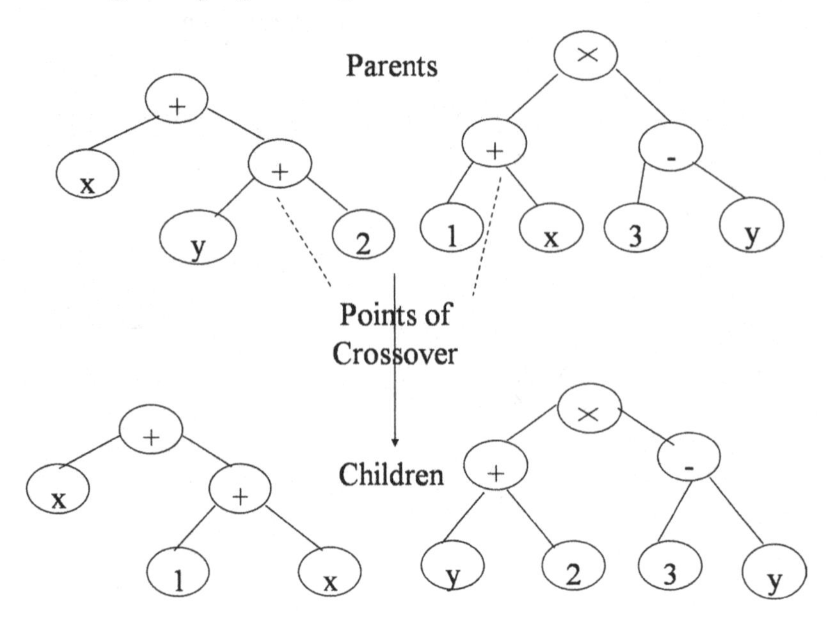

which the operators would be applied to the new representation scheme. The first major operator of use is crossover. Here two or more parents are mixed to form two or more children. In case of a tree-based representation, we first select points of crossover in the two trees, which are nodes of the tree. Then the entire sub-trees in the two individuals rooted by these nodes are exchanged. The first sub-tree is transferred to the second individual to replace the second sub-tree and vice versa. This generates two children from two parents. The crossover process is summarized in Figure 10.

Multi-parent/multi-child crossover may be done in similar manner. The crossover operation may at times lead to wrong syntax in many representation scenarios. Hence, the children need to be operated by some repair operation, or accordingly constraints may be put on the crossover scheme. Further, the crossover operation may have the children being produced of size greater than the allowed size. This may further be restricted before carrying out the crossover operation.

Mutation

The other major operator of use for evolution is mutation. Unlike genetic algorithms, mutation in case of genetic programming has two roles to play. The first role is the optimization of the structure of the tree. This deals with the addition, deletion, and amendments to the number of chidden for a node. Nodes may be added to the leaf nodes, some nodes may be deleted. In all the entire tree-structure needs to be modified. This task is performed by the application of structural mutation operator. This operator must though check for the syntax and the maximum length of the tree. This operation is shown in Figure 11.

The other task to be performed is the alteration of the genes. This means changing the operators at the non-leaf nodes and the operands at the leaf nodes. The change again needs to be syntactically correct; else, a repair operation would be required. This process is similar to the application of mutation to single gene is a conventional genetic algorithm. This operation is shown in Figure 12.

Figure 11. Structural mutation

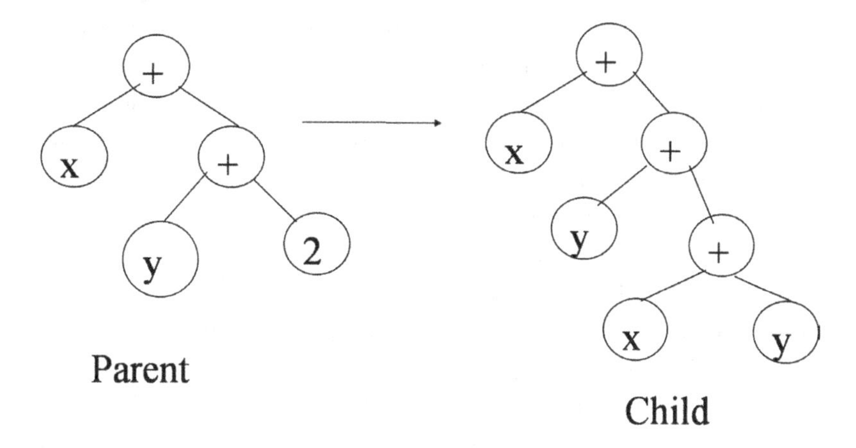

Other Operators

The evolution may not be able to take place effectively by the stated operators along. The evolution is further enhanced by a variety of other operators. Just like genetic algorithms, elite is used to preserve the individual with the best fitness value from dying or being modified in the evolutionary process. Replication is another operator used. This operator simply creates a carbon copy of an individual with high fitness value. Since multiple operators are used to work over the different individuals, the two like individuals soon change their formats. This operator creates a better chance for the fit individual to be worked over by a variety of genetic operators.

We have already discussed the working of the repair operator that takes as an input an individual, which does not follow the syntax or has other infeasibility which may not be represented in the syntax. This operator removes all these inconsistencies and returns a feasible individual. The other operator we discussed was used to analyze the individual and reduce it in its base form, which was especially practical in scenarios like expression evolution, where redundancy may be very high. A large number of problem modeling scenarios are prone to these problems.

Figure 12. Parametric mutation

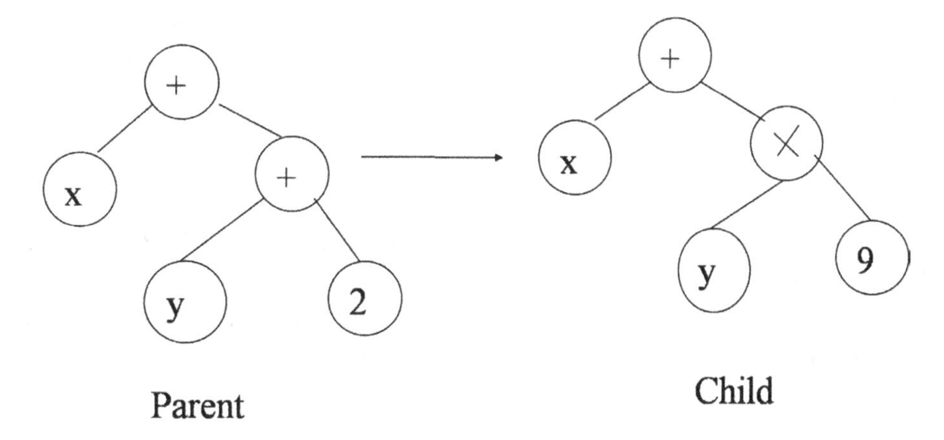

Fitness Evaluation

The next task is to make a fitness function that computes the fitness of any individual. Like the genetic algorithm, the fitness function in this case also is highly dependent on the problem being solved. In fact, the task is usually to convert the genotype into the phenotype and use it to solve the problem. The indicators of the problem indicate the effectiveness of the solution and hence the fitness value. Consider the problem of evolving an expression that we have been taking as a synthetic example. The problem may be to regress a set of values by this expression. Hence, we may use the tree-based representation and convert it into a mathematical expression. The entire set of values may be passed by this expression to compute the regression error. This regression error may give us the fitness cost (or the negative of fitness value), which needs to be minimized by the algorithm.

However, we may further have a preference for smaller sized trees. The size may be measured by the maximum height of the tree or the number of nodes it has. This is because solutions to most of the problems are usually simple and can be obtained by a set of good program modules or sub-trees. Very long trees may be able to get lesser cost or error, but make the evolution un-necessarily complex. This further has the problem of excessive dimensionality, which must be avoided. Hence, we may penalize the large tress by some penalty factor. The penalty is proportional to the size of the tree. This method of preferring smaller trees is called as parsimony. The fitness may hence be given by Equation 10:

$$Fit(I) = f(I) - \alpha S(I) \quad (10)$$

Here, f(I) is the true fitness of the tree or individual, S(I) is the size of the tree, α is the penalty constant.

Initially in the evolutionary process when the trees are small, increase of size results in a major change to fitness. Hence, the penalty added may be neglected in lieu of added goodness of solution. However, excessively large trees result in minor increase in fitness, but as compared, the increase in penalty may be high. Hence, the over-large trees are avoided by this approach.

Analysis

The analysis of genetic algorithm with a binary representation of the individual was carried out with the help of a schema theory. Here a schema was taken of length equal to the length of the genotype individual. Each gene contained a 0, 1, or * depending whether all individuals of the population have 0 or 1 at the specific gene, or some have 0s while the others have 1. We follow a similar mechanism to analyze and explain the convergence, exploration, and exploitation in genetic programming.

A schema of genetic programming contains a sub-tree. All the nodes of the sub-tree are marked by their nodal values, if all the individuals of the population have the same value at the specific node, as well as have the same architecture. In other case a * is marked at the same location. Initially, all the individuals are random and hence the schema just consists of a * as the root. After a few generations, all the individuals start to have the same root, and at later stages of time, the same set of children. Subsequent evolution may again result in the individuals taking specific architecture and node values at various locations. All these are marked in the schema. An example of the schema and the optimization at various stages of time is shown in Figure 13.

GRAMMATICAL EVOLUTION

In the previous section, we discussed about one of the major evolutionary paradigms that as Genetic Programming. This technique is able to solve

Figure 13. Figure genetic programming schemes at various stages

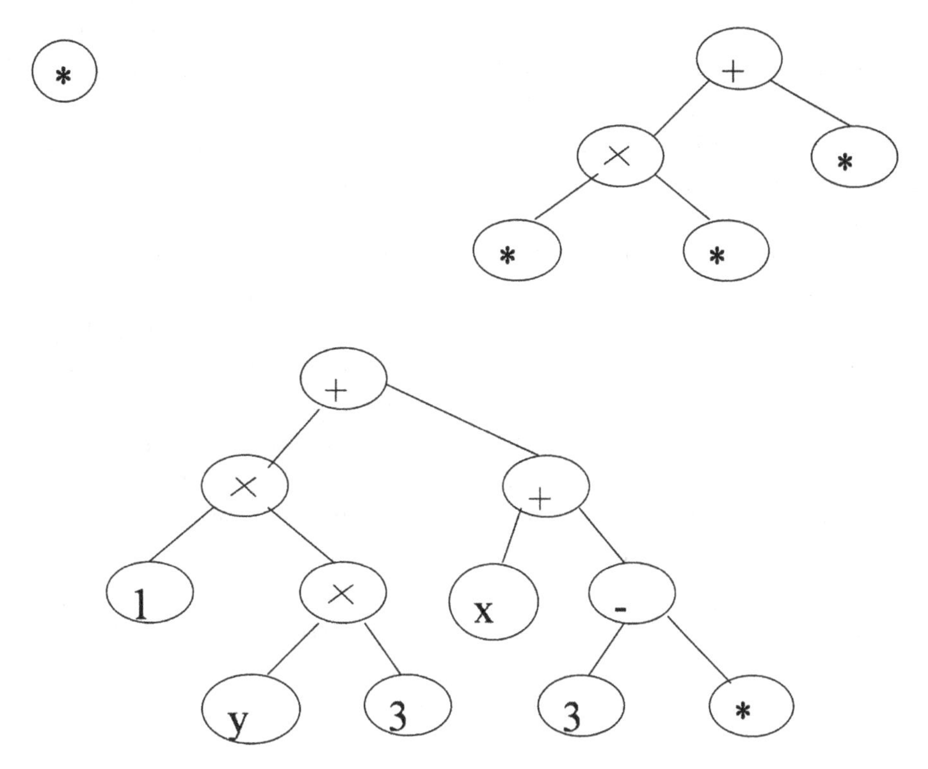

problems reasonably well, especially in situations where the solution may be easily represented in a program-like architecture, which is a solution to the problem being solved. The GP with its designed evolutionary operators may be easily able to evolve the optimal individual subjected to the constraints of computation time. This technique, however, has a limitation of a complex individual representation technique, in case the genetic individual is represented into a tree format. The tree-like structure is difficult to handle and further the implementation of the evolutionary operators may not be very simple. This motivates the use of some technique in which the genetic individual of the GP may be represented in a linear form. This is done by the use of Grammatical Evolution (GE).

This evolutionary technique uses the fundamental principle of Grammar for effectively representing an individual into a linear format. The Grammar is a part of the central individual representation scheme that is common to all individuals in the population, and all individuals that are evolved or generated as the algorithm proceeds. This gives a base or scope for information specification outside the genetic individuals. Since some part of information, in form of grammar, is prevalent outside the individual, this may sometimes lead to requirement of smaller individuals as compared to the GP counterparts.

The smaller individuals may further mean that optimization is easy to carry, does not suffer from the problems of curse of dimensionality, as well as requires lesser time. Excluding the individual representation technique, the rest of the algorithm is similar to that used in Genetic Algorithms and further follows similar principles of evolution and operator usage. In the subsequent sub-sections, we discuss about the various aspects and the general working of this evolutionary technique. For original texts over the topic and related research, refer to O'Neil and Brabazon (2001), O'Neil and Ryan (2001, 2003), and Rosca (1997).

Grammar

The major underlying principle behind the GE is the grammar, which forms the language of the solution specification or the individual representation. The grammar is defined as a collection of Terminating Symbols, Non-Terminating Symbols, Start Symbols, and a set of Production Rules. This may be given by Equation 11:

$$G = \{T, N, S, P\} \quad (11)$$

The grammar is specified in the Backus-Naur Form (BNF), which gives an easy mechanism of representation of each of its components, serving the cause to ease the individual representation technique. The purpose of the evolution technique is to ultimately evolve a program, executing of which would give the extent to which it solves the desired problem, or the fitness value. The program is ultimately seen as a collection of symbols. The symbols may have their own meaning from the problem context or from the context of the phenotype. All the symbols that may appear in the final genetic individual that convey some end meaning as per the problem domain are known as terminating symbols (T). The grammar specification further uses some additional symbols called as the non-terminating symbols (N). These symbols may not be present in the end program or solution, but may be used in the intermediary stages of the generation of the phenotype from the genotype. The non-terminating symbols as well usually convey a meaning as per the problem domain, but they require a replacement by a terminating symbol, before being used as a valid phenotype for fitness evaluation or as a solution to the problem.

The entire usage of Grammar when dealing with this evolutionary technique comes in the individual representation technique. In discussion over the past evolutionary approaches, we saw that the evolutionary algorithm is carried out over a genotype, which is required to be converted into an equivalent phenotypic representation so that the fitness may be evaluated. The phenotypic individuals have individuals represented in a usable manner, which may be directly used as a solution to the problem, and compute the problem solving effectiveness or the fitness value. The conversion of the genotype to the phenotype in this technique is governed by the Grammar. The conversion is carried out in multiple steps or iterations.

The conversion starts with a start symbol, which is a non-terminating symbol that is fixed for a problem. At every step, we replace a non-terminating symbol by one or more than one terminating and non-terminating symbols. The terminating symbols cannot be replaced by any other symbols in the subsequent steps. However, the non-terminating symbols compulsorily need to be replaced by other symbols before the conversion process exits. We discussed about the importance of syntax in the GP, which assures that the program that so generates is syntactically correct and can be used for fitness evaluations. Similarly, the replacement steps in this algorithm have to ensure that the general program architecture so far is that ultimately leads to a syntactically correct program. This is done by the use of production rules in the grammar specification.

The production rules are a collection of all rules that indicate which symbol may be replaced by what sequence of terminating and non-terminating symbols to generate the individual phenotype. Each rule has an antecedent part and a consequent part. The antecedent part indicates the non-terminating symbol, which if found can be referred to this rule for replacement. The consequent part is a collection of terminating and non-terminating symbols. In case the symbol, as specified in the antecedent of the rule is found, and this rule is chosen for replacement, the entire consequent part replaces the selected symbol in the phenotypic individual being prepared. The rules are written in a manner so as to assure that the syntax is always maintained.

The concept of grammar and rules, and the use of Backus-Naur form come from the natural languages. Consider the language English. Any of the words into the dictionary like 'the,' 'is,' etc. form the set of terminating symbols. These words come as it is in the final sentence that we read and write. The terms like 'verbs,' 'nouns,' etc. have a special relevance to the language. These do not come in the end language we read and write; however, they are used for specifying the grammatical rules. The grammatical rules themselves act as the basis to judge the correctness of a sentence. Ideally, only grammatically correct sentences may be used in conversation. These rules act as guiding principles for the generation of the sentence.

Consider the problem of generation of a Boolean expression which woks over the Boolean constants and operators. Here we try to specify a grammar base that would enable us to solve the problem of generation of the expression. The grammatical base must be extensive so as to allow the generation of any type and form of expression by its usage. Consider the following grammar:

```
G = { T, N, S, P}
T = { and, or, xor, nand, not, true,
false, (,) }
N = { <expr>, <biop>, <uop>, <bool> }
S = { <expr> }
P = {
1. <expr>::= (<expr> <biop> <expr>)
| <uop> <expr>
| <bool>
2. <biop>::= and
| or
| xor
| nand
3. <uop>::= not
4. <bool>::= true
| false
}
```

Here, the set of terminating symbols contain all the operators and constants (true and false) that the expression may contain. Brackets are used to separate operations in the final expressions. The non-terminating symbol <expr> is used which denotes any general expression and is also the start symbol. The non-terminating set contains symbols for the binary operators (biop), unary operators (uop), and the Boolean constants (bool). Each of these has a production rule associated to it. The production rules for every non-terminating symbols contains the entire manner in which it may be possibly used or replaced, separated by an 'or' (|) sign.

Individual Representation

Once the grammatical base has been formulated, the next task is to have an individual representation technique. As discussed in the previous section, the main task that the algorithm demands is to be able to convert a genotype into a syntactically correct phenotype. We have the genotype individual in hand for the same purpose. In this technique, the genotype is a set of integers. The total length of the genotype individual is fixed to a constant value, which may not necessarily have a relation to the length of the program. These integers guide the conversion process from the genotype to the phenotype. Therefore, for the conversion, we have in hand an array of integers or the genotype and the grammar. We place a pointer in the genotypic array, which points to the next gene or integer to be read from the array.

The integers are read linearly from one end to the other, each read value guiding the conversion process. Similarly, starting from the start symbol, the conversion process has the phenotypic individual containing a set of terminating and non-terminating symbols. At the first iteration, the individual has only one non-terminating symbol, which is the start symbol. As the iterations proceed, the replacement governed by the production rules leaves the individual with terminating and non-terminating symbols at different locations in the individual.

Consider any iteration in the conversion process. The first job is to spot the first non-terminating symbol in the program so far. This is the symbol that this iteration would attempt to replace. Now the set of production rules are considered. For the particular symbol being attempted to replace, there would be some options available, which may replace the symbol by the rule consequent part. Let us say that the total number of rules having the selected symbol as the antecedent is c in number. Now one of these rules needs to be selected for replacement. For selection of this rule, we take help of the genotype. Let the next integer to be read is x. The kth rule out of the available set of rules is selected where k is given by Equation 12:

$$k = x \bmod c \tag{12}$$

After reading the integer, the pointer is moved to the next integer in the genotype. The selected symbol is replaced by the consequent part of the selected rule. This process goes on along with iterations. If at any time no non-terminating symbol is present, or in other words the program contains only terminating symbols, the conversion process stops and the resultant program is retuned as the phenotype. The integers not read from the genotype are simply ignored. Similarly, if all the integers have been read from the genotype and the program still has non-terminating symbols left, the left non-terminating symbols may be replaced by default replacements containing only terminating symbols. In some implementations, the pointer may be reset to point to first integer once more in order to make the individual terminate. If it still does not terminate, the default values may be used.

Consider the problem of generation of the Boolean expression, the grammar for which was described in the previous sub-section. Consider the genotype to be as given by Equation 13:

$$X = \{ 10, 15, 5, ... 12\} \tag{13}$$

The start symbol is <expr> which may be replaced by any of the 3 rules that is <expr> <biop> <expr> | <uop> <expr> | <bool>. We select the rule number 10 mod 3 since the first gene is 10. Hence, the second rule is selected (1st rule if indices start from 0). The expression hence becomes <uop><expr>. Now the first non-terminating symbol is <uop> which may be replaced by any of the 1 rules that is not. We select the 15 mod 1 or the first rule and replace it with not. The program becomes not<expr>. Now the first non-terminating symbol is <expr> which may again be replaced by any of the 3 rules, for which 5 mod 3 or 3rd rule is selected making the expression not<bool>. Similarly, we may proceed in same manner.

Genetic Operators

Based on the concept of grammar and linear individual representation, we are able to convert the genetic individual into a linear genetic individual or the genotype. Now the conventional genetic operators may be easily used for the evolution. The use of crossover, mutation, elite, and other operators may be done in exactly the same manner as the genetic algorithm. For the same reasons the analysis of the GE follows similar trends of GA and GP considering the genotype and the phenotype respectively. Based on the same principles the specialized genetic operators may be designed as per the problem requirements.

Similar is the nature of the fitness evaluation as well. The GE individual may be easily converted into a phenotype for fitness evaluation purposes. As a phenotype, it represents a potential solution or a program, which may be conventionally used for the task of evaluation of its goodness or computing the fitness value.

PATH PLANNING BY GENETIC PROGRAMMING

The next task is to use the studied concepts of Genetic Programming and Grammatical Evolution to solve the problem of motion planning of robots. In this case, we take a maze-like map. The walls are all at 90 degrees to each other. Some paths are allowed, whereas the others are blocked. The robot needs to find a path from the source to the goal in such a scenario. There might be multiple paths possible, but the robot must choose the shortest one possible. Unlike the other approaches, the robot is restricted to move straight in any direction or turn in multiples of 90 degrees. This eases the problem to be solved to an extent, but the harder path is that since it is a maze, the route to goal may not necessarily be straight or almost straight. Here, we make use of Grammatical Evolution for the problem solving. The following sub-sections explain the solution formulated, and give the simulation results.

Individual Representation

The first and major task to solve the problem is to have an individual representation strategy. In this case, the individual is a program, which upon execution gives the phenotype or a robotic path. The individual or the program is hence a collection of commands given to the robot. From the problem, it is known that all motions of the robot are either horizontal or vertical, since the robot is operating in a strict maze-like environment. Hence, once the robot is at a path, it can only opt to go straight. However, whenever there is any block where multiple paths meet, the robot needs to be able to decide the path, which it must follow. It is clear that it cannot go back, as moving forward and going back would mean returning back which could have been avoided by not picking the path at the first place. This would naturally not happen in the optimal path. Hence, the robot needs to chose from the available choices and move accordingly. The instruction set or the program simply dictates the path to be selected to the robot. The robot follows these instructions and simply moves forward. As per the Grammatical Evolution principles, the path needs to be a set of integers.

In this individual representation technique, each integer represents an instruction to the robot regarding the turn to make. The robot continues to move in the direction it is moving, until it comes to a point where there is more than one possible way to proceed, or in other words, where two paths meet. In such a situation, the algorithm looks at the next integer in the genotype and uses it to decide the move. Suppose that there are c possible moves, which does not include the move, which takes the robot back on the path it was coming from. Let the integer read from the genotype is x. The robot selects the x mod c path to proceed. It would continue to walk straight, until it arrives at another similar situation. In case the robot reaches the goal, the conversion is stopped and the other genes are left unread. In case the robot lands at a situation where it cannot move forward unless it turns back, the location of the robot is taken as final position and rest of the genes are not read. In case the robot does not reach the goal and the genotype ends, the genotype may be reset to a maximum of one time. Henceforth, the position of the robot is taken as the final position that it could reach.

Fitness Evaluation

The fitness of the individual is simply the length of path tracked by the robot. Every block of the maze is assumed to have a unit block length. However in case the robot fails to reach the goal, an extra penalty is added proportional to the Manhattan distance between the goal and the last point touched by the robot. Note that in this case the smallest distance between any points is their Manhattan distance since the robot can only travel in multiple of 90 degrees. The fitness may hence be given by Equation 14:

$$Fit(I) = l + \alpha \| X - G \| \quad (14)$$

Here, $\| x \|$ is the Manhattan norm. G is the goal. X is the last point touched by the robot. l is the length of the path. α is the penalty constant.

Results

The algorithm as discussed was developed in MATLAB. The map was made in a very low resolution with 1 pixel representing a block and loaded as a bmp file in MATLAB. Various maze-like maps were used for testing purpose. The number of individuals was 100. The optimization was done until 50 generations. Crossover rate was 0.7. Gaussian mutation, stochastic uniform selection, and rank-based scaling were used. The genotype had 100 as the genomic length. The results to some of the scenarios presented are given in Figure 14.

Figure 14. Experimental results for grammatical evolution

CONCLUSION

This chapter further explored the potential of the evolutionary computing paradigm. The chapter was basically structured into two similar parts. The first part covered the swarm intelligence domain. In this section, we saw the inspirations we took from the various natural phenomena and used it for the design of artificial systems for problem solving. The individual algorithms we considered in this domain included the particle swarm algorithm, ant colony algorithm, probability-based incremental learning algorithm, and artificial bees colony algorithm. Each of these algorithms could carry out the task of some kind of optimization. The same led us to use them for the task of planning the path of the mobile robot.

The other part of the chapter concentrated upon the use of genetic programming. Here, we used an individual as a program that could solve the problem being considered. This formed one of the major classifications of evolutionary computing.

The tree based individual representation was used. The genetic operators, typically the crossover and mutation operators, were modified to work over this individual representation technique. We further discussed about a linear mechanism by which an individual could be represented. This was by using the concept of grammar and the corresponding evolution technique was known as grammatical evolution. The major task here was to formulate a mechanism to convert the individual in genotype form into corresponding phenotype form. This technique was also used for the task of planning for the mobile robots. Here, a special form of individual representation had to be designed for the specification of the robot path.

REFERENCES

Altenberg, L. (1994). The evolution of evolvability in genetic programming . In *Advances in Genetic Programming* (pp. 47–74). Cambridge, MA: MIT Press.

Baluja, S. (1994). *Population-based incremental learning: A method for integrating genetic search based function optimization and competitive learning. Technical Report: CS-94-163*. Pittsburgh, PA: Carnegie Mellon University.

Banzhaf, W., Nordin, P., Keller, R. E., & Francone, F. D. (1998a). *Genetic programming: An introduction on the automatic evolution of computer programs and its applications*. San Francisco, CA: Morgan Kaufmann.

Banzhaf, W., Nordin, P., Keller, R. E., & Francone, F. D. (1998b). *Genetic programming: An introduction*. San Mateo, CA: Morgan Kaufmann.

Banzhaf, W., Poli, R., Schoenauer, M., & Fogarty, T. C. (1998c). Genetic programming. *Lecture Notes in Computer Science*, *1391*, 97. doi:10.1007/BFb0055923

Clerc, M., & Kennedy, J. (2002). The particle swarm-explosion, stability, and convergence in a multidimensional complex space. *IEEE Transactions on Evolutionary Computation*, *6*, 58–73. doi:10.1109/4235.985692

Dorigo, M., & Caro, G. D. (1999). The ant colony optimization metaheuristic . In Corne, D. (Eds.), *New Ideas in Optimization* (pp. 11–32). London, UK: McGraw Hill.

Dorigo, M., Maniezzo, V., & Colorni, A. (1996). Ant system: Optimization by a colony of cooperating agents. *IEEE Transitions on System, Man, and Cybernetics – Part B*, *26*(1), 29–41. doi:10.1109/3477.484436

Eberhart, R. C., & Shi, Y. (2001). Tracking and optimizing dynamic systems with particle swarms. In *Proceedings of the IEEE Congress Evolutionary Computation*, (pp. 94-97). Seoul, South Korea: IEEE Press.

Hohfeld, M., & Rudolph, G. (1997). Towards a theory of population-based incremental learning. In *Proceedings of the 4th IEEE Conference on Evolutionary Computation*, (pp. 1–5). IEEE Press.

Karaboga, D. (2005). *An idea based on honey bee swarm for numerical optimization. Technical Report-Tr06*. Kayseri, Turkey: Erciyes University.

Karaboga, D., & Basturk, B. (2007). A powerful and efficient algorithm for numerical function optimization: Artificial bee colony (ABC) algorithm. *Journal of Global Optimization*, *39*(3), 459–471. doi:10.1007/s10898-007-9149-x

Kennedy, J., & Eberhart, R. C. (1995). Particle swarm optimization. In *Proceedings of IEEE International Conference on Neural Networks*, (pp. 1942-1948). Perth, Australia: IEEE Press.

Langdon, W. B. (1998). *Genetic programming and data structures: Genetic programming + data structures = automatic programming!* Dordrecht, The Netherlands: Kluwer.

Luke, S., & Spector, L. (1997). A comparison of crossover and mutation in genetic programming. In *Proceedings of the Second Annual Conference on Genetic Programming*, (pp. 240–248). IEEE.

Maniezzo, V., Gambardella, L. M., & Luigi, F. D. (2004). Ant colony optimization . In Onwubolu, G. C., & Babu, B. V. (Eds.), *Optimization Techniques in Engineering* (pp. 101–117). Heidelberg, Germany: Springer.

Maurya, R., & Shukla, A. (2010). Generalized and modified ant algorithm for solving robot path planning problem. In *Proceedings of the 3rd IEEE International Conference on* Computer *Science and Information Technology,* (vol. 1), (pp. 643-646). IEEE Press.

O'Neill, M., & Brabazon, A. (2004). Grammatical swarm. *Lecture Notes in Computer Science, 3102*, 163–174. doi:10.1007/978-3-540-24854-5_15

O'Neill, M., & Ryan, C. (2001). Grammatical evolution. *IEEE Transactions on Evolutionary Computation, 5*, 349–358. doi:10.1109/4235.942529

O'Neill, M., & Ryan, C. (2003). *Grammatical evolution*. Boston, MA: Kluwer.

Parpinelli, R. S., Lopes, H. S., & Freitas, A. A. (2002). Data mining with an ant colony optimization algorithm. *IEEE Transactions on Evolutionary Computation, 6*(4), 321–332. doi:10.1109/TEVC.2002.802452

Pham, D. T., Eldukhri, E. E., & Soroka, A. J. (2006). The bees algorithm – A novel tool for complex optimisation problems. In *Proceedings of the Intelligent Production Machines and Syst*ems *Conference*, (pp. 454–459). IEEE.

Rosca, J. P. (1997). Analysis of complexity drift in genetic programming. In *Proceedings of the Second Annual Conference on Genetic Programming*, (pp. 286–294). Morgan Kaufmann.

Shi, Y. (2004). Particle swarm optimization. *IEEE Neural Networks Society Bulletin,* 8–13.

Southey, F., & Karray, F. (1999). Approaching evolutionary robotics through population-based incremental learning. In *Proceedings of the IEEE International Conference on Systems, Man, and Cybernetics,* (vol. 2), (pp. 710–715). IEEE Press.

Stutzle, T., & Dorigo, M. (1999). ACO algorithms for the traveling salesman problem . In Miettinen, K., & Neittaanmaki, P. (Eds.), *Evolutionary Algorithms in Engineering and Computer Science* (pp. 160–184). West Sussex, UK: Wiley.

Stutzle, T., & Hoos, H. (1997). Improvements on the ant system: Introducing MAX–MIN ant system. In *Proceedings of the International Conference on Artificial Neural Networks and Genetic Algorithms*, (pp. 245–249). Springer.

Tackett, W. A. (1993). Genetic programming for feature discovery and image discrimination. In *Proceedings of the Fifth International Conference on Genetic Algorithms*, (pp. 303–309). San Mateo, CA: Morgan Kaufmann.

Yang, S. (2005). Population-based incremental learning with memory scheme for changing environments. In *Proceedings of the 2005 Conference on Genetic and Evolutionary Computation*, (pp. 711–718). IEEE.

Yang, S., & Yao, X. (2003). Dual population-based incremental learning for problem optimization in dynamic environments. In *Proceedings of the 7th Asia Pacific Symposium on Intelligent and Evolutionary Systems*, (pp. 49–56). IEEE.

Yang, S., & Yao, X. (2005). Experimental study on population-based incremental learning algorithms for dynamic optimization problems. *Soft Computing, 9*(11), 815–834. doi:10.1007/s00500-004-0422-3

Yang, S., & Yao, X. (2008). Population-based incremental learning with associative memory for dynamic environments. *IEEE Transactions on Evolutionary Computation, 12*(5), 542–561. doi:10.1109/TEVC.2007.913070

KEY TERMS AND DEFINITIONS

Genetic Programming: Evolutionary technique where the intent is to evolve a program whose execution solves the problem in hand. A tree-based representation whose construction is determined by the program syntax is commonly used.

Grammar: A set of terminating symbols, non-terminating symbols, start symbols, and production rules for a program. Each program is a collection of terminating symbols, which is regarded as syntactically correct only if it follows the rules of grammar. The construction of the program is usually traced with a start symbol being replaced by terminating and non-terminating symbols, each non-terminating symbol again replaced by terminating and non-terminating symbols and so on, until the program is obtained. Every step must be consistent with production rules (or grammatical rules).

Grammatical Evolution: A type of genetic programming technique where the individual is represented as a linear array of integers based on which the program may be generated with the help of grammar.

Grammatical Evolution-Based Path Planning: Assuming map as a map with all paths at multiples of 90 degrees, the robot is made to start from the source. Any point where multiple ways are possible is decided by looking at the next gene of the genotype. The path terminates if the robot reaches the goal, gets trapped, or all genes in genotype are read.

Swarm Intelligence: Class of algorithms inspired by the collective intelligence found in natural swarms where multiple individuals collectively get some work done as a result of individual performance and coordination between themselves. Many algorithms similar to genetic algorithm are designed.

Chapter 6
Behavioral Path Planning

ABSTRACT

A large part of our everyday motion is governed by behaviors. We necessarily do not look at the entire map and formulate the best way out, but rather take instinctive actions regarding our motion. We naturally reach the desired locations fairly easily and near optimally. With the same inspirations in mind, in this chapter, the authors explore the behavioral systems for the task of motion of the mobile robot. In this chapter, they study two different algorithms, fuzzy inference systems and artificial neural networks. The fuzzy systems are governed by a set of rules, which determine the behavior of the system, for any applied input. The major task involves the use of fuzzy sets for the output computations. As per the theory of these sets, every input belongs to every set by a varying degree called as the membership degree. The authors use this concept of fuzzy-based inference to design a system for the motion of the mobile robot. They further introduce the neural networks paradigm, which is an inspiration from the human brain for problem solving. Neural networks process applied input layer wise by unit processing centers known as artificial neurons. These systems may be trained by a training database that is particular to a problem. The authors use both these algorithms to design systems for behavioral path planning of mobile robots.

INTRODUCTION

The manners in which we react to situations, take everyday common decisions, and carry our daily jobs constitute our behavior. Behavior is developed over the time and may or may not have reasoning associated with it. It may usually be very difficult to justify every big and small action we just made, or are making. However, it would be fairly right to say that the taken actions were governed by whatever we have learnt in life, experienced, whatever believes and values we possess, and

DOI: 10.4018/978-1-4666-2074-2.ch006

whatever thinking capability we have. By this virtue of behavior, we can take our own decisions in life for every big and small problem. In fact, to some extent behaviors characterize our personalities and abilities in a world full of individuals. It is, hence, very necessary to develop behaviors so as to succeed in accomplishing any task.

Another important characteristic associated with behaviors is that they are instinctive. We do not necessarily think out a lot or attempt to foresee long consequences when making a decision. In most scenarios, we let our instincts make decisions, which depend upon the individual behaviors we possess. It would be interesting to realize the kind of information processing that governs our actions based on these behaviors or instincts. We naturally do not carry intensive computations or attempt to analyze each and every aspect of the problem in our decision-making. Rather, we are more likely to see a part of it and use the summarized form of our experiences to quickly come up with the action plan. This form of decision-making is quick, and based on our behaviors may be effective.

Based on the natural behaviors, there is a sure urge to make behaviors for the artificial systems that govern their actions to the problems that they are given. The aim is to replicate the same problem solving methodology in these systems, as the humans showcase in their daily problem solving. By this mechanism, we are likely to give the systems the ability to quickly decide and act upon problems, rather than making a full-analysis and deciding on the same basis. A complete analysis on many of the problems may not be possible owing to uncertainties in inputs, non-understanding of the problem domain, lack of availability of information, etc. Many other times there may be serious computational constraints, considering the fact that complexity drastically increases with small additions in the modeling scenario, which has a very adverse effect on the computational requirements. Hence, it is advisable, and in many problems the only solution, to design behavioral systems for decision-making.

The problem of path planning is a similar problem. Consider your everyday motion where you move from one spot to another in the entire room or city at large. In regular motion, we hardly consider the obstacles that lie or are likely to lie too far off. Similarly, we are likely not to consider the complexities of the route. Rather, we normally walk, avoiding possible collisions with any of the obstacles we may find on our way. Everyday walking is a proof of us being able to navigate long distances with ease and without any collision with any of the obstacles.

Based on the same principles, there is a motivation to construct a planning algorithm for the navigation of the mobile robot. This system, in place of looking at the entire map and computing all complexities, may only concentrate on its surroundings and try to steer itself through obstacles, moving towards the goal position. This is somewhat like the manner in which we walk. A clear advantage of this system comes in cases of dynamically changing environment. While walking, if we encounter any obstacle that may be moving or may have suddenly appeared, we do not get struck. In turn, we keep walking, ensuring that we do not collide with the newly observed obstacle. This is a very important characteristic in planning knowing operations are in a place where uncertainties are high and anything may suddenly appear, or where map is fairly unknown.

The behavioral systems, owing to their rapid planning cycles, can as well naturally overcome these obstacles. As soon as some new obstacle is seen that lies close to the vicinity so as to affect the motion plan, the input is quickly modified and given to the planning algorithm, which uses its instinctive behavioral approach to compute the robot action. The advantage of the quick planning, however, does come with the cost of completeness. The behavioral systems are not complete. Hence, it is possible that there exists a path from the source to goal; however, these systems are unable to find one. Further, it is possible that these systems make the robot move to a place where subsequent motion may not be possible. This may further make the robot wander about a set of points, not able to reach the goal. This may happen in everyday walking as well. Consider the situation where we walk over a road, unknowing that its other end is temporarily blocked. In such a situation, we may use a higher level of planning to escape, a capability that purely behavioral systems may not have.

In the upcoming sections, we study the use of neural networks for solving the problem of motion planning of robots. These are models inspired by the human brain. An interesting experimentation reports implanting the brain of a rat into the processing unit of a mobile robot (Reading, 2011). The work involves interfacing of a biological brain with a mobile robot. Experiments were done on a miabot robot, which is equipped with 16 sonar sensors on all its sides. The sensor readings were taken, which were fed into the biological brain. The brain produced stimulus, which were taken from electrodes. Based on these outputs, the robot was turned to avoid obstacle. The task of robot was to move in a collision free manner, avoiding all obstacles. After being developed and trained, the biological brain could control the mobile robot reasonably well.

In this chapter, we attempt to engineer these behavioral systems by two approaches. The first approach makes use of fuzzy inference systems. Here the modeled inputs are computed and given to a fuzzy-based inference system, which decides the robotic action. To understand how the system solves the problem of motion planning, we first describe the concept and working of the fuzzy systems in brief. The other system we deal with is the use of neural networks for solving the problem. Here also the inputs are modeled and fed into a multi-layer perceptron model of neural network. We discuss the neural networks and the problem solving by these networks in the subsequent sections.

FUZZY CONCEPTS

The first major algorithm of study is the fuzzy inference system. Here, the main task is the use of an inference system, which attempts to take some inputs and process them into some outputs based on some inferences, which are fed to the system in the form of production rules. The manner of working of this system is fuzzy in nature, a concept that differs from the normal rule- and state-based systems that we encounter. We study the fuzzy sets and the related concepts, before having a broader look at the fuzzy inference system. For greater details over the working and methodology, refer to books, original works, and some general texts by Cox (1994), Dubois and Prade (1980, 1985), Kandel (1991), Kasabov (1995, 1998), Klir et al. (1997), and Yager and Zadeh (1992).

Fuzzy Sets

Sets are commonly used mathematical principles, where a set denotes a collection of numerals, objects, or other entities. These sets have the property that every entity can only be present in a set or not present in the set. In other words, either an entity is completely a part of a set or it is not at all part of the set. This is because the conventional mathematical theory of sets is completely discrete in nature. The fuzzy sets are sets where every entity is a part of every set. Hence, every set holds all entities and all entities are present in all sets. The association of an entity to a set may however be of varying degrees. Hence, some entities may have a significantly large association to some set, whereas others may have a fairly poor association. The degree to which an entity belongs to a set is known as its membership value.

The fuzzy sets form a more realistic modeling of the system as compared to the discrete set (Zadeh, 1965, 1968, 1979). Consider a system that forms sets of distance of an object from the goal. Let us say that the modeling considers two sets being close and far. The discrete mechanism of set modeling would state that all points having the distance less than, say 30 meters, belong to the set close, while the other sets belonging to the set far. Now, we know the actual system would process the inputs from the two sets differently by different sets of principles or rules. Such a modeling of the system means that two points lying very close to each other at each side of the 30 meters mark are processed differently. This however does not happen in everyday life where we do act on distant and far objects differently, but there is no major classification of this sort, classifying the inputs to two completely different sets.

In the actual world, we make a smooth transition in our action, as the inputs change. Hence, the discrete system of classification may not be realistic. A better way to model the input would be to use a fuzzy-based approach. Here, we state that every input belongs to every set (both sets in this case). The degree of membership may be different for the different sets. Hence, in the above problem an input of 10 meters may be fairly within the set of close, say to a membership value of 0.8. However, the same input may also belong to the set far to a smaller value, say 0.2. As we move across points, further from the goal, the membership value of sets close may decrease and that of sets far may increase.

Membership Functions

The degree of association of an input to a particular set is known as the membership degree. It may be impossible to denote a membership value for every input to every set. This task is performed by use of functions. Consider the input x whose membership value to the set u needs to be found. We model this by a membership function, which may be denoted by $\mu_u(x)$. The complete specification of any number x belonging to a set u by degree $\mu_u(x)$ may hence be conveniently be represented by $x/\mu_u(x)$. This notation of fuzzy numbers is used multiple times for calculation purposes. It is common to use some standard functions as the membership functions. Commonly used membership functions include triangular, Gaussian, etc. Some of these membership functions are given in Figure 1.

Figure 1. Some membership functions: (a) Gaussian membership function, (b) triangular membership function, (c) sigmodial membership function

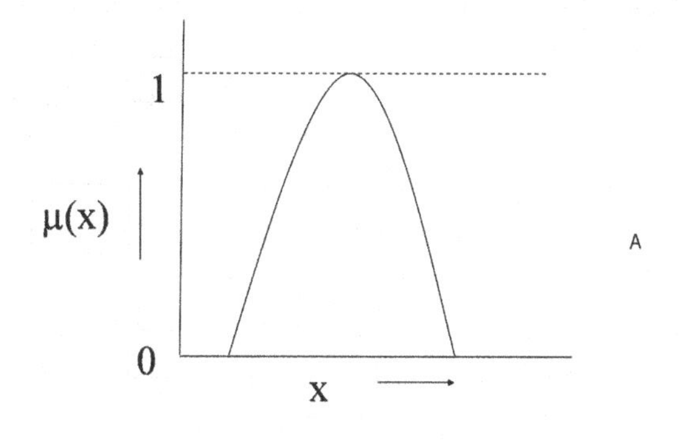

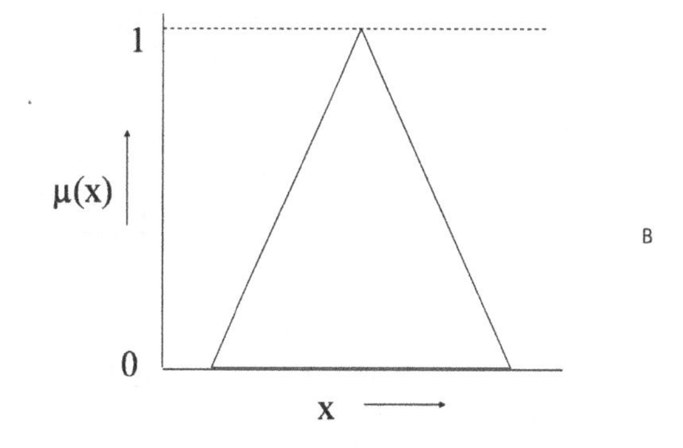

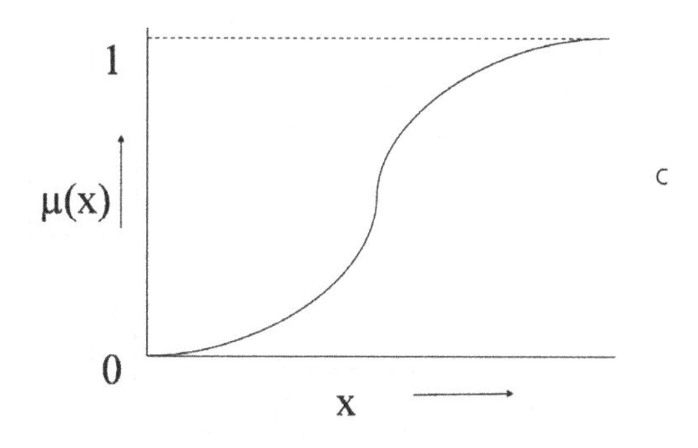

Consider the problem of modeling of an input distance, which we break into 3 sets, say close, medium, and far. For each of these functions, we use a membership function curve that best describes the change in the membership degree as we change the inputs. Consider the curve for these three variables as shown in Figure 2.

Here we have considered the fact that the inputs are bounded from a minimal value of 0 to a maximal value of 100. Consider any input within this range, say 80. This input has a membership value for each of the sets formulated. Looking at the figure we may easily compute the membership values which would come out to be $\mu_{close}(80)=0$, $\mu_{medium}(80)=0.4$, and $\mu_{far}(80)=0.6$. Further, looking at the figure we can easily see that the close membership curve gets a membership value of almost 0 to inputs that lie far from it. Similar discussion may be done for the other sets as well. Hence, there is some region where every set gives a membership degree of 1. Around this region, the membership degree falls until it reaches a value of 0 denoting no association of the input to the set.

FUZZY OPERATORS

The fuzzy formalism of problem is different from the conventional formalism of the problem. The difference comes from the added concept of fuzziness, which changes the manner of work. The conventional system of working with sets and using production rules for problem solving over these sets is simple. Since every input either belongs to a set or not the computation is easy. Rule specification normally consists of conditions joined by unary or binary Boolean operators. The validity or trueness and non-validity or falseness of these rules may be easily determined from the knowledge of the truth table of the individual operators. If the resultant condition evaluates to be true, the rule fires, else it does not fire. The rules that fire determine the output. Hence, knowing the sets to which the input belongs makes the computation an easy task.

The problem that comes with the fuzzy sets is the added fuzziness. Now the inputs belong to the different sets by different degrees. Considering the same rule set, it is important for the different operators to account for this fuzziness in their computation. This calls for a migration from the

Figure 2. Various membership functions for an input

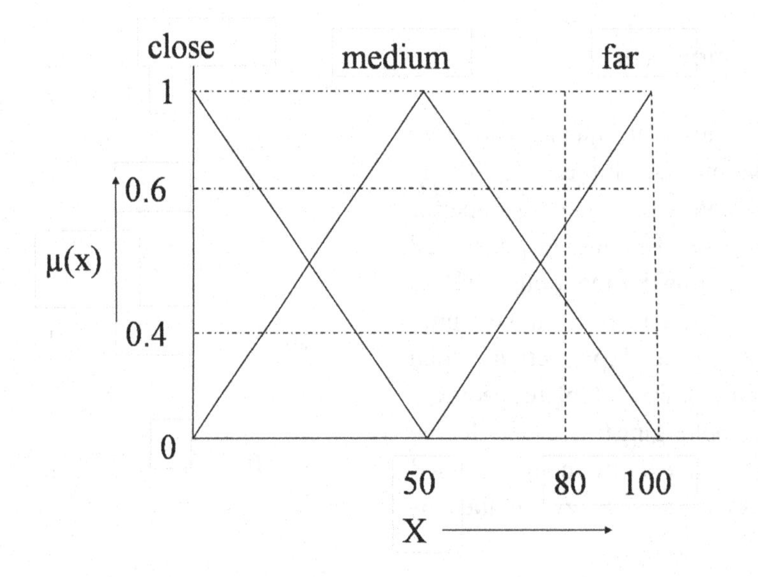

normal Boolean operators to the fuzzy equivalents that work over fuzzy numbers (Mamdani, 1977; Sugeno, 1974, 1985; Tagaki & Sugeno, 1985). In the subsequent sub-sections, we discuss about the fuzzy operators. The basic methodology would be to first look at the individual operators and then look at the entire fuzzy inference system at large.

AND

The first major operator of use is the AND operator. In Boolean terms, this operator returns a true value whenever all the inputs are true, in all other cases it returns a false value. In fuzzy terms, since all inputs belong to all sets, the question of absolute belongingness of the resultant obtained after the AND operation is irrelevant. This operator in the fuzzy domain takes in 2 membership function values denoting the degree to which the inputs belong to the participating sets in the rule. The task of the AND operator is to return a resulting membership value, which denotes the degree to which the inputs belong to the condition represented by the AND operator.

There are two ways to model the AND operator. The first method is known as the min method, which simply states the resultant membership value to be the minimal of the two input membership values. The other method is known as the product method, which states the output membership value to be the product of the input membership values. Hence if c_1 denotes the membership degree of the first input set x and c_2 denotes the membership degree of the second input set y, the resultant membership degree of the output x AND y may be given by Equation 1:

$$x \text{ AND } y = \begin{cases} \min(c_1, c_2) & \text{For min method} \\ c_1.c_2 & \text{For product method} \end{cases} \quad (1)$$

The relevance of these two methods may be easily realized. Consider the situation with a task to be performed which has two related sub-tasks, both of which must be completed to fully complete the given task. In such a case, it may be assumed that the person who can clear the task which has lesser chance of being completed (or which requires higher skill), may be easily able to complete the total task. Hence, the capability requirements of the tasks are the capability requirements of the most difficult task. In other words, the chance of the task being completed is equivalent of the lesser probabilistic sub-task being completed, which leads us to the min realization. For realization of the product, we look into the scenario from the point of view of probability theory. If the probability of the first sub-task being completed is c_1 and that f second sub-task being completed is c_2, the probability of the resultant task being completed is $c_1.c_2$, which is the product realization.

OR

The other very important operator used in Boolean arithmetic is OR, which we hence attempt to modify for use in the fuzzy arithmetic. In the Boolean arithmetic, where the output to this operator may only be a true or a false, this operator gives a false only when all the inputs are false. If any of the applied input is true, this operator gives the output as true. In the fuzzy equivalent, we have the same problem that the inputs belong to the sets to varying degrees. Let the applied input be such that its membership with the input set x participating in the condition is c_1 and membership with the input set y participating in the condition is c_2. Similar to the AND operator, the OR operator may be realized by either a max or a probabilistic or. This is given by Equation 2:

$$x \text{ OR } y = \begin{cases} \max(c_1, c_2) & \text{For max method} \\ c_1+c_2-c_1.c_2 & \text{For probabilistic or method} \end{cases} \quad (2)$$

Similar to the AND operator, the OR operator may be realized easily. Let the task contain two sub-tasks such that either of them needs to be done in order to complete the main task. Now a person is likely to select the task that is easier to do or is more probabilistic to happen. Hence, if a person has capabilities to complete the task which has a higher probability of happening he may be able to complete the entire task. This gives the max realization. Similarly, if c_1 and c_2 denote the probabilities of the two sub-tasks happening, the probability that either of these tasks may happen may be given by probabilities that either task may happen minus probability that both tasks happen as per the probability theory. This is given by the probabilistic OR.

NOT

The next major operator that is commonly used in Boolean operations is NOT. This operator is a unary operator that takes a single input and simply returns the opposite as its output. The input true returns false and vice versa. Similar is its operation in the fuzzy arithmetic, where this operator takes a single input and returns the opposite. Hence, let c_1 be the membership value of the input set x. The output value, which denotes the negation simply given by Equation 3:

$$\text{NOT } x = 1 - c_1 \tag{3}$$

The realization of this scheme may be given by the probability theory. If the probability of an event happening is c_1, the probability of the event not happening is $1-c_1$.

Implication

Consider any general rule. It has two parts, the antecedent and the consequent. The antecedent part consists of all the input membership functions that participate in a rule separated by the operators. The consequent part of the rule consists of the output membership functions that participate in the rule. Hence, the general form of the rule may be given as:

```
If x is MF2(x) AND y is MF1(y) AND z
is NOT MF3(z) then p is MF2(p) and q
is MF3(q)
```

For simplicity of working this may be easily converted to the form:

```
If (MF2(x) AND MF1(y)) AND
(NOT(MF3(z)) → MF2(p) and MF3(q)
```

The brackets denote the operator precedence, which has been taken as it is from the Boolean arithmetic. Since we know the inputs, their membership degrees may be easily obtained. Using the operator arithmetic discussed in the previous sub-sections, we may easily replace the antecedent part by a membership value, which denotes the degree to which the rule fires in deciding the output. It should be noted here that the AND used in the antecedent part is different from the 'and' used in the consequent part. In antecedent, the AND is a binary operator, whereas in consequent, it is just a separator.

The next operator is implication that maps the antecedent to the consequent. Unlike other operators that were used between two membership values, this operator is used between a membership value, which is given by the antecedent, and a membership function curve given in the consequent. In case multiple outputs are there in the consequent, this operator may be applied separately for all the outputs separated by the 'and' separator. The output of this operator denotes the output as computed by this particular rule. Since the arithmetic is fuzzy, we do not get a single output but rather a set of outputs, each output being indicated by a membership degree denoting the likeliness of the output being correct as per modeling.

The implication operator may be performed in two ways. These are min and product. Consider x to be the membership value of the rule as computed by the earlier steps and Y be the output membership curve. The output of the implication operator, which is again a curve, may be given by Equation 4:

$$x \rightarrow Y = \min(x, Y) \text{ For min method } \; x.Y \text{ For product method} \quad (4)$$

The two techniques may be graphically represented as given in Figure 3.

The min method is often known as the slicing method, since it slices the output curve Y along the membership value of x. The product method is also known as the compression method, since it compresses the output membership degree curve Y by a magnitude of x.

Aggregation

The rule base of the system under consideration may consist of multiple rules, all of which are fired to varying degree of extend and contribute towards deciding the output. The problem comes in how to combine the individual outputs of each rule and present it as a unanimous output for every output variable. This task is performed by the aggregation operator. For every output variable, this operator simply aggregates the individual rule outputs to formulate the resultant output. The output is in the form of a membership curve denoting the possibilities of the various outputs to be the final outputs. This function takes as inputs multiple membership curves and in turn produces as output a membership curve. The aggregation is done point by point, iterating thorough all the possible output values within the spectrum of the output variable. The corresponding membership values are worked over to give the membership value of the particular output, from the membership values of the individual rule outputs or the aggregation operator input. The aggregation may either be performed by a sum operator, or by max operator, or by probabilistic or method. The aggregation over two membership curves X and Y give the membership function Z given by Equation 5:

$$\begin{aligned} Z = {} & \max(x, y) \text{ for max method, } x \in X, y \in Y \\ & x + y \text{ for sum method, } x \in X, y \in Y \\ & x + y - x.y \text{ for probabilistic or method, } x \in X, y \in Y \end{aligned} \quad (5)$$

Figure 3. Implication by min and product

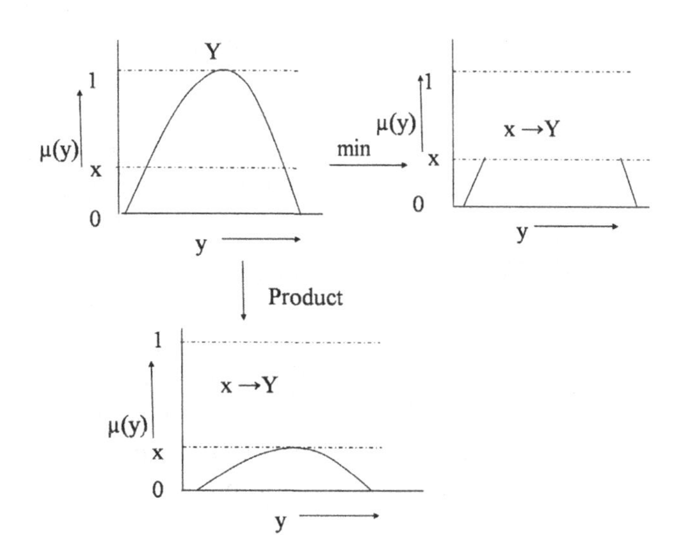

Here, x and y are the corresponding membership degrees of the output variable in the two inputs. The three methods of aggregation are shown by Figure 4.

Defuzzification

The last operator we discuss is defuzzification. The aggregation operator gives the output of every output variable. However, the output is still fuzzy in nature. This means that in place of a crisp output having some definite value, we have a broad range of values, each value having some membership degree denoting its possibility of being the discrete output as per the system modeling. The system, however, needs as output a crisp value corresponding to the crisp numbers that were applied as the inputs. This task is done by the defuzzification operator. This operator analyzes the entire output membership curve and gives as output a crisp number, which forms the output of the complete system. A number of methods are used for the task of defuzzification like centroid, bisector, mean of maximum, minimum of maximum, largest of maximum, etc. The centroid method is one of the most common methods of use. This method simply takes a weighted mean of all the possible output values, where the individual membership degrees act as weights. This may be given by Equation 6:

$$output = \frac{\int_i x_i \mu(x_i)dx}{\int_i \mu(x_i)dx} \quad (6)$$

FUZZY INFERENCE SYSTEMS

The purpose of introduction of fuzzy arithmetic and the fuzzy concepts is to be able to use the same concepts for designing of inference systems. The inference systems are the ones that take certain inputs, process them to give the desired outputs. The conversion of inputs to outputs is based on some inference build within the system. The fuzzy inference systems are based upon the concepts of fuzzy sets and fuzzy arithmetic for the computation of the outputs (Furuhashi, et al., 1993; Mizumoto & Zimmermann, 1982; Plonka & Mrozek, 1995). The major concept behind these systems is the use of a rule base or a collection of rules that denote the logic behind the system. This logic specified by a collection of rules is specified in the design time.

The entire task of designing a fuzzy inference system involves a series of steps, which when done may be used for the specific purpose for which the system was designed for. For every input that is applied henceforth, the usual manner of working as briefly explained in the previous steps works. In this section, we look at these steps that govern the design of the fuzzy inference system with an emphasis on the working of the system for computation of the output to any applied input. The entire process of computation of output may be summarized in Figure 5.

Determination of Input and Output

One of the first steps to be performed in any system is to identify the inputs and outputs explicitly. The entire system being designed may be using the fuzzy inference system as one of the sub-systems, in which situation it becomes even more important to clearly identify the inputs and outputs. These must be known at the first place, along with the range of values that they take.

Membership Functions

Each input and output needs to be broken down into a number of membership functions. Hence, we need to decide the number of membership functions as well as the type and parameter of each membership functions. The optimal number and parameters may not be set at the first try, and the entire procedure may take intensive experimenta-

Figure 4. Aggregation methods

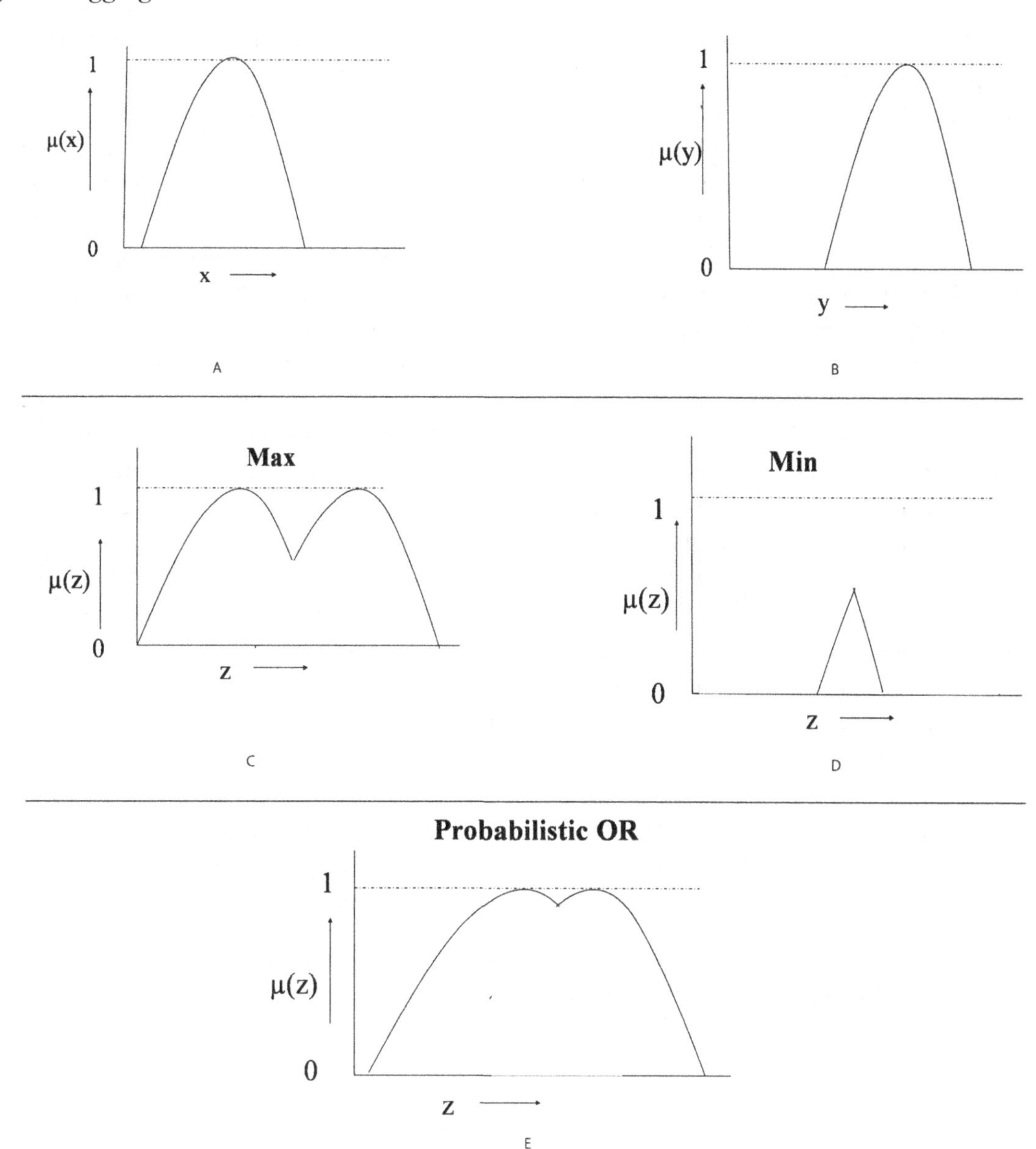

Figure 5. Fuzzy inference process

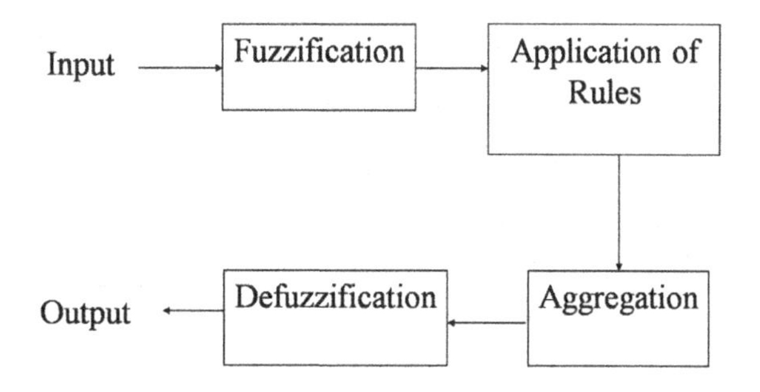

tion in determining the optimal set of membership functions. Having as much understanding of the system as possible eases this procedure. Usually it is better to keep the number of membership functions low. Lower number of membership functions denotes a smooth change in system behavior as the inputs change. Further, this creates scope for lesser number of rules, which means that creation of rules would be simple without necessarily going into the complexities of interactions between different inputs with different membership functions.

However, many times the system may have a rather complex manner in which the outputs change to change of inputs, for which a larger number of membership functions would be a necessity. It is again worth noting that the membership functions may not be uniformly distributed or of the same types and numbers across inputs and outputs. Inputs and regions where system shows complex behaviors may be fitted with more membership functions, the density being sparse at other places. Figure 6 shows a distribution of a sample inputs into membership functions. It is a normally used technique to vary a membership function maxima to minima, and to have another membership function maxima at the minima at both extremes. Here, maxima means a membership value of 1 and minima means a membership value of 0. This allows functions to smoothly transcend from one function to another. It may be noted that any two inputs with same value of membership value for all membership functions would give same outputs. Hence, regions not covered by membership functions, regions covered by the same membership function with the same membership value may be different sides of extremity, etc. are all likely errors.

Whenever an input is applied to the system to compute the output, these membership functions act upon them to convert a crisp input into membership values, with a unit membership value for each membership function. In this manner, every input is mapped into all membership functions that the input was broken down into. This process is known as fuzzification.

Rules

The entire performance of the fuzzy inference system is based upon rules. Good rules that narrate the actual system behavior lead to a good system, whereas poor rules lead to system giving

Figure 6. Diversion of input into membership function

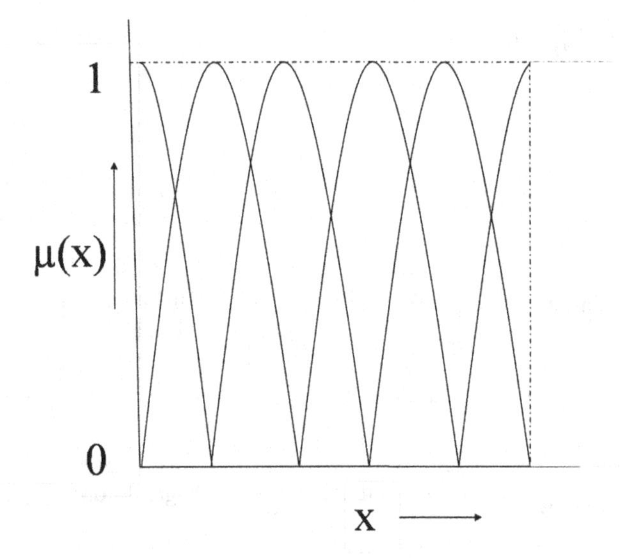

erroneous outputs. The rules in a fuzzy system are designed by a domain expert who attempts to link every input or combination of inputs to the output. Like the membership function breakup, these rules may not necessarily be correct at the very first attempt. The system may require intensive experimentation before a set of rules may be finalized. Each rule in a fuzzy system broadly covers the performance of the system for some kind of input combinations. Lesser number of rules is normally preferred as they are easy to vary to notice the changing system performance. Further, they denote simplicity of the overall system and its understandability. If the system, however, is complex by its behavior, having a larger number of rules is necessary.

Every rule in the system has an antecedent and a consequent. The antecedent lists all the participating inputs and the membership functions by which they participate in the rule. These are separated by Boolean operators. The consequent as well contains the participating outputs in the rule, and the membership functions by which these participate. The discussed set of operations using fuzzy arithmetic is carried out to compute the output by the membership degrees, which have already been computed in the fuzzification process.

Inference

The individual rules outputs may be easily aggregated to give a unanimous output. The output is however fuzzy and needs to be defuzzified to give a crisp output over which the system works.

This completes the entire system design. One of the main aspects of the fuzzy inference system is the ability to understand the conversion of the inputs to the outputs. Every applied input effectively belongs to a single or a couple of membership functions, with significant membership value. It is possible to trace the rules that are effectively being fired with significant degree. Mostly for every input combination a couple of rules would be effective, which effectively determine the output. Hence, understanding of the system is high when dealing with fuzzy systems. It was stated repeatedly that the initial designed system may work for some inputs, but may not work for the other inputs. Here the understanding of the rules and systems comes into play, where amendments may be made to the rules and membership functions to get the needed outputs. First attempts must be made to alter the rules, failing which the membership functions may be altered. It is better to increase as less rules as possible, to retain the simplicity and understandability of the system. If still the desired behavior is not met, the membership functions may be increased. Repeating a few times the desired fuzzy system may be effectively designed.

MOTION PLANNING BY FUZZY INFERENCE SYSTEM

In this section, we develop a model for the motion of a mobile robot using a Fuzzy Inference System (FIS)-based planner. The task is to plan the path of a mobile robot from source to goal in a given map using fuzzy-based planner. It is assumed that the source, goal, and map are already available. The complete mechanism of designing and testing the fuzzy system is dealt with in this section. The FIS is used for the movement of the robot. The FIS was generated by hit and trial method. The FIS is supposed to guide the robot to reach the goal position. The FIS tries to find out the optimal next move of the robot. In this manner, the environment of the robot is assessed and the designed fuzzy inputs are derived. These inputs are given to a FIS, which gives as its output the next move of the robot. This move is executed. The execution of move places the robot in a new environment. Now again the inputs to the fuzzy system are derived from this environment.

In this manner, the procedure goes on, until the robot reaches the goal. The purpose is to design a FIS that given any position of the robot, gives an output the near-optimal next move of the robot. This section is reprinted with few modifications from Kala et al. (2010). With kind permission from Springer Science+Business Media: Artificial Intelligence Review, Fusion of probabilistic A* algorithm and fuzzy inference system for robotic path planning, volume 33, 2010, 275-306, Kala, R., Shukla, A., Tiwari, R.

For more related works in robotic planning, please refer to Aguirre and Gonzalez (2000), Antonelli and Chiaverini (2004), Baturone et al. (2004), Maaref and Barret (2000), Yang et al. (2003), and Zhang and Bohner (1993).

Inputs and Outputs

The first major task in FIS design is to decide the inputs and outputs. The input is the interpretation of the environment by the system. Considering the design limitations, we need to reduce the complete environment in which the robot is into a few numerical inputs in our FIS design. To decide the inputs we again take an analogy from the manner in which we walk and move. The aim is to always align ourselves into direction of goal, go straight ahead, and avoid any obstacle on the way by either moving from its left or right. The designed FIS takes 4 inputs. There are angle to goal (α), distance from goal (d_g), distance from obstacle (d_o), and turn to avoid obstacle (t_o).

The angle to goal is the angle (α, measured along with sign) that the robot must turn in order to face the goal. This is measured by taking the difference in current angle of the obstacle (φ) and the angle of the robot (θ). The result is always between -180 degrees and 180 degrees. This is shown in Figure 7.

Figure 7. The angle to goal

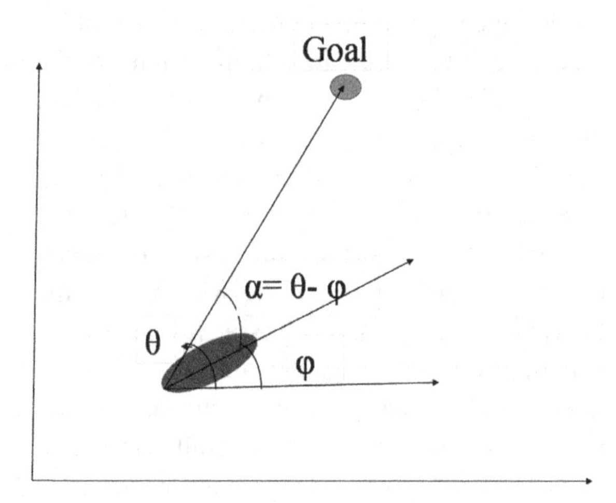

The distance from goal (d_g) is the distance between the robot and the goal position. This distance is normalized to lie between 0 and 1 by multiplying by a constant. Similarly, the distance from obstacle (d_o) is the distance between the robot and the nearest obstacle found in the direction in which the robot is currently moving. This is also normalized to lie between 0 and 1.

The turn to avoid obstacle (t_o) is a discrete input that is either 'left,' 'no,' or 'right.' These stand for counter-clockwise turn, no turn, and clockwise turn respectively. This parameter represents the turn that the robot must make in order to avoid the closest obstacle. This input is measured by measuring the distance between the robot and the obstacle at three different angles. The first distance is the distance between the obstacle and the robot measured at the angle at which the robot is fac-

Figure 8. The determination of turn to avoid obstacle

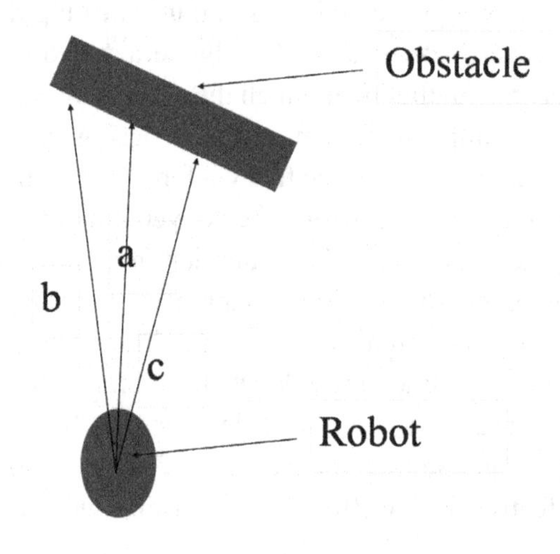

ing (a). The second distance is the same distance measured at an angle of dθ more (c). The third distance is measured at an angle dθ less (b). These three distances are given in Figure 8.

Various cases are now possible. The first case is c>b. This means that the obstacle was turned in such a way that turning in the clockwise direction made it even further. In this case, the preferred turn is clockwise with an output of 'right.' The second case is b>c. This means that the obstacle was turned in such a way that turning in the anti-clockwise direction made it even furtherer. In this case, the preferred turn is anti-clockwise with an output of 'left.' The third case is b=c. This is the case when the robot is vertically ahead of the robot. In such a case, we follow a 'left preferred' rule and take a 'left' turn.

Similar to the inputs, a major task is determination of the output of the robot. In this system, we are simply concerned about determining the path of the robot, which means figuring out all its turns and movements. Hence, there is a single output that measures the angle (β) that the robot should turn, along with direction.

Once the inputs and outputs are known, we need to divide them between membership functions. It is natural that perfect distribution of membership functions cannot be obtained in a go, and there needs to be constant experimentation of the complete system to get desirable combinations of membership functions. After some experimentation, the membership functions of the different inputs and outputs that we get are given in Figure 9.

Rules

Rules are the driving force for the FIS. Based on the inputs, we frame rules for the FIS to follow. The rules relate the inputs to the outputs. Each rule has a weight attached to it. Further some inputs have been applied with the NOT operator as well. All this makes it possible to frame the rules based on the system understanding.

Similar to the mechanism of determination of inputs and outputs, the design of rules is done in analogy with the mechanism of natural walking of humans. The rules can be classified into two major categories. The first category of rules tries to drive the robot towards the goal. The second category of rules tries to save the robot from obstacles. If an obstacle is very near, the second category of rules become very dominant. In all other cases, the first category of rules gains importance and pushes the robot towards the goal. It is again evident that ideal set of weights may only be obtained after a lot of experimentation. Tracing the rules fired for every scenario and tuning them for better performance is the step that needs to be performed multiple times. There may be necessity to alter rules, delete them, or add them during experimentation. With all this in mind, 11 rules are identified. These are given in Figure 10. The numbers in brackets denote the weights.

Experimentation

The discussed model was developed in MATLAB. The maps were input as a bmp file. The fuzzy inference system was developed using the Fuzzy Logic toolbox in MATLAB. The complete simulator design was done by a separate code which moved the robot using the developed system as well as displayed the path. The simulator was also a source to display the various scenarios and look into the loopholes in the system. This was used to modify the membership functions and rules, and further experiment the system. The final testing of the system was done on some entirely new set of maps. The results to some of these maps are given in Figure 11.

NEURAL NETWORKS

Neural networks are an inspiration from the manner of working of the human brain. The human brain is capable of accomplishing all the

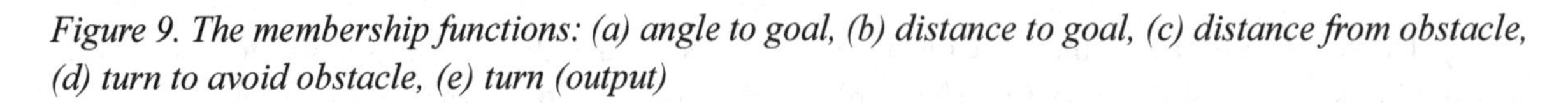

Figure 9. The membership functions: (a) angle to goal, (b) distance to goal, (c) distance from obstacle, (d) turn to avoid obstacle, (e) turn (output)

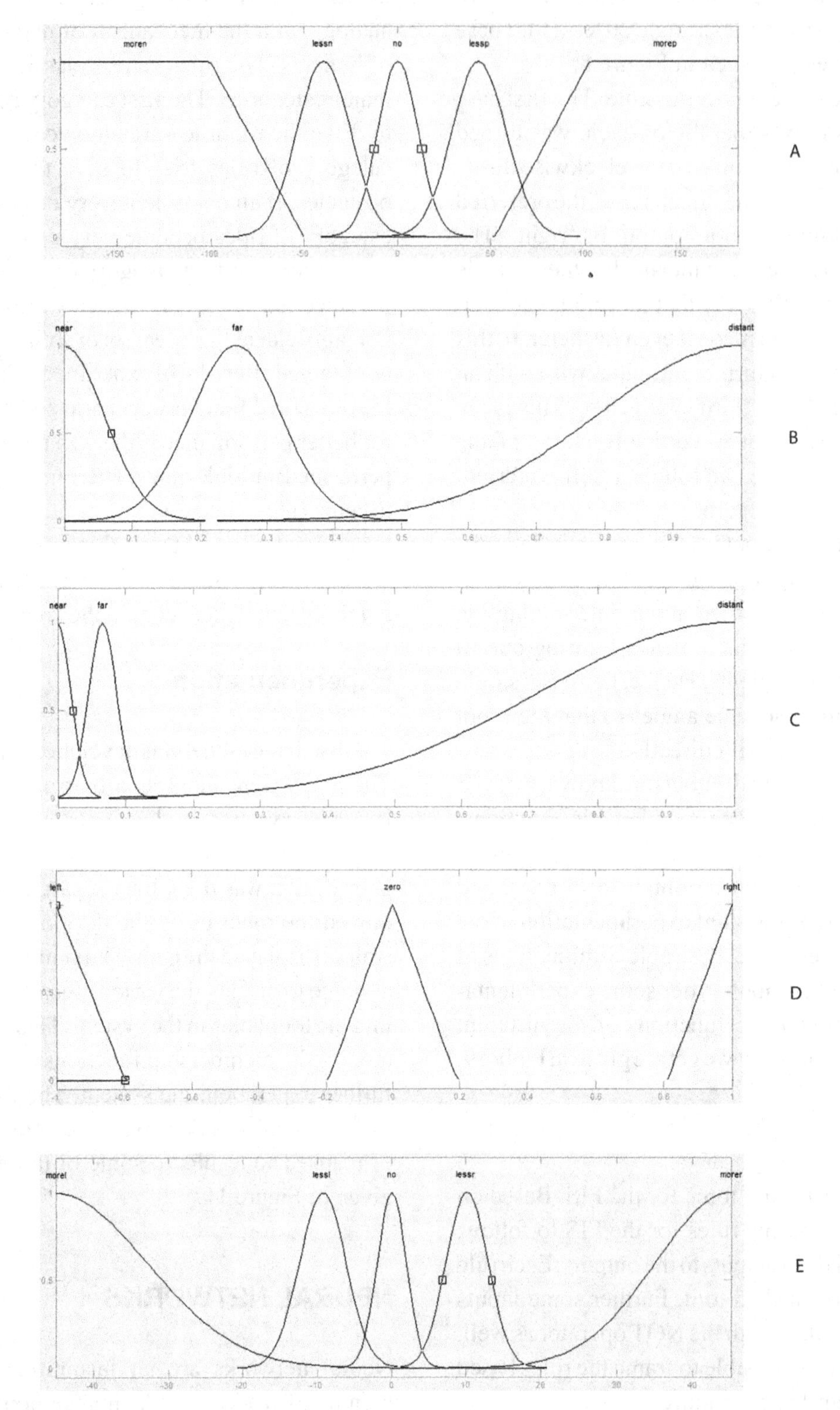

Figure 10. The FIS rules

Rule1: If (α is less_positive) and (d_o is not near) then (β is less_right) (1)
Rule2: If (α is zero) and (d_o is not near) then (β is no_turn) (1)
Rule3: If (α is less_negative) and (d_o is not near) then (β is less_left) (1)
Rule4: If (α is more_positive) and (d_o is not near) then (β is more_right) (1)
Rule5: If (α is more_negative) and (d_o is not near) then (β is more_left) (1)
Rule6: If (d_o is near) and (t_o is left) then (β is more_right) (1)
Rule7: If (d_o is near) and (t_o is right) then (β is more_left) (1)
Rule8: If (d_o is far) and (t_o is left) then (β is less_right) (1)
Rule9: If (d_o is far) and (t_o is right) then (β is less_left) (1)
Rule10: If (α is more_positive) and (d_o is near) and (t_o is no_turn) then (β is less_right) (0.5)
Rule11: If (α is more_negative) and (d_o is near) and (t_o is no_turn) then (β is less_left) (0.5)

wonderful tasks that we see in our daily lives. While some of the tasks may look trivial, a close analysis would reveal their complexities. This fact is further explained when we attempt to engineer machines that perform the same or similar tasks. The beauty of the human brain is still the source of inspiration for numerous people, who attempt to learn from the manner in which the human brain works, and attempt to add the same functionality in the intelligent machines. The human brain is supposed to contain biological neurons, which are small processing elements. The brain

Figure 11. Experimental results with FIS

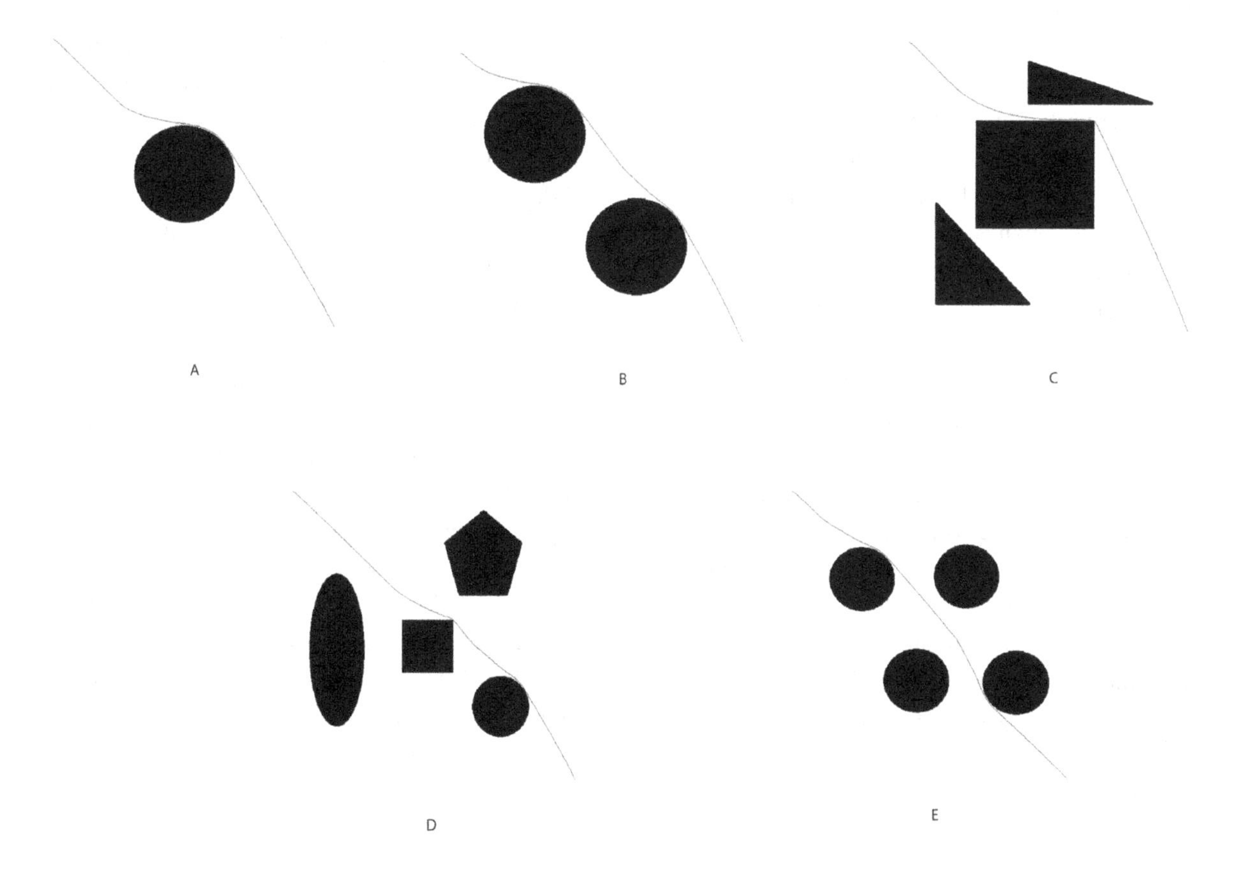

has about 10^{11} neurons, which are all connected to each other making the number of connections as 10^{22}. All these neurons perform computation in parallel and exchange information. This leads to all the simple to complex information processing in the human brain, which gives the humans their intelligence.

With the inspiration from the human brain, there have been numerous attempts to make systems on the same lines. The artificial neural networks are a collection of artificial neurons that are connected to each other. Similar to the human brain, the artificial neurons carry small amount of computations. Their results are given to other neurons, which further work upon the computed information, until the information is fully processed and can be given as the output to the system. The processing of the information does takes place in a layer-by-layer mechanism, with the initial layers working on almost the raw input data, and the latter layers working on more processed data, that lies in form closer to the output.

One of the most important aspects associated with the neural networks is their ability to learn. These networks can be given a data set containing collection of data items with inputs and outputs. These networks would themselves find the patterns and associations between the inputs and the outputs. Hence, these networks are able to mine the data set and extract out some algorithm that maps the inputs to the outputs. This means that next time whenever any of the similar inputs would be given to the network again, it would be able to give the correct or nearly correct output. This depends upon the learning of the network.

Further, these networks are known to have a high degree of generalizing ability or generalization. This means that if any new input is given to the system that is related to the inputs originally in the data set, the system should be able to generalize the mapping rules to the newly applied input, and give possibly the correct output. This capability of generalization forms an excellent feature for their use in multiple domains. Domains where humans possess no understanding of the system outputs, or the underlying algorithm is too complex to state, have neural networks as a reasonable option. The notion that little domain expertise is required in the use of neural networks has let to their widespread use within wide variety of applications.

The manner in which information processing takes place is similar in human brain and artificial neural networks. The major difference between the two lies in the number of neurons and connections. The number of neurons (typically of the order of 10s) and connections (typically of the order of 100s) in an artificial neural network are naturally much less than that in human brain (10^{11} and 10^{22} respectively). A single neuron of human brain is believed to perform slower than a single neuron of artificial neural network when involved using modern electronics and computation. However, in the brain all neurons process in parallel at the same time in a pipelined architecture. In artificial neural networks, the processing is neuron by neuron and layer by layer. This makes them slow.

Learning in human brain is continuous and lifelong, as an infant matures to young age and until he/she dies. However, in artificial neural networks learning is from a data set, which should be available beforehand and should be dimensionally not very large. Besides, the two models differ in regard to plasticity or forgetting old facts. Humans tend to forget old information while the artificial neural networks retain old information until some new trends are found which leads to forgetting old trends. From a stochastic point of view as well, the models differ. Human brain may give different outputs to same inputs, which is usually not true with an artificial neural network.

The neural networks are essentially used for problems of functional approximations and classification. The functional approximation problems have a data set of inputs and outputs, where the outputs lie within some continuous domain. The task is to regress the data set and to come up with a simple or sometimes very complex formula that

Figure 12. Perceptron

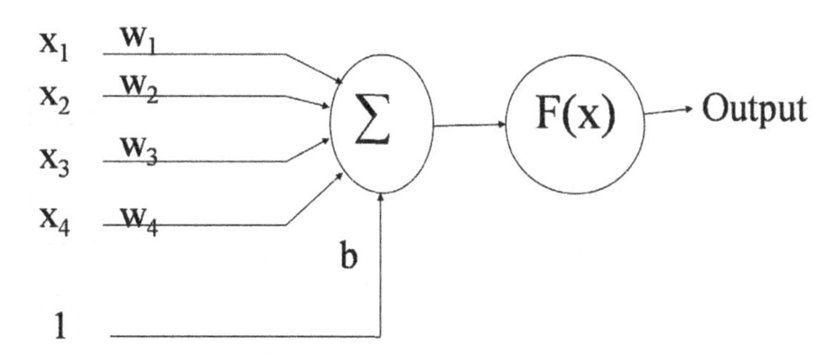

maps the inputs to outputs. The classification problems ask the neural networks to select a class out of a possible set of classes to which the input potentially belongs to. There are a large number of models of neural networks that are widely used. These include Multi-Layer Perceptron, Radial basis Function Networks, Learning Vector Quantization, Self Organizing Maps, etc. Some of these are only used for classification problems. In this chapter we only deal with a model that we would be later using for the problem of motion planning. This is the multi-layer perception. For more detailed discussion over the various aspects and concepts, please refer to the books and papers by Graupe (1999), Kasabov (1998), Konar (1999), Rosenblatt (1968, 1961), Rumelhart and McClelland (1986), Windrow (1962), Windrow and Hoff (1960), and Windrow and Winter (1988).

Perceptron

The simplest artificial neuron is the perceptron, which consists of a set of inputs that give information to the neuron and a unit output which denotes the information obtained after the neuron's processing. The inputs go to the perceptron via connections. The connections have weights, which get multiplied to the data being given to the perceptron. An extra input called the bias is usually applied to the perceptron, which always has a numeric equivalent of 1 and has a weight b associated with it. This input enables to control the overall output of the perceptron. This data passes through a non-linear activation function. This gives the ability to the neuron to work over non-linear problems, which is a property prevalent in almost all problems of any data sets. The basic architecture of the perceptron is given in Figure 12.

Consider the inputs to the peceptron to be x_i and the corresponding weights be w_i. The weighted addition produces the sum which goes to the non-linear layer. This sum is given by Equation 7:

$$y = \sum_i w_i x_i + b \qquad (7)$$

This sum when passed over the non-linear activation function gives the output. Let the non-linear function be given by f(.). Some of the commonly used activation functions are tan sigmoidal, exponential, identity, etc. The resultant output of the neuron hence becomes as given in Equation 8:

$$o = f(y) = f\left(\sum_i w_i x_i + b\right) \qquad (8)$$

Multi-Layer Perceptron

The perceptron in itself is unable to perform well and solve any significant degree problem. For most commonly found problems, we do not use a single perceptron, but rather a large number of perceptrons. Each of these artificial neurons takes

Figure 13. Multi layer perceptron

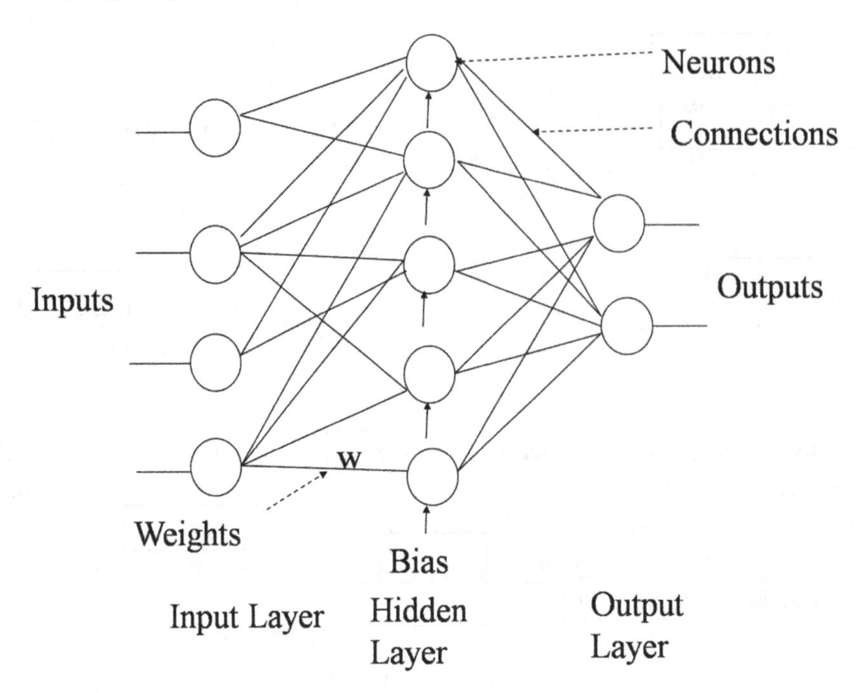

input, processes, and further gives the computed output to other neurons input or as output. These neurons are usually connected in a layered architecture, hence the name Multi-Layer Perceptron (MLP). The general architecture of the MLP is given in Figure 13.

Every MLP has at least two layers. The first compulsory layer is the input layer. This is the layer that has the same number of neurons as the inputs in the system. Each neuron denotes an input. Unlike other layers and neurons, this layer does not do any information processing. It simply collects the inputs to the MLP and transmits them to the next layer and the neurons it is connected to. If the MLP is of an all connected architecture, which means that all neurons of the preceding layer are connected to all the neurons of the next layer, a single input may be given to multiple neurons. All connections from a particular neuron of the input layer to any neuron of the subsequent layer have exactly the same input, which is equivalent to the input applied to the particular input neuron. The last layer is the output layer, which has exactly the same number of neurons as the number of outputs in the system. The neurons of this layer are perceptrons and hence do exactly the same computations as described in the previous sub-section. The neuron has weights for every incoming connection and a bias for every output neuron.

Between the input layer and the output layer there may be none or any number of hidden layers. Each of these hidden layers may have a different number of neurons. Each neuron is a perceptron. A neuron of any layer is connected to some or all neurons of the next layer. Presence of a connection means that the output computed by this neuron would go to all the neurons to which it is connected. This becomes the input to those neurons. Each connection has a weight and every neuron has a bias. This does not apply to the input layer, which is a passive layer.

Mostly problems give decent solutions when fixed with a single hidden layer and few neurons in this layer. As we increase the number of hidden layers or the number of neurons in the hidden layer, the curve imitated by the MLP gets more complex. This should be avoided as it leads to a loss of generalization. However, it may not be possible to have the MLP give good solutions to problems with very less neurons, in cases where the problem by behavior has complex behavior. This is similar to the number of rules and membership functions in fuzzy inference systems.

Learning

The MLP architecture and working as presented in the previous sub-section is capable of giving some outputs to inputs, which are applied at the input layer. The output, however, is governed by the weights of all connections as well as the biases of the individual neurons. The task comes in optimal setting of the weights and the biases so that the complete neural network behaves as per the expected behavior. This means it should give desired outputs to the inputs. This setting of the weights and biases to their desired values is done with the help of a training database and the procedure known as learning. During learning, the neural network attempts to extract the patterns and relations between the inputs and outputs and stores them in form of weights and biases. In other words, it attempts to adjust the values of its weights and biases so that the output is close to the output stored in the data set.

One of the commonly used learning algorithms for the MLP is the back propagation algorithm. This algorithm is a supervised training algorithm. Here the training algorithm is given the training inputs along with the training outputs to train the network. For every input, the simple procedure follows to apply the input and attain the output as per the present set of weights and biases. This stage is known as the feedforward stage as the processing of the network is from back to forward or from input layer to the output layer. The error is measured as the difference between the output obtained by the system and the actual desired output.

This error is then used to compute the corrections in weights and biases. The corrections are applied so that next time the same input appears, the output would be closer to the desired output. The corrections are applied in feedback direction starting from output layer to the input layer. The algorithm is called a back propagation algorithm as the errors are propagated backwards. We do not discuss this algorithm in detail due to the shortage of space, but only study the applications of the algorithm for the motion planning of mobile robots.

PATH PLANNING USING NEURAL NETWORKS

The next major algorithm we use for solving the problem of motion planning of mobile robots is neural networks. Similar to the fuzzy inference systems, this is a behavioral approach. One of the approaches, which may hence be thought of, is to replace the fuzzy inference system discussed in the previous sections with a neural network. The neural network may have the same inputs and outputs as the fuzzy system. Unlike the fuzzy system, the neural network may itself make its own rules. Learning of the neural network may take place in two ways. The first method is to make a training data set consisting of desirable outputs to some set of outputs. This may be manually created using human intelligence, or other algorithms like fuzzy planner may be used to compute desirable output to a set of inputs. A similar approach would be discussed in the next few sub-sections.

The other method may be to start with a random set of neural networks for planning, and assign them fitness values based on the quality of generated path. Genetic Algorithm may be used to extend this notion and use it for learning. A similar approach would be discussed in the subsequent chapters where we use Genetic Algorithm for optimization of the Fuzzy Planner. We provide a different kind of approach in this section to give a new flavor of use of neural network. Excerpts of text and figures are taken from (Kala et al., 2009).

Problem Modeling

The first task we deal with is to discuss the modeling scenario. Unlike the previously discussed approaches, this approach is implemented over a dynamic environment, which means that the various obstacles can move. For simplicity, we take only dynamic obstacles for experimentation, and each of the obstacles of a unit size, which moves at a constant speed but varying direction. We assume that the information of every obstacle is available and is being constantly updated in the threshold time (η), which is the unit time assumed by the algorithm.

We take the following movements of the robots as valid that can be performed in a minimum threshold time. The movements have been quantized for algorithmic purposes: Move forward (unit step), Move at an angle of 45 degrees forward (unit step) from the current direction, Turn clockwise/ anti-clockwise (45 degrees), Turn clockwise/ anti-clockwise (90 degrees). It is assumed that the robot only rotates/moves in angles of a multiple of 45 degrees. Hence, we can only move the robot in the directions (by rotations and forward move) North, North-East, North-West, South, South-East, South-West. For algorithmic purposes, the directions have been denoted as N by 0, NE by 1, E by 2, SE by 3, S by 4, and so on. This is given in Figure 14. Hence, if the robot is at direction 5, it has the following possible moves: No motion; Change direction to 4, 3, 6, or 7; Move forward in direction 5; Move forward in direction 6 or 4. The modeling may resemble the graph approaches where the robotic moves were quantized to a small sub-set of value.

Figure 14. The various angles possible for the robot

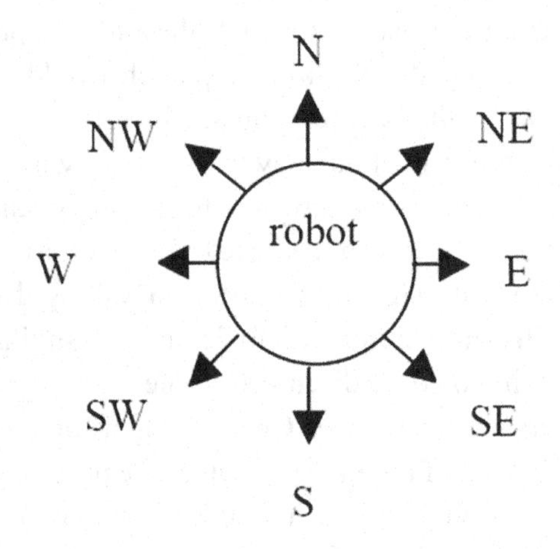

Input Modeling

The first task that needs to be done in the system is to identify the inputs of the system. Again based on the analogy of human motion, we intend to orient the robot towards the goal and walk straight. For avoidance of obstacle, ideally the robot should know the complete map. Since neural networks can take in only a small amount of inputs, we segment out a small part of the entire map around the robot and give it to the neural networks as input. There are a total of 26 inputs to the neural network (I_0-I_{25}). I_0 and I_1 denote the rotations needed for the robot to rotate from its present direction to the direction in which the goal is present. We know that the rotation can be either -2 or +2 (180 degrees left or 180 degrees right), -1 (90 degrees left), 0 (no rotation), +1 (90 degrees right). These are represented by these 2 inputs as follows:

$(I_0, I_1) = (0,0)$ represents +2 or -2
$(I_0, I_1) = (0,1)$ represents +1
$(I_0, I_1) = (1,1)$ represents 0
$(I_0, I_1) = (1,0)$ represents -1

It should be noted that these are numbered according to gray codes so that even if one of the bits is corrupted by noise, the effect on the neural network is the minimum.

The other bits (I_2 to I_{25}) represent the condition of the map. Considering the whole map as a grid of (M x N), we take a small portion of it (5 x 5), with the robot at the center of the map. Each coordinate of this map is marked as 1 if the robot can move to this point in the next step, or 0 if the point does not exist or the robot cannot move to this point because of some obstacle. Therefore, we have taken a small part of the graph. If (x,y) be the coordinate of the robot, we take (x-2,y-2)...... (x+2,y+2) as the input. These are 25 points. These points, excluding the center of the graph (where our robot stands) are fed into the neural network.

Output Modeling

There are total 3 outputs (O_0,O_1,O_2), each can be either 0 or 1. We can make 8 combinations of outputs using these. The meanings of these outputs are as under:

(O_0,O_1,O_2)=(0,0,0) Rotate left 90 degrees.
(O_0,O_1,O_2)=(0,0,1) Rotate left 45 degrees.
(O_0,O_1,O_2)=(0,1,0) Move forward.
(O_0,O_1,O_2)=(0,1,1) Rotate left 45 degrees and move forward.
(O_0,O_1,O_2)=(1,0,0) Do not move.
(O_0,O_1,O_2)=(1,0,1) Rotate right 90 degrees.
(O_0,O_1,O_2)=(1,1,0) Rotate right 45 degrees and move forward.
(O_0,O_1,O_2)=(1,1,1) Move right 45 degrees.

It should be noted that this sequence is in gray code as well, if we arrange it in gray code sequence, the order of the moves will be 90 degrees left rotate, 45 degree left rotate, 45 degree left rotate and move forward, move forward, 45 degrees rotate right and move forward, 45 degrees rotate right, 90 degrees rotate right. Hence, there is a transition in series. This ensures that the effect of noise is minimum.

Special Constraints put in the algorithm

There were a few limitations of the algorithm, these were removed by applying some constraints to the algorithm, which ensured the smooth working of the algorithm. Both these constraints are important from the point of view of modeling scenario.

It may happen that the robot does not move or keeps turning its position continuously. Since there is no feedback in test mode, this is likely to continue indefinitely. For this we apply the constraint that if the robot sits idle for more than 5 time stamps and does not make any move (excluding rotations), we take a random move.

It may happen that the robot develops affinity to walk straight in a direction; hence, if it is situated in the opposite direction, it may go farther and farther from the destination. For this, we limit that if the robot makes 5 consecutive moves (excluding rotations) which increase its distance from the destination, we make the next move only which decreases its distance (excluding rotations).

Procedure

The algorithm first generated test cases, so that the established neural network can be trained. These are done by generating input conditions and solving them using A* algorithm and getting the inputs and outputs. The neural network is established and trained using these test cases. The training algorithm used is the conventional back propagation algorithm. We know that if we train the robot how to act in various obstacle situations, the robot can learn the various actions to be taken. We lay more stress on the rotational parameters of the robot that is the capability of the robot to rotate from its present direction to the goal direction. After the training is over, we enter into the test case. Looking at the condition of the map, the input cases are generated and are fed into the neural network. The output is fetched and decoded. The same is followed and robot is moved.

Figure 15. Experimental results with ANN

Results

For the testing of the algorithm, we made a program using Java Applets that generated n number of obstacles each controlled by an independent thread. These obstacles were moved completely randomly independent of each other. These were also moved on the lines of collision free movement. Another thread was started that moved our robot from the source to the goal. This thread used the algorithm discussed to move the robot in its path. All the obstacles and our robot were displayed in the applets. The directions were displayed as the head pointing towards the direction. An open figure denotes obstacle and a closed denotes robot. The applet also collected the complete path followed by the robot to give a path trace of the robot. The algorithms were tested using this technique. In all the movement of the robot at each step was recorded.

In this algorithm, we use a total of 26 inputs, 2 hidden layers, and 3 output neurons. The number of neurons in the first hidden layer is 20 and that in the second hidden layer is 15. The activation functions used for the two hidden layers and the one output layer are respectively tansig, purelin, and purelin. The inputs and outputs as per design can be either 0 or 1.

The map taken for experimentation was of size 100 x 100. We used total 500 obstacles, which all could move a unit step at any unit time. The threshold time η was fixed to be 1 seconds. We moved the robot from position (1,1) to the goal position (99,99). The path traced by the robot for multiple runs is given in Figure 15. A random stage during the simulation is shown in Figure 16. Here each particle denotes a moving obstacle with head denoting the direction of motion. The direction of motion of all obstacles was changed at every time step.

PATH PLANNING USING NEURAL DYNAMICS

The last model we briefly touch upon is based upon neural networks. In this model, we embed the entire map with neurons. The various neurons are connected to each other and pass the activations to each other. Each neuron is located at some physical region of the map. Each neuron

Figure 16. Random state in experimentation with ANN

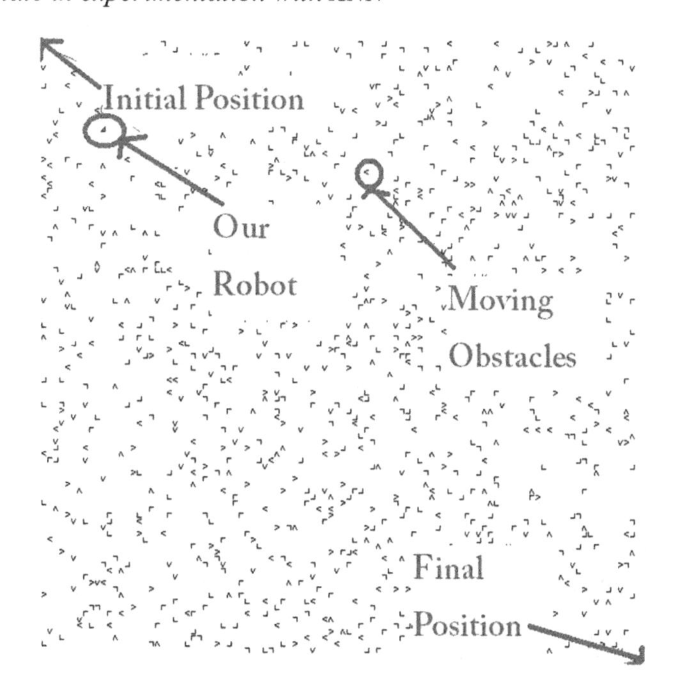

does some basic processing. The purpose is to use all these basic processing by the neurons for computing the robot path. The processing of each neuron constitutes its activity or its dynamics. Each neuron transfers its activity to other neurons physically located to its neighboring positions to which it is physically connected by connections having weights. In this manner, there is a lot of exchange of information between the neurons that leads to a massive amount of computation. Since the computation is spread across neurons, it may be done in parallel. This gives the capability to this form of planning to be very fast and hence may be used in real time environments for dynamic obstacles. It may further be used in scenarios where there is a sudden emergence of obstacles.

The motion of the robot is based upon two fundamentals. The first states that it must move towards the goal. This is a global fundamental, which the robot caters to in its entire journey. The second states that it must avoid all obstacles. This is more of a local fundamental, where the robot attempts to avoid obstacles in the close vicinity, and by repeatedly doing this, it ensures that no collision takes place. Based on these fundamentals the processing of neurons is done. Every neuron activity may have an exhibitory and inhibitory part. The exhibitory part attracts the robot towards the goal. The neuron at the goal possesses this global attraction or activity, which is communicated to the other neurons by the corresponding connections. The inhibitory part repels the robot from the obstacles. This factor is dominant at neurons near the obstacles.

The activity of every neuron is shared to another neuron. The Euclidian distance between neurons acts as a basis of considering the effect of any neuron's activity over other neuron's activity. For every neuron, the activity is the resultant of inhibitory and exhibitory terms, which resemble the shunting equations used in cellular neural dynamics. Since the environment is dynamic and may even change suddenly, the neural landscape constituting the activities of all neurons is rapidly changing. In case there is no change in environment, the neural activity may stabilize. The robot moves as per the activity of the neurons until it reaches the goal position. Further details of the model and the algorithm may be seen in Yang and Meng (2003). For background and related works, refer to Grossberg (1973), Wang et al. (1994), Yang and Meng (2000), and Zelinsky (1994).

CONCLUSION

The focus in this chapter was to explore the behavioral mechanism of planning a robot. Here, the entire paradigm of planning was modeled as a robot behavior. The approach was to adopt a representation of this behavior, and to later develop a system for the motion. The robot was expected to use this behavior for its motion. For the representation of the robot behavior, two approaches were presented, fuzzy inference systems and neural networks. The first approach was the fuzzy inference system. Here, the basic notion was to use fuzziness in system modeling. The method of working of the system constituted the division of the entire range of inputs and outputs into fuzzy sets. Rules were constituted for various combinations of sets. Whenever an input is applied, the fuzzification gives the membership values of various inputs to the various membership functions or sets. Application of rules gives a fuzzy output, which is aggregated for different rules and defuzzified to give final system output.

This approach was used for the planning of a mobile robot. The inputs were the distance and angles of the robot from obstacles and goals. The output was the turn that the robot should perform to avoid obstacle and reach the goal. The other method used was neural networks. Here, the basic motivation was imitation of the human brain. Machine learning was the interest, where a system learns behavior based on past patterns. The method was applied for the problem of path planning. Direction to goal and a small segment of map was fed as input. The output was the next move of the robot. Further planning based on neural activity was also presented.

REFERENCES

Aguirre, E., & Gonzalez, A. (2000). Fuzzy behaviors for mobile robot navigation: Design, coordination and fusion. *International Journal of Approximate Reasoning*, *25*(3), 255–289. doi:10.1016/S0888-613X(00)00056-6

Antonelli, G., & Chiaverini, S. (2004). Experiments of fuzzy lane following for mobile robots. In *Proceedings of the IEEE Conference on American Control,* (vol 2), (pp. 1079 2004). IEEE Press.

Baturone, I., Moreno-Velo, F. J., Sanchez-Solano, S., & Ollero, A. (2004). Automatic design of fuzzy controllers for car-like autonomous robots. *IEEE Transactions on Fuzzy Systems*, *12*(4), 447–465. doi:10.1109/TFUZZ.2004.832532

Cox, E. (1994). *The fuzzy systems handbook.* London, UK: Academic Press.

Dubois, D., & Prade, H. (1985). A review of fuzzy sets and aggregation connectives. *Information Sciences*, *36*, 85–121. doi:10.1016/0020-0255(85)90027-1

Furuhashi, T., Hasegawa, T., Horikawa, S., et al. (1993). An adaptive fuzzy controller using fuzzy neural networks. In *Proceedings of IEEE Fifth International Fuzzy Systems Association World Congress*, (pp. 769–772). IEEE Press.

Graupe, D. (1999). *Principles of artificial neural networks*. Singapore, Singapore: World Scientific.

Grossberg, S. (1973). Contour enhancement, short term memory, and constancies in reverberating neural networks. *Studies in Applied Mathematics*, *52*, 217–257.

Kala, R., Shukla, A., & Tiwari, R. (2010). Fusion of probabilistic A* algorithm and fuzzy inference system for robotic path planning. *Artificial Intelligence Review*, *33*(4), 275–306. doi:10.1007/s10462-010-9157-y

Kala, R., Shukla, A., Tiwari, R., Roongta, S., & Janghel, R. R. (2009). Mobile robot navigation control in moving obstacle environment using genetic algorithm, artificial neural networks and A* algorithm. In *Proceedings of the IEEE World Congress on Computer Science and Information Engineering, CSIE 2009*, (pp. 705-713). Los Angeles, CA: IEEE Press.

Kandel, A. (1991). *Fuzzy expert systems*. Boca Raton, FL: CRC Press.

Kasabov, N. K. (1995). Hybrid connectionist fuzzy production systems: Toward building comprehensive AI. *Intelligent Automation and Soft Computing*, *1*(4), 351–360.

Kasabov, N. K. (1998). Foundations of neural networks. In *Fuzzy Systems, and Knowledge Engineering*. Cambridge, MA: MIT Press.

Klir, G. J., Clair, U. H., & Yuan, B. (1997). *Fuzzy set theory: Foundations and applications*. Upper Saddle River, NJ: Prentice-Hall.

Konar, A. (1999). *Artificial intelligence and soft computing: Behavioral and cognitive modeling of the human brain*. Boca Raton, FL: CRC Press. doi:10.1201/9781420049138

Maaref, H., & Barret, C. (2000). Sensor-based fuzzy navigation of an autonomous mobile robot in an indoor environment. *Control Engineering Practice*, *8*(7), 757–768. doi:10.1016/S0967-0661(99)00200-2

Mamdani, E. (1977). Application of fuzzy logic to approximate reasoning using linguistic synthesis. *IEEE Transactions on Computers*, *26*, 1182–1191. doi:10.1109/TC.1977.1674779

Mizumoto, M., & Zimmermann, H. (1982). Comparison of fuzzy reasoning methods. *Fuzzy Sets and Systems*, *18*, 253–283. doi:10.1016/S0165-0114(82)80004-3

Plonka, L., & Mrozek, A. (1995). Rule based stabilization of the inverted pendulum. *Computational Intelligence: An International Journal*, *11*(2), 348–356.

Rosenblatt, F. (1958). The perceptron, a probabilistic model for information storage and organization in the brain. *Psychological Review*, *65*, 386–408. doi:10.1037/h0042519

Rosenblatt, F. (1961). Principles of neurodynamics. In *Perceptrons and the Theory of Brain Mechanisms*. Washington, DC: Spartan Press.

Rumelhart, D. E., & McClelland, J. L. (1986). *Parallel distributed processing: Exploring in the microstructure of cognition*. Cambridge, MA: MIT Press.

Sugeno, M. (1974). *Theory of fuzzy integral and its applications*. (PhD Thesis). Tokyo Institute of Technology. Tokyo, Japan.

Sugeno, M. (1985). An introductory survey of fuzzy control. *Information Sciences*, *36*, 59–83. doi:10.1016/0020-0255(85)90026-X

Takagi, T., & Sugeno, M. (1985). Fuzzy identification of systems and its application to modeling and control. *IEEE Transactions on Systems, Man, and Cybernetics*, *15*(1), 116–132.

University of Reading. (2012). Biological interfaces with computer systems. *University of Reading*. Retrieved January, 2012, from http://www.prospectus.rdg.ac.uk/cirg/cirg-manmachine.aspx

Wang, Y., Linnett, J. A., & Roberts, J. (1994). Kinematics, kinematic constraints and path planning for wheeled mobile robots. *Robotica*, *12*(5), 391–400. doi:10.1017/S026357470001794X

Widrow, B. (1962). Generalization and information storage in networks of ADALINE neurons. In Yovits, M. C., Jacobi, G. T., & Goldstein, G. D. (Eds.), *Self-Organizing Systems* (pp. 435–461). New York, NY: Pergamon.

Widrow, B., & Hoff, M. E. (1960). Adaptive switching circuits. In *Proceedings of the 1960 IREWESCON Convention,* (pp. 96–104). IREWESCON.

Widrow, B., & Winter, R. (1988). Neural nets for adaptive filtering and adaptive pattern recognition. *Computer*, *21*, 25–39. doi:10.1109/2.29

Yager, R., & Zadeh, L. (1992). *An introduction to fuzzy logic applications in intelligent systems*. Boston, MA: Kluwer Academic.

Yang, S. X., & Meng, M. (2000). An efficient neural network approach to dynamic robot motion planning. *Neural Networks*, *13*(2), 143–148. doi:10.1016/S0893-6080(99)00103-3

Yang, S. X., & Meng, M. Q.-H. (2003). Real-time collision-free motion planning of a mobile robot using a neural dynamics-based approach. *IEEE Transactions on Neural Networks*, *14*(6), 1541–1552. doi:10.1109/TNN.2003.820618

Yang, X., Moallen, M., & Patel, R. (2003). An improved fuzzy logic based navigation system for mobile robots. In *Proceedings of the IEEE Conference on Intelligent Robots and Systems*, (pp. 1709-2003). IEEE Press.

Zadeh, L. (1968). Probability measures of fuzzy events. *Journal of Mathematical Analysis and Applications*, *22*, 421–427. doi:10.1016/0022-247X(68)90078-4

Zadeh, L. (1979). A theory of approximate reasoning. In Hayes, M. M. (Ed.), *Machine Intelligence* (*Vol. 9*, pp. 149–194). New York, NY: Elsevier.

Zadeh, L. A. (1965). Fuzzy sets. *Information and Control*, *8*, 338–353. doi:10.1016/S0019-9958(65)90241-X

Zelinsky, A. (1994). Using path transforms to guide the search for findpath in 2D. *The International Journal of Robotics Research*, *13*(4), 315–325. doi:10.1177/027836499401300403

Zhang, J., & Bohner, P. (1993). A fuzzy control approach for executing subgoal guided motion of a mobile robot in a partially-known environment. In *Proceedings of the IEEE Conference on Robotics and Automation,* (vol 2), (pp. 545 -550). IEEE Press.

KEY TERMS AND DEFINITIONS

Artificial Neural Networks: A network structure consisting of individual neurons connected to each other by connections, each neuron doing some basic processing which enables the network to return some output for an applied input.

Behavioral Planning: Planning by developing a behavior of the robot, which decides its instantaneous actions to avoid nearby obstacles, and move towards goal. Planning is in multiple cycles, at each cycle a basic move is made based on nearby surroundings.

Fuzzy Interference System: A mechanism to determine the output when applied with some inputs based on logic or rule set. The mapping is based on fuzzy algebra, where multiple rules may simultaneously fire by varying degrees to decide overall outputs. Planning these systems demands feeding the current goal and nearby obstacle information as inputs, writing rules characterizing robot behavior, and getting outputs to inputs which are used for the robot's current move.

Learning: The process of tuning the parameters of a system, so that the system generates desired outputs to the known inputs. For the planning task, learning enables robot behavior. If robot's actions to some sets of inputs are known and can be used for learning, the robot may self-learn from its past motions.

Planning with Neural Dynamics: Planning technique where neurons are embedded into the robotic map, each neuron propagating activity based on activity of nearby neurons, nearby obstacles, and the goal.

Chapter 7
Hybrid Graph-Based Methods

ABSTRACT

The basic methods of planning for the mobile robot were the point of discussion in the previous few chapters. In each method, the authors observed that there were some advantages and some disadvantages. The disadvantages of the method restricted their use. In this chapter, the authors explore the possibilities of hybridizing different methods to solve the same problem, such that the advantages are added up and the disadvantages are eliminated as much as possible. They study two methods, which have their bases in graph techniques, in this chapter. In the first part, the authors make use of the Multi Neuron Heuristic Search (MNHS) algorithm. The algorithm is implemented in a hierarchical manner, where each generation of the algorithm gives a more detailed path that has a higher reaching probability. The map used for this purpose is based on a probabilistic approach where they measure the probability of collision with an obstacle while traveling inside the cell. As cells decompose, the cell size reduces and the probability starts to touch 0 or 1, depending upon the presence or absence of obstacles in the cell. In this approach, it is not compulsory to run the entire algorithm. One may rather break after a certain degree of certainty has been achieved. In the second approach, the authors hybridize the MNHS algorithm and Evolutionary Algorithms (EA). MNHS is slow but gives good paths for problems, with the added advantages of ensuring completeness. On the other hand, the EA gives results in finite time, but the optimality or completeness cannot be guaranteed. Here, the authors propose to mix these two techniques to get the added benefits of both these algorithms. The MNHS improves the performance of the algorithm while the EA does the task of time optimization in case of complex graphs. The EA carries forward the task of selection of points in the robotic map. These points are checked for feasibility and then converted into a traversable graph. The same is used by MNHS to find the optimal path from source to destination. In this way, the algorithm finds the best path without robotic collision.

DOI: 10.4018/978-1-4666-2074-2.ch007

INTRODUCTION

There exist a significantly large number of algorithms used for solving the problem of robotic motion planning. Many of these algorithms were discussed in the previous chapters. Each of the algorithms had its own way of modeling the problem and then solving it. This made the individual algorithms have a number of advantages and a number of disadvantages. The advantages enabled the solutions return results to the problem, whereas the disadvantages implied some assumptions that must hold in the problem for the algorithm to find a result. The reason all the algorithms seemed to give fairly good results to the presented map was the fact that the input map had all the assumptions holding good in which the algorithms give results. As for example, we never gave a high-resolution map as the input to the graph-based search algorithm. Similarly, we never gave a map to an evolutionary algorithm that had a maze-like structure or a narrow corridor that make the resultant solution. If given as input, these algorithms would have completely failed, exposing the limitations of the individual algorithms.

Since limitations exist in all the algorithms of the study, it is important to formulate any mechanism to minimize the assumptions. In reality, we need an algorithm that computes very fast, gives results in real time, works for all kinds of obstacles including maze and narrow corridor, works with a high resolution map, etc. However, in reality no single algorithm may be able to cater to all the needs; hence, an attempt needs to be given to reduce as many limitations as possible, to as high a degree as possible. Humans are extremely good in solving maps, or puzzles as an analogy, and figuring out paths in all means. Humans can adapt rapidly to the type of scenario and adjust themselves automatically. In order to make the machines have this capability, or party have the capability, we cannot stick to a single algorithm.

We start with the task of selecting an algorithm and to attempt to reduce as many problems associated with the algorithm, as possible. This should lead to increased use of the algorithm, with maybe some added assumptions, which are more realistic. The task is hone by hybridizing the algorithms with other algorithms. This must be done in a manner that leads to increased advantages of both the algorithms and reduced disadvantages. Coupling two individual algorithms in any architecture must lead to better performances as the individual algorithms. This leads to a wide range of possibilities where we can attempt to mix any algorithm with any other algorithm or itself and study the resultant system. Some of the different possibilities would be discussed in this chapter and chapters to come.

Almost any algorithm can be fused with any other algorithm or even itself. This gives a fascinating playground to work with. The manner of hybridizing the maps is important to understand. A large number of models may be made that make use of multiple algorithms to solve the problem. It is not necessary that every two algorithms hybrid each other, but a large number of algorithms can be used for hybridization with each other. Even an algorithm can be fused with itself to create altogether a different algorithm. Some of these methods would be seen in this chapter and the couple of chapters to come.

We broadly present three fundamental techniques of hybridization of algorithms for problem solving. In the first technique, the map is broken down into multiple levels. Each of the levels has a different resolution (as we shall see later in this chapter). Different levels have different resolutions and hence difference computation times and demands for solving at the particular level. We may have different algorithms (or different instances of same algorithm) working at different levels or resolutions of the map. In this manner, the entire problem may be solved. An intelligent breakup technique may be used. All methods of hybridization we study in this book will fall under this category.

The second category is when the algorithms may work parallel to each other with different algorithms (or different instances of the same problem) working over different parts of the map. Hence, the entire map is broken down into regions. Each region has its own algorithm, which is only concerned with the problem of generating a path within that region. The results may be combined by some integration technique.

The last category of hybridization also makes the algorithms work parallel to each other. However, here different algorithms work in the same entire map. Each of these algorithms computes some path. Different algorithms may compute paths of different goodness. If we look inside each path, some part of the path for some region would be optimal, whereas another part of the path in some other region would be sub-optimal. It is possible that path computed by some other algorithm for this region is optimal, whereas for other region is sub-optimal. Hence, different paths computed by different algorithms are optimal for some regions or sections. Some integration techniques may be used to combine the best parts of the paths returned by all the algorithms to ensure that the overall path is optimal.

In chapter 1, we introduced a large variety of tasks that the robots perform with the help of a manipulation hand. This is especially true with industrial robots, or robots working in production lines. Planning these robots is an interesting task that becomes difficult when the robotic manipulator or robotic hand has many degrees of freedoms. It denotes many possible actions that the robot may take from its current state. In terms of graph, it means every node has a large number of edges, while the total number of vertices is naturally high. In such a state, it becomes important to design an effective planning algorithm that gets the task done. A natural possibility is to first design a vague plan of action for the robotic hand for the task in hand. This vague plan may be later tuned to produce a finer plan. This is the essence of multi-layer planning.

In this chapter, we take the base algorithm that we try to improve as Multi Neuron Heuristic Search (MNHS) for the purpose of solving the path planning problem (Shukla & Kala, 2008). This algorithm is an improvement over the standard A* algorithm. This algorithm was found to perform better in situations where the heuristic function may fluctuate very quickly between adjacent nodes in the graph. The MNHS improves performance in situations where the heuristic function behavior is uncertain by expanding multiple heuristic nodes simultaneously, in place of the node with the best heuristic as was the case with conventional A* algorithm. It is important to analyze the various advantages and disadvantages of graph search-based approaches before we proceed with the chapter. These are summarized in Table 1.

We discuss two approaches in this chapter. In the first approach, we hybridize MNHS with itself, with the algorithm executed in hierarchies. Here, we discuss our work reported in Kala et al. (2011). Discussions in the introduction, conclusions and sections covering the entire algorithm are reprinted with few modifications from Kala, Shukla, and Tiwari (2011). Reprint is done on the basis of retained author rights.

The map in this approach is modeled in a way similar to the quad tree approach (Kambhampati & Davis, 1986). In this model, the graph is divided into a set of nodes of varying sizes. All nodes are arranged in a layered manner from top to bottom, each with different sized nodes, and different

Table 1. Advantages and disadvantages of graph search algorithms

S. No.	Advantages	Disadvantages
1.	Complete	High computation time, work only in small resolution maps
2.	Optimal	Non-iterative

Figure 1. Hierarchical MNHS algorithm

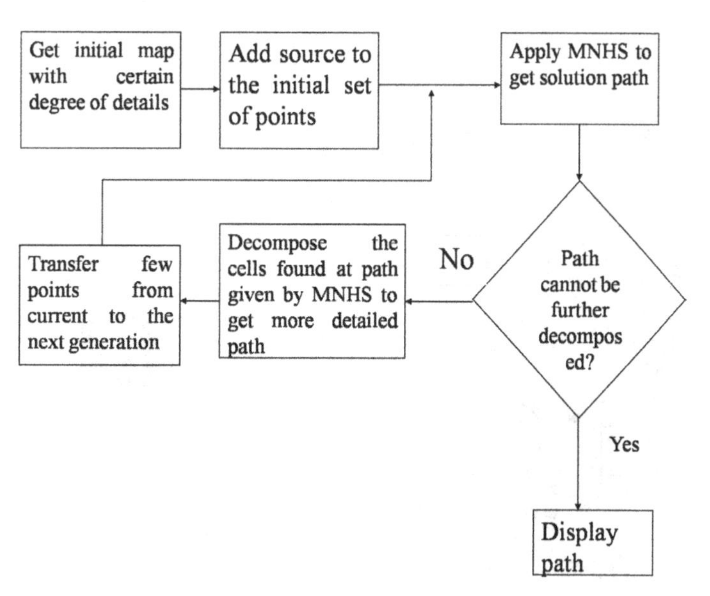

levels of uncertainty. Here at the top, the nodes have a high degree of uncertainty regarding the presence or absence of the obstacles. This uncertainty is measured in terms of grayness of the cell. A completely white cell denotes the absence of any kind of obstacle from the entire region. On the other hand, full black color denotes the confirmation of presence of obstacles in the cell. The grey denotes the intermediate values with the intensity denoting the probability of the cell being free from obstacles.

In practical problems, the map may be too large for the standard MNHS algorithm to perform. This problem is solved by making a hierarchical solution to the problem. The first little iterations are supposed to find the initial vague solutions with lesser details. The probability of collision with obstacles is undetermined to a large extent in the first little iteration. The size of the cells is quite large. Towards the later iterations of the algorithm, the cells lying on the possibly optimal path are decomposed. As a result, the path keeps adding details. In addition, the presence or absence of the obstacles keeps becoming clearer.

In the second approach, we aim to hybridize MNHS with evolutionary algorithms. Discussions in the introduction, conclusions, and sections covering the entire algorithm are reprinted with a few modifications from Kala et al. (2009a).

The MNHS algorithm gives good results when used in the problem of robotic path planning. Completeness of the solution is another added advantage. However, the MNHS algorithm is known to be computationally expensive, especially in the case of high-resolution maps. The motivation behind the use of MNHS over A* algorithm is to keep backup paths ready and explore them also from time to time, so that if some region completely fails to give a correct solution, the other paths are already explored to a good extent. If the second backup path also fails, then the third backup path is also explored to a fair extent. This is possible due to the inherent nature of the MNHS that equally respects the various heuristic values from bad to good and expands all of them. This is done as it is possible that the bad heuristics may suddenly turn good and vice versa.

As the entire algorithm can be time consuming, there needs to be some mechanism to reduce the time of execution. This is done with the help of

Figure 2. (a) The map at any general point of time, (b) the equivalent graph

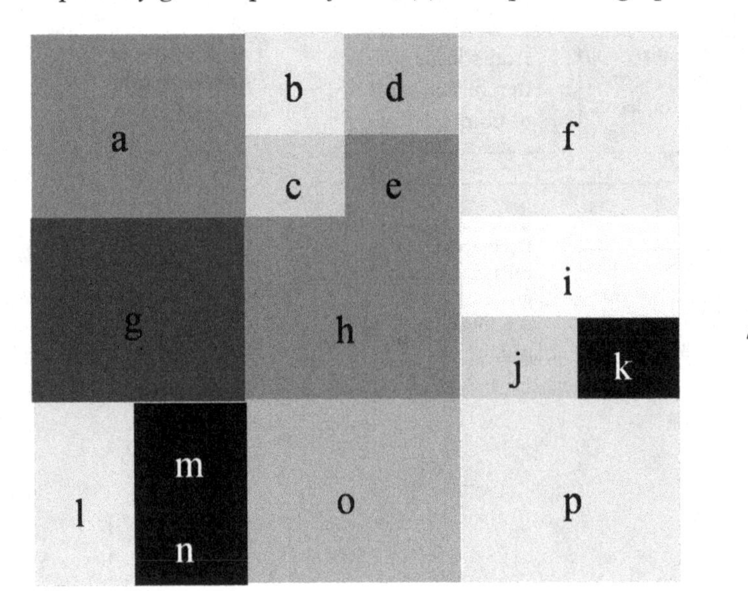

A

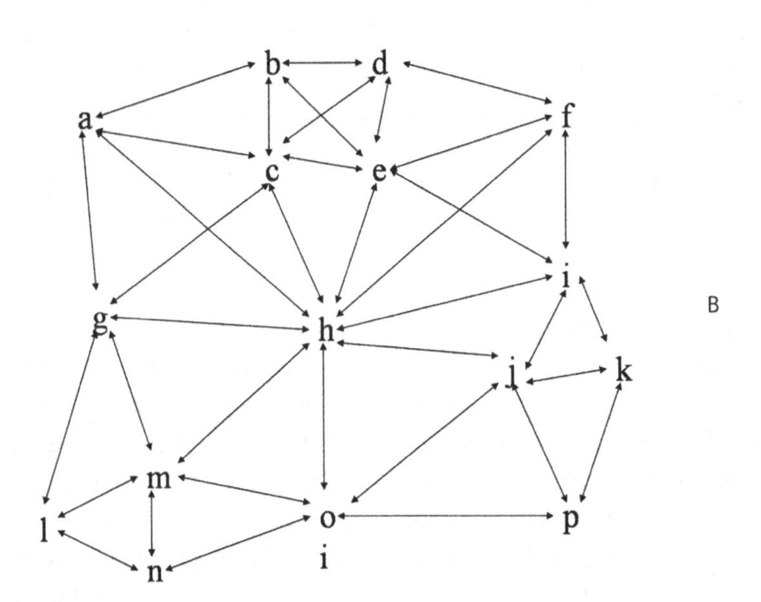

B

Evolutionary Algorithms (EA). Rather than giving the entire map to the MNHS for the purpose of path planning, only a small set of points chosen by the EA are given to the MNHS for the task of finding the smallest route. This greatly limits the search space of the MNHS and results in a big computational optimization. The location of the various points may be optimized by the EA as the algorithm runs and proceeds for convergence.

Since we have studied the basic methods of dealing with the problem, before we enter into deep discussion over the methods of hybrid planning

Figure 3. The MNHS path exploration

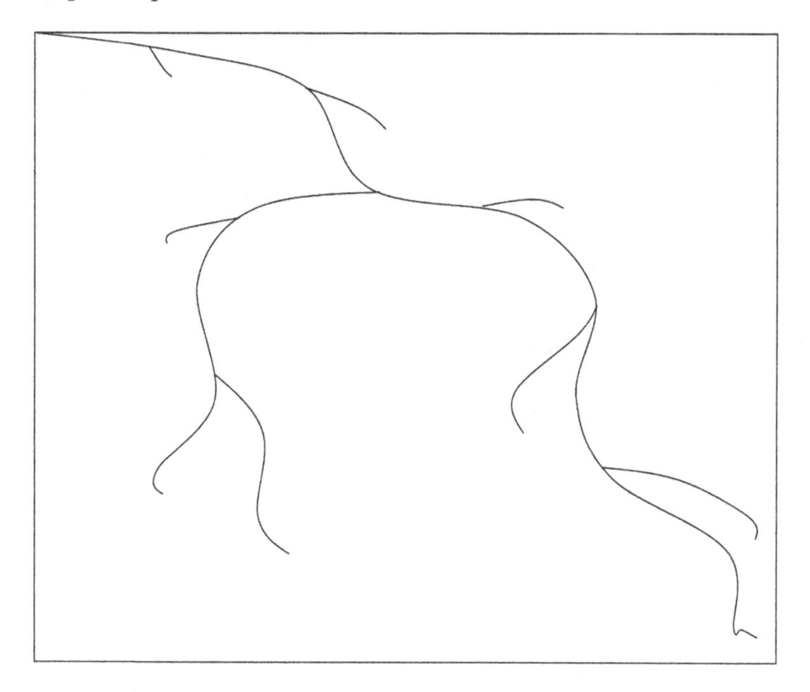

intended for the chapter, it would be wise to discuss the literature in brief. Without loss of continuity, readers may even prefer to go directly into discussion over the two hybrid methods discussed from the next section onwards. The problem of path planning has been a very active area of research, especially during the last decade. The problem has seen numerous methods and means that solve the prevalent research issues to varying extents.

Pozna et al. (2009) solved the problem using a potential field approach for obstacle avoidance. Other potential field methods include Tsai et al. (2001). Hui and Pratihar (2009) gave a comparison between the Potential Field and the soft computing solutions. Various statistical approaches have also been used. This includes the work of Jolly et al. (2009) who used Bezier Curve for path planning. Goel (1994) solved the problem for dynamic obstacles using an adaptive strategy. Quad Tree (Kambhampati & Davis, 1986), Mesh (Hwang, et al., 2003), Pyramid (Urdiales, et al., 1998) are representations that have been tried for better performance. Another good amount of work exists using the soft computing approaches especially Genetic Algorithms (Alvarez, et al., 2004; Juidette & Youlal, 2000; Xiao, et al., 1997), A* Algorithms (Shukla, et al., 2008), and Artificial Neural Networks (ANN) (Kala, et al., 2009b). Shibata et al. (1993) used Fuzzy Logic for fitness evaluation of the paths generated. Various other approaches (Camilo, et al., 2008; Cortes, et al., 2008) have also been proposed.

Zhu and Latombe (1991) used the concepts of cell decomposition and hierarchal planning. Here, they represented the cells in a similar concept of grayness denoting the obstacles. Urdiales et al. (1998) used a multi-level probability based pyramid for solving the path-planning problem. Hierarchical planning can also be found in the work of Lai et al. (2007) and Shibata et al. (1993). Along with the problem of path planning, the researchers have also studied the degrees of freedom and dimensionality as they have a deep impact on the problem. Jan et al. (2008) presented his work for solving in 3 degrees of freedom.

Chen and Chiang (2003) made an adaptive intelligent system and implemented using a Neuro-Fuzzy Controller and Performance Evaluator. Their system explored new actions using GA and generated new rules. In the field of multi-robot systems, Carpin and Pagello (2009) used an approximation algorithm to solve the problem of robotic coordination using the space-time data structures. They showed a compromise between speed and quality. Pradhan et al. (2009) solved a similar problem for unknown environments using Fuzzy Logic. Peasgood et al. (2008) solved the multi-robot planning problem ensuring completeness using Spanning Trees. Hazon and Kaminka (2008) analyzed the completeness of the multi-robot coverage problem. O'Hara et al. (2008) gave the idea of using embedded networks as sensors for solving the problem.

We first discuss the hierarchical MNHS algorithm and then the hybridized MNHS and evolutionary algorithm.

HIERARCHICAL MNHS ALGORITHM

In this section, we give a general outline to the hierarchical MNHS algorithm that we have developed for solving the problem of robotic path planning. The concept of probability based fitness has also been introduced. The general structure of the algorithm is given in Figure 1. This section and subsequent sub-sections covering the entire algorithm are reprinted with few modifications from Kala, Shukla, & Tiwari (2011). Reprint is done on the basis of retained author rights.

The algorithm starts by taking as input the initial graph. This graph may be approximately built by the map-building algorithm. This means that we do not know for sure whether obstacle lies at some cell or not (Kambhampati & Davis, 1996). The MNHS starts the search by processing its open list, which initially comprises just the source node. On successful termination, the MNHS algorithm returns a path. The path has a certain probability of collision associated with it depending upon the probability of the occurrence of obstacles at the cells (Urdiales, et al., 1998). We then decompose all the cells that were traversed in this path. The decomposition may be defined as adding more details by the map algorithm or the breaking up of cells into smaller cells for more accuracy in the prediction of presence or absence

Figure 4. Decomposing gives finer paths: (a) the original path, (b) path after decomposition

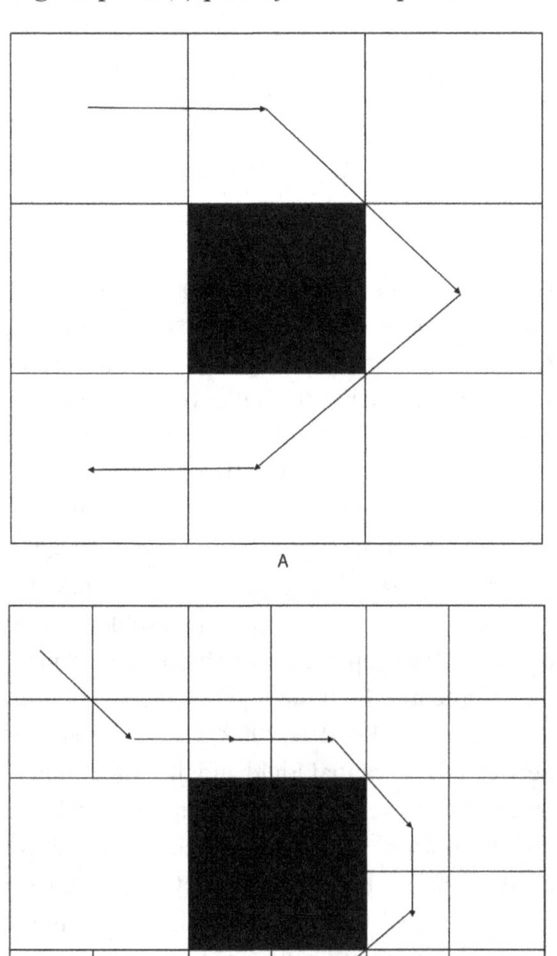

Figure 5. The decomposition of cells

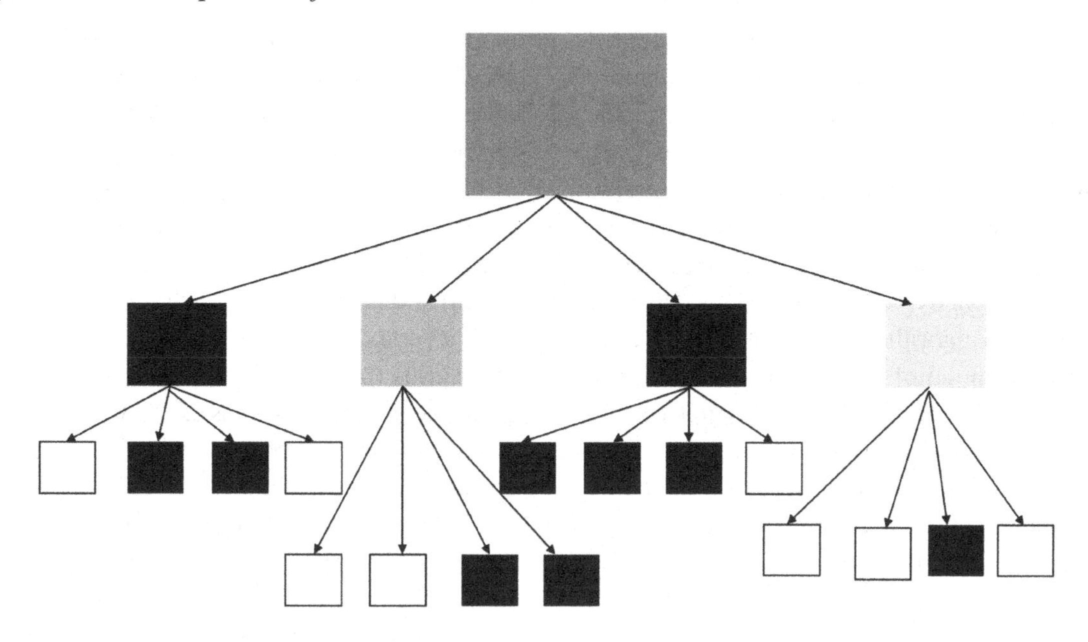

of obstacles. If the cells cannot be further broken down, the algorithm would terminate.

Another concept introduced in the algorithm was the transfer of points from one generation to the other. It would be a waste of computation to start the MNHS algorithm from scratch the next time, since a lot of information is already available from the previous run. Rather than starting the algorithm from scratch, we set some initial points in the open list and closed list of the MNHS

Figure 6. The passing of nodes between generations

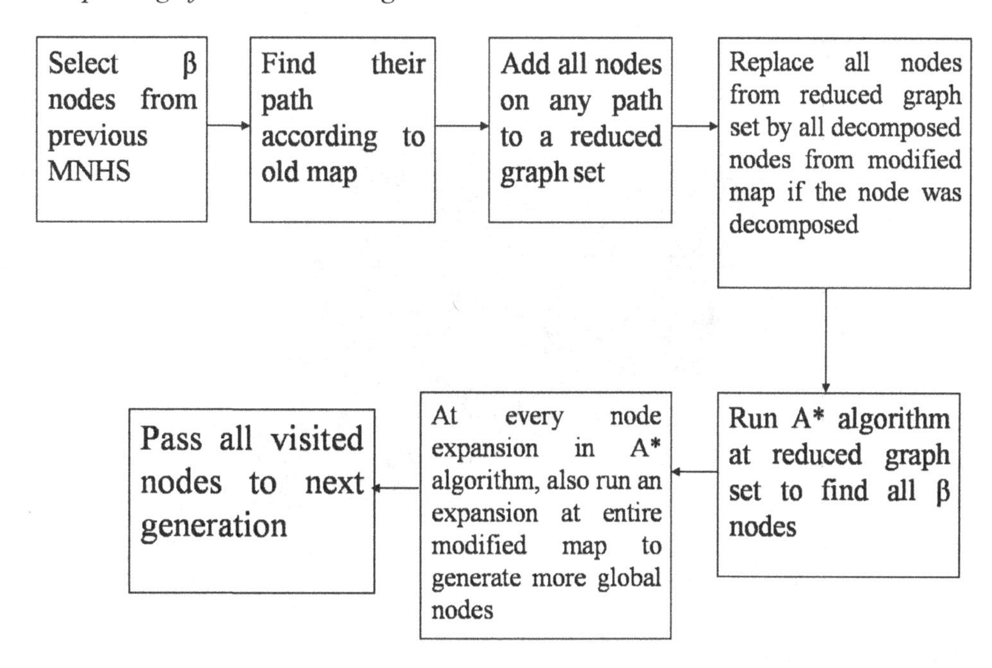

algorithm. These points are taken from the open list of the previous MNHS run. They are recomputed according to the present modified graph before transferring them to the next generation. After the transfer of points to the next generation is over, the MNHS is made to run again on a finer graph. This transfer is motivated by Genetic Algorithms.

The algorithm also has a provision for external termination. Many times, due to the real time nature of the algorithm, the result is required at any specific point of time (Shukla, et al., 2008; Kala, et al., 2009b). Here, we may not be in a position to wait for the algorithm to continue. In such a case, this algorithm may be externally terminated and the robot can be moved in the best path generated so far.

We discuss the various steps of the algorithm one by one.

Map

Map is the representation of the robotic world. The path-planning algorithm uses map to get information about the obstacles and paths that are accessible. Here, we have used a probabilistic approach to represent a map. The map is represented in the form of grids of variable sizes. Each grid is marked with a color from white to black through grey. The color denotes the probability of the obstacles lying in that region. White means no obstacle and black means a confirmed presence of obstacle. The search operation operates on a graph, and it is necessary to convert the map into a graph. The graph is a collection of vertices (or nodes) and edges. In the graph, each grid represents a vertex, and all grids that are accessible by other grids (irrespective of presence or absence of obstacles) are marked as edges. One such representation of the map is given in Figure 2a. The corresponding graph may be given by Figure 2b.

The grayness is the measure of the obstacles in the cell. A gray value of 0 means the cell is fully covered by obstacles and vice versa. Please note that this is the convention followed in the algorithm, with lower numeric value denoting a higher obstacle count. The grayness of any cell c may be calculated by using Equation 1:

$$Grey(c) = 1 - \frac{\text{total area covered by obstacles}}{\text{total area of the cell}} \quad (1)$$

Figure 7. The plot of fitness function

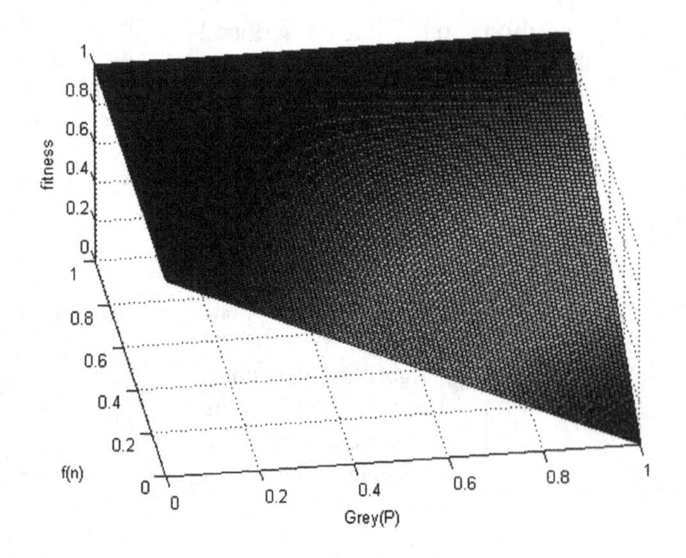

Figure 8. (a) The solution to maze generated by MNHS, (b) the solution to maze generated by A algorithm*

01	02	04	06	08	09	11	13	15	17
03	00	00	00	00	00	00	00	00	19
05	10	26	28	35	40	00	00	00	21
07	00	00	00	00	00	00	22	00	00
12	32	00	00	00	00	00	00	00	23
14	00	00	00	00	00	00	00	00	25
16	24	38	00	00	00	00	00	00	27
18	00	00	00	00	00	00	00	00	00
20	29	30	31	33	34	26	37	39	41

A

01	02	03	04	05	06	07	08	09	10
17	00	00	00	00	00	00	00	00	11
18	19	20	21	22	23	24	25	00	12
30	00	00	00	00	00	00	26	00	13
31	32	33	34	35	36	00	27	00	14
39	00	00	00	00	37	00	28	00	15
40	41	42	43	00	38	07	29	00	16
44	00	00	00	00	00	00	00	00	00
45	46	47	48	49	50	51	52	53	54

B

Suppose the whole map is built on a unit map consisting of unit cells. The formula given by Equation 1 in such a case converts into Equation 2:

$$Grey(c) = 1 - \frac{\text{total unit cells occupied by the obstacles}}{\text{total unit cells covered by c}} \quad (2)$$

It is natural that grayness of any cell lies between 0 and 1 and denotes the probability of collision of the robot in that cell.

Initially, we start with a graph of size M grids X N grids. A grid represents the smallest possible division in the graph. As the algorithm proceeds, we keep decomposing the cells into smaller cells. This is repeated numerable number of times, unless it is not possible to decompose any grid of the path given by the algorithm. The shape given in Figure 2 is formed as a result of multiple decompositions of multiple grids over time.

MNHS Algorithm

We discussed the principles and working of the MNHS in chapter 2. In this section, we again briefly discuss the basic motivation behind the use of

Figure 9. The path traced with no obstacles

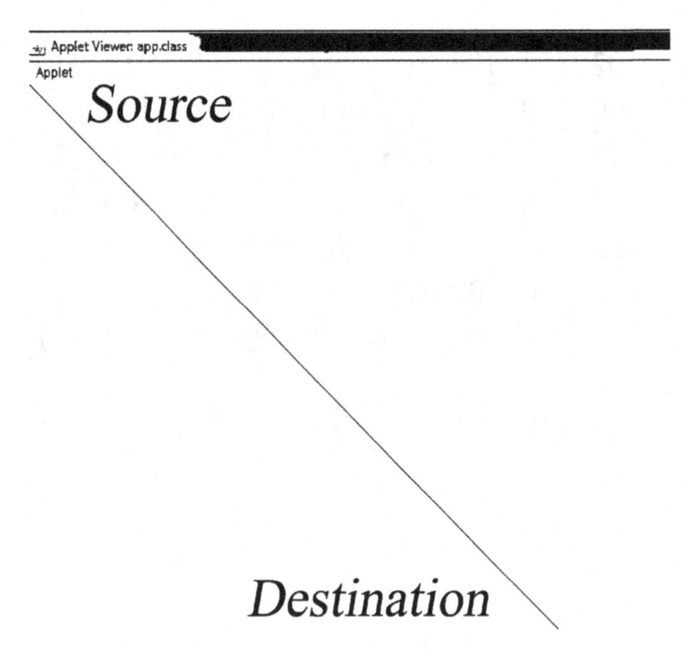

MNHS and the key features of the algorithm that we use for solving the problem of path planning.

The research so far has been using path planning for relatively simple paths. Researchers try to place obstacles in the way and try to see the behavior of the robots. In practical life, it can never be assumed that the path would be so simple. The reason is the numerous possibilities of obstacles in a variety of ways. Consider a robot cleaning a house. There would be multiple paths possible with numerous obstacles of varying sizes. The robot is supposed to avoid all of them and reach the destination. The scalability of algorithms is quite limited in nature. Planning with complex maps can, hence, be difficult. An example would be the maze-like structure where a robot has to find its way out of the maze.

The MNHS algorithm takes care of these problems by trying to exploit each and every path possible. This has a multiplying effect on the computational time, but in return is an assurance in case of maps with rapid change in heuristics. The extra exploration by the algorithm would come to the rescue if the robot reaches quite near to the goal only to find that there is no way to reach it.

Along with the possibilities of early discovery of goal for complex maps, the benefits of MNHS in this algorithm further extend to maintenance of backup paths between iterations. Since we are denoting each path by a probability measure at each stage, it is possible that after certain levels of decomposition, we realize that the path is infeasible due to the presence of some obstacle. This is a very common problem in such probability based algorithms. In such a situation, starting right from scratch would be wasteful. Even if we start again from scratch, the same results might come again where it is impossible to reach the goal. The MNHS is an improvement as it always generates many paths of varying lengths. Some of these paths are almost near to the goal whereas some are just near the source as the algorithm proceeds. Even if the best path fails, due to the sudden exposure of some obstacle, we can always continue with the second best path, which must have been expanded to a good degree.

Figure 10. The condition of the grid at (a) start, (b) generation 3, (c) generation 6

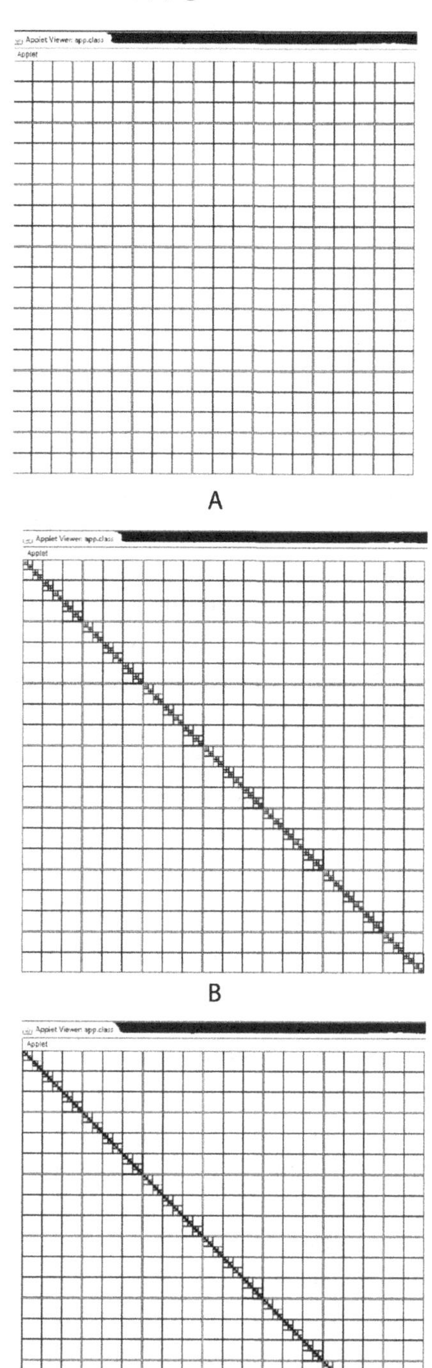

This concept is shown in Figure 3. Here, the best path has almost reached the goal. At the same time, the other paths have been expanded to a reasonably good degree that are ready to provide a backup. The obstacles have not been shown in the figure. The figure is just meant to explain the general concept of the algorithm.

Decomposition of Cells

The algorithm is advancement over the existing algorithm due to the hierarchical decomposition of cells that the algorithm follows. The decomposition of any cell gives a set of finer cells. These cells are more detailed and smaller in size as compared to the original cell. These cells carry more information regarding the presence or absence of the obstacles.

It is natural that a cell that is already of unit size cannot be further decomposed. In such a case, the color of the cell would be either black or white. Black denotes the presence of some obstacle at the position and white denotes the absence of obstacle.

In addition, it is natural that we would not be interested in a cell if the probability of the obstacle avoidance (or its color) is almost 1. This is because we can be pretty sure that walking on this cell would most probably not result in a collision. Similarly, we would not be interested in a cell decomposition if its probability (or its color) is almost 0. This is because of the fact that we can be pretty sure that walking on this cell would most probably result in an error. The only purpose for decomposing these cells is to get finer paths that are more realistic to the shortest path. This is given in Figure 4.

In this algorithm, we decompose all cells that lie on a successful path found by the MNHS. This decomposition divides any cell into 4 cells of equal size. The color is recalculated for all the cells. If again these cells form part of a successful path, they are again broken down. This is given in Figure 5.

Figure 11. The various costs between generations

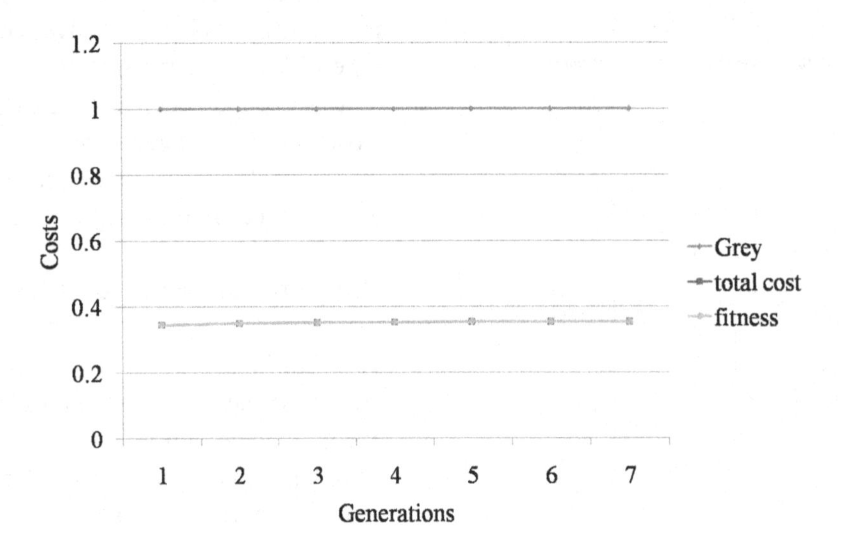

Generations

In this algorithm, we use the concept of generations similar to that used in Genetic Algorithms (GA). Here, the algorithm operates in generations and the solution keeps improving between generations. The characteristics are passed from one generation to the other. The newer generations are found to adapt well to the changing environment or map.

Unlike the GA, we do not pass the entire population with all characteristics from one generation to the other. The reason behind this is that the environment changes between generations. We cannot recalculate the entire population set according to the new environment. That would make the whole algorithm very expensive. We can only recalculate the effect of change in environment on some of the populations.

Population in this case refers to the open list and the closed list of the MNHS. It is natural that with the change in environment, the entire costs would change. The vertices in MNHS that earlier had poor cost values may now have good cost values and vice versa.

In total, we first select β points from the current open list of the MNHS. Since we cannot say with guarantee what would be the effect of the changing environment on the points, we select these β populations of varying costs from the best to the worst. The advantages of the MNHS are also applicable to this approach.

Then we need to recalculate the new costs of these β points. Therefore, we trace their path according to the old map. All the nodes visited are added in a set called as the reduced graph set. If any point in any of the paths in the reduced graph set was decomposed by the decomposition algorithm, we replace the point with all the decomposed points. This gives us modified reduced graph set with all new nodes. We then run a local A* search on the points in the reduced graph set to calculate the new paths of the selected β nodes. This gives us the new paths that lead us to these β nodes.

As we run this A* algorithm to find the new paths of the selected nodes, we also expand the nodes in the entire graph. This gives us more points whose costs are known, and we can directly pass to the next generation. This multiplies the number of points that we pass to the next generation without much computational overhead.

All these β nodes along with the other generated nodes are added in the initial set of points

Figure 12. The passing of nodes between generation (a) 1st to 2nd generation, (b) 2nd to 3rd generation, (c) 3rd to 4th generation

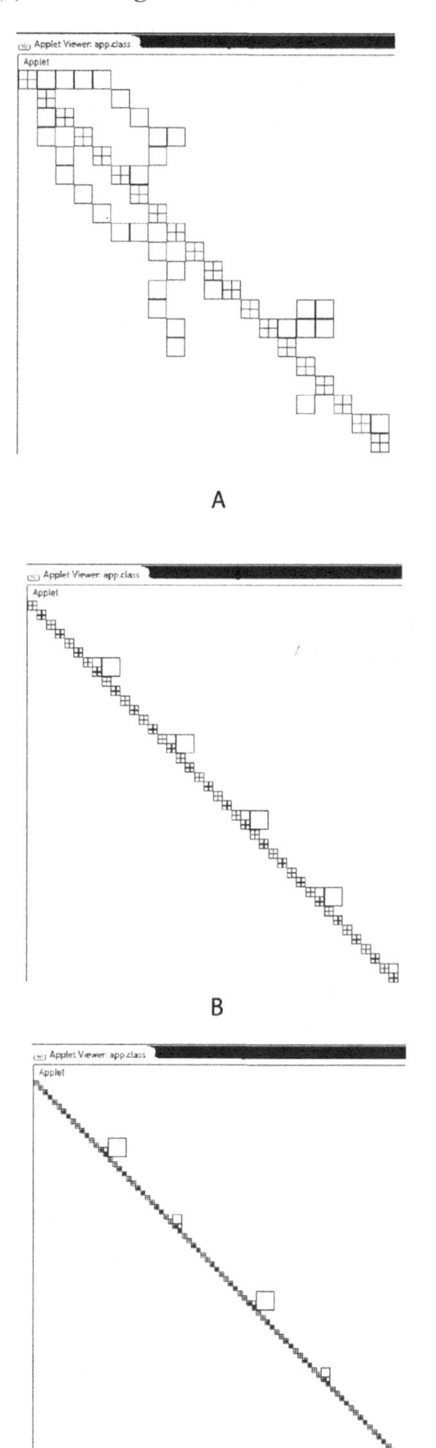

for the next run of the MNHS. If these were in the closed list, they are added in the closed list of next MNHS and the same in the case of open lists. These are passed from the previous generation to the next generation. If any of these points corresponded to the goal position, it is not passed to the next generation.

This algorithm is given in Figure 6.

THE HIERARCHICAL APPROACH

We have already discussed some of the hierarchical concepts of the algorithm in terms of generation, map decomposition, etc. In this section, we study these concepts and the way they go with each other along with the motivation behind the approach. The basic approach used in the problem is that of the MNHS. The whole map has been generalized to reduce number of cells by introducing the probabilistic approach. The search on the reduced graph can be easily performed. In the previous sections, we saw how the various fragments of the graph can be broken down or decomposed depending upon the need. We also saw how the information was being passed from one generation to the other.

Since we are following a probabilistic approach, as we shall see in the next section, every solution generated by the MNHS would be having decent probability of avoidance of obstacles. For this reason, we decompose exactly this path with the hope that the collisions would be avoided. If there are no collisions in the path, then we can expect the algorithm breaking this path down or decomposing this path again and again. The end result would be that we get the entire path. Nevertheless, in reality, the path may face obstacles at any generation. It may also be possible that no other path in close vicinity is possible. The MNHS approach becomes a very useful approach in such cases. In such situations, we already have

Figure 13. Time required by algorithm v/s generations

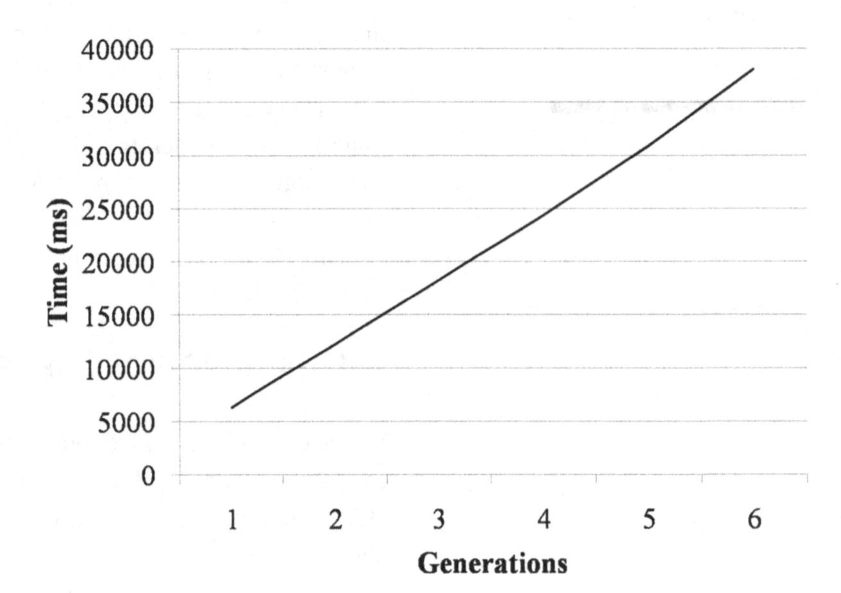

a backup path ready. If this path also fails, another path is ready.

Another important concept that comes from the same problem is that we need proper means to run the MNHS over the solution again and again. This would ideally not be possible as the map changes between generations and all the costs are naturally changed. We saw in the previous section how we pass on the needed information to the next generation so that the MNHS works.

The basic motive behind this approach is that we run a local A* algorithm over a reduced node set. Since this has limited number of nodes in the search domain, it does not take much time. As a result, we can get modified values of good number of vertices in smaller amount of time. Re-calculating all values or running the MNHS again would have taken a large amount of time. While we recalculate the values of some points, we somewhere hope that the final path would be using these nodes that the A* algorithm works on and passes to the next generation. If this is true, a large part of the computation of the MNHS is already done. As discussed earlier, if the path being decomposed has no obstacles on further decomposition, the entire path is generated by this A* algorithm in negligible time and there is virtually no need for MNHS to run again. On the contrary, if we suddenly see an obstacle in this path, then another path needs to be selected. The expansion by the A* algorithm gives a fair idea to the MNHS regarding the points.

The motivation here is to take β points and pass it to the MNHS. If suppose β gets very large, we would end up in recalculating the entire map that was left by the previous MNHS. After recalculation of the entire map, the next iteration of MNHS would continue. The reduced graph set in this case would become very heavy comprising of too many elements making the A* algorithm computationally expensive. However, at the same time, the computational load from the MNHS would reduce. If suppose β gets too small. In such a case the A* algorithm would hardly give points. The simulation would be similar to a fresh start of MNHS provided that the previously decomposed graph contains obstacles.

Figure 14. The path traced with one obstacle at (a) 1st generation, (b) 2nd generation, (c) higher generations

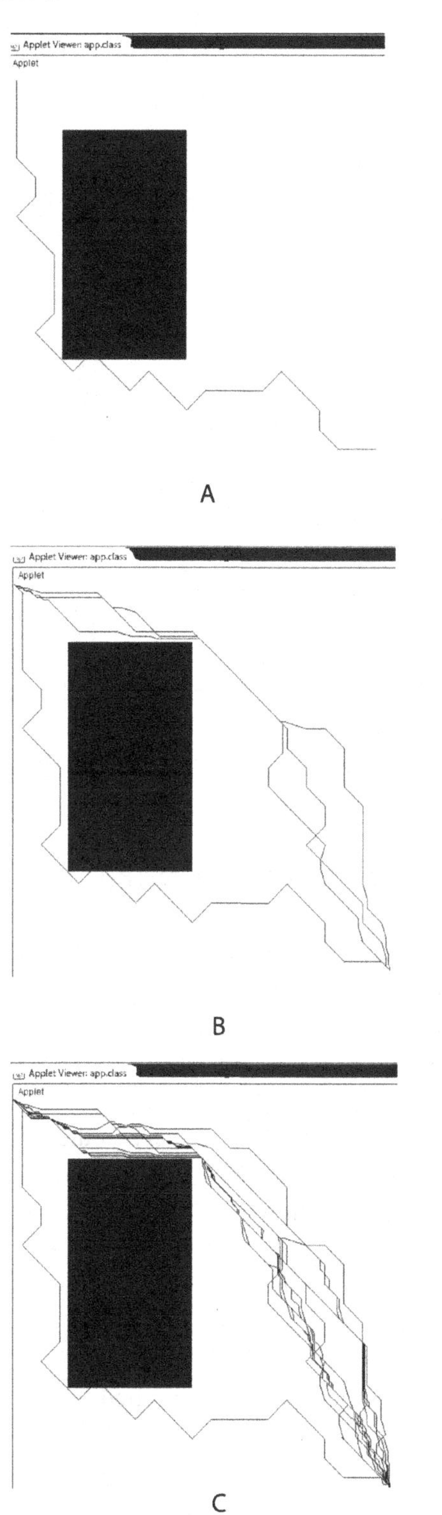

Figure 15. The condition of the grid at (a) generation 2, (b) generation 4, (c) higher generations

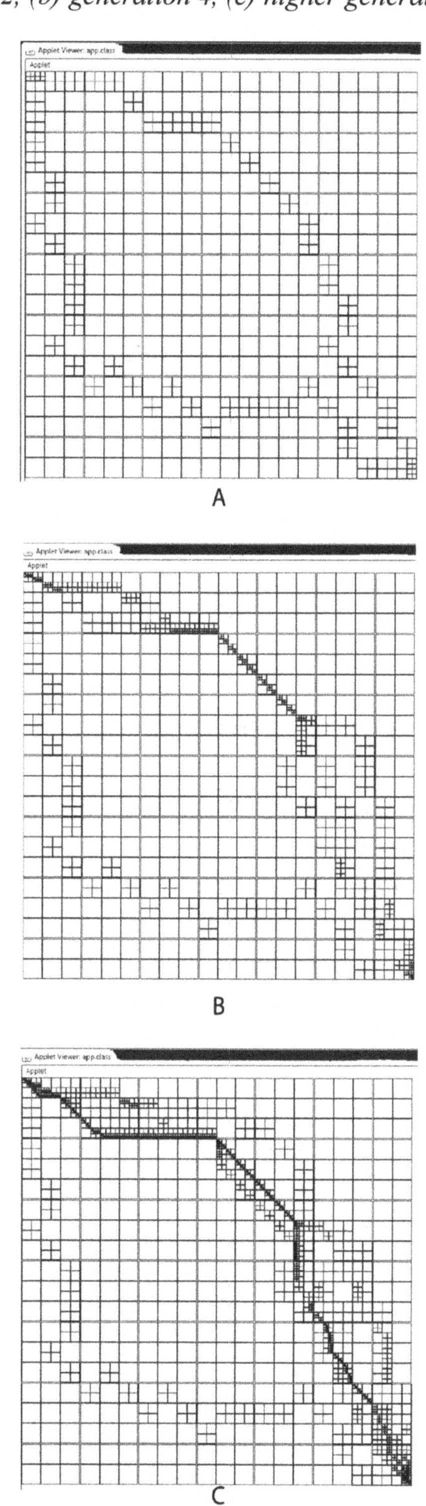

Figure 16. The various costs between generations

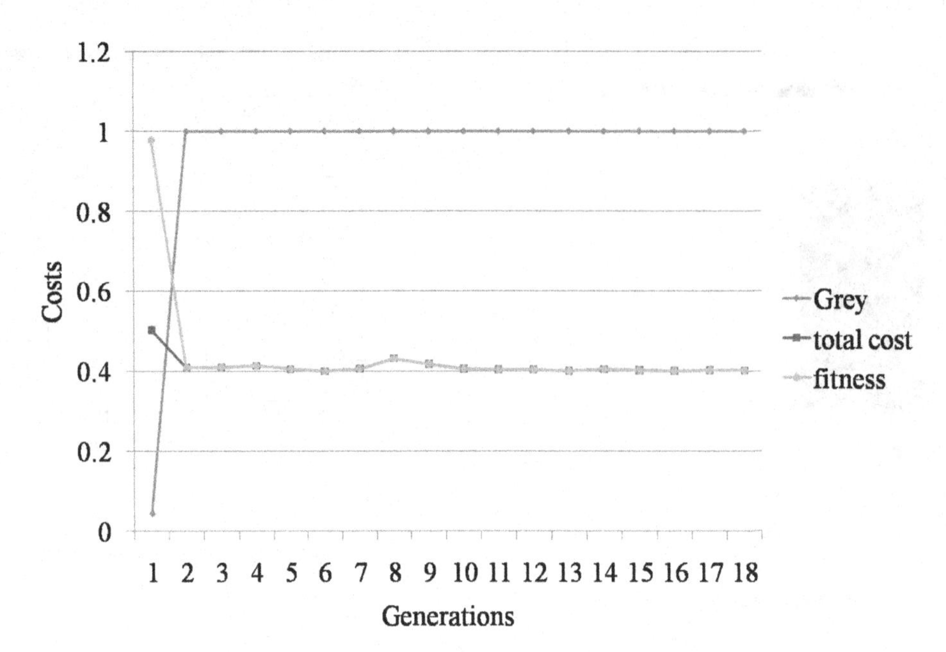

PROBABILITY-BASED FITNESS

The MNHS uses the fitness (also called as cost) for its functioning. The cost decides the goodness of the solution. The better solutions have lower costs. Hence, the approach is generally to find smaller costs and expand them further.

This algorithm is based on probability that is denoted by the probability of finding obstacles in the cell. Hence, each path we traverse has some probability of success associated with it. This probability for a path is given by Equation 3.

$$Grey(P) = \prod_i Grey(p_i) \quad (3)$$

Here p_i are the consecutive nodes that make up the total path P.

The other costs that the algorithm uses are Heuristic Cost h(n) and the Historical Cost g(n). For the MNHS, the total cost f(n) is given by the sum of Heuristic Cost h(n) and Historical Cost g(n). In this algorithm, we use the historic cost as distance of the center of the node from the source and the heuristic cost as the cost of the center of node to the goal.

The total fitness C(n) in this case is the probability based total cost. This is given by Equation 4:

$$C(n) = f(n)\ Grey(P) + (1-Grey(P)) \quad (4)$$

Here P is the path of traversal till node n.

f(n) has been normalized to lie between 0 and 1.

It may be noted here that Equation (4) has been derived keeping in mind the following points:

- If Grey(P) is 0, it means that the path is not feasible. The fitness in this case must have the maximum possible value, i.e. 1.
- If Grey(P) is 1, it means that the path is fully feasible. The fitness in this case must generalize to the normal total cost value, i.e. f(n).
- All other cases are intermediate.

The graph of this function is as shown in Figure 7.

Figure 17. The passing of nodes between (a) 1st to 2nd generation, (b) 3rd to 4th generation, (c) between higher generations

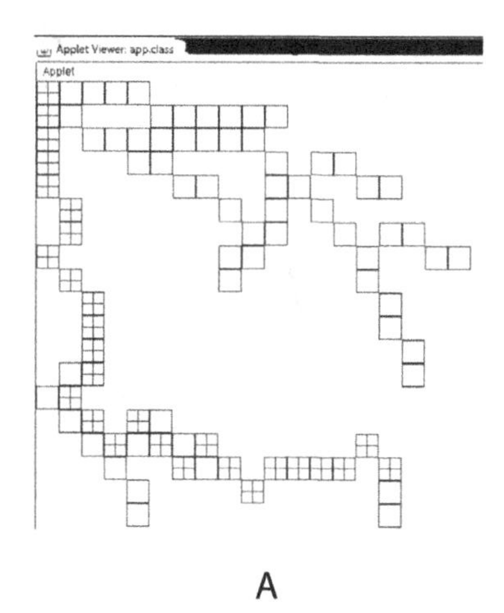

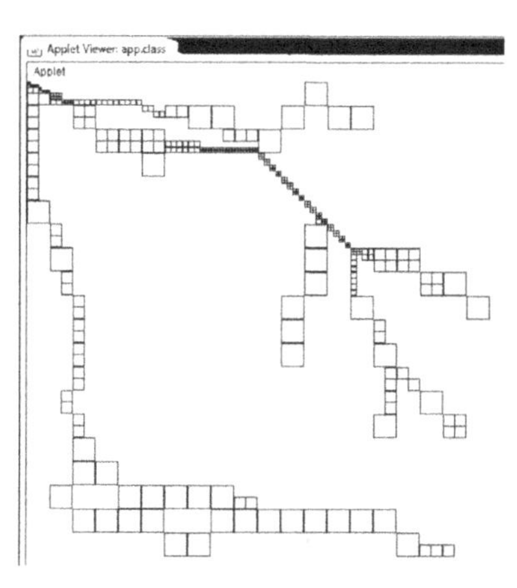

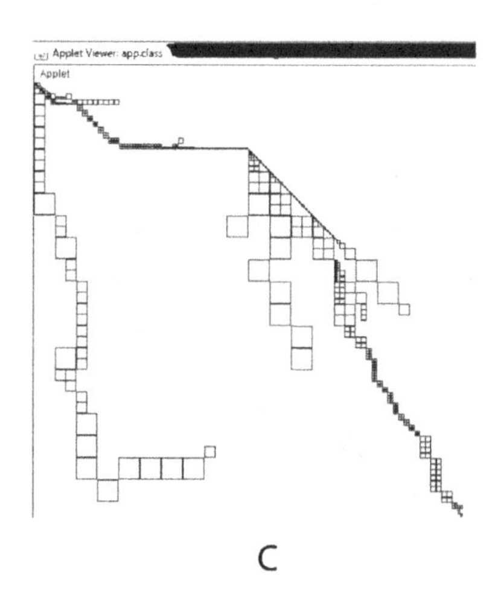

SIMULATION AND RESULTS

In order to test the working of the algorithm, we made our own simulation engine. Every attempt was made to ensure that the simulation engine behaves in a way similar to actual robot. This would ensure that the algorithm can be easily deployed to the real robot for the purpose of path planning. All simulations were done on a 2.0 dual core system with 1 GB RAM using the self generated simulation engine. The simulation engine that was built had the functionality of on-demand cell decomposition, graph generation, time monitoring. Various views were generated during simulation to have a multi-dimensional view of the solution and its creation. The initial graph was taken as an input in form of an image. The image depicted the obstacles as black regions and the path as the white region.

We first provide an experimental setup to prove the supremacy of the MNHS over the conventional A* algorithm. We take an arbitrary maze-like map as shown in Figure 8a. Here, each cell denotes a vertex of the graph used by the graph search algorithm. We give the same map to both the MNHS as well as the A* algorithm. The time any vertex is explored by the corresponding algorithms is noted and displayed in the solutions. It is natural that the time of exploration of the goal needs to be as less as possible for any search algorithm. Figure 8a shows the solution generated by the MNHS and Figure 8b shows the solution generated by A* algorithm. It can be easily seen that MNHS explored fewer vertices and reached the goal earlier as compared to the A* algorithm. This shows that MNHS is a betterment over the A* algorithm.

We conducted 4 tests to see the functionality of the algorithm. (1) The first test was on a path with no obstacles. This was done to ensure the robot performs well in situations where there are no obstacles. (2) The second test was using a single obstacle. This tested the capability of the robot to surpass simple obstacles. (3) The third

Figure 18. Time required by algorithm v/s generations

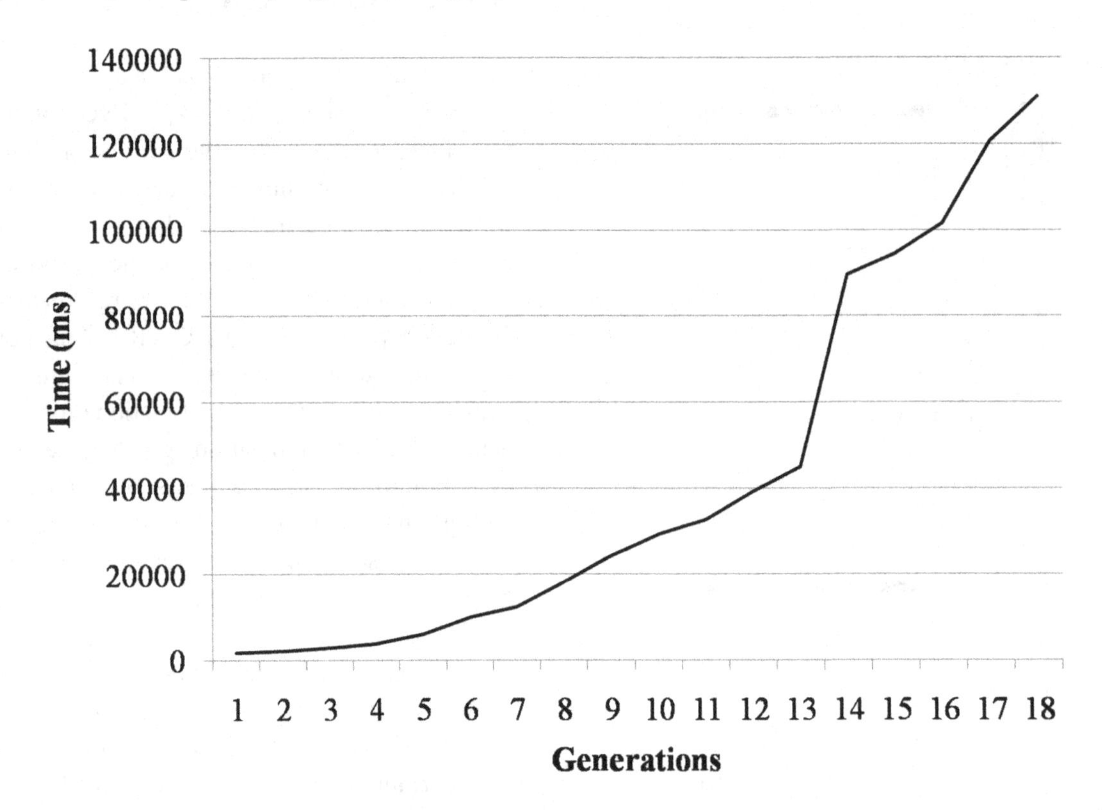

test was the scalability test where the robot was given good number of obstacles. This ensured the capability of the algorithm to solve complex problems. (4) The last was the scattered obstacle test. Here the algorithm was given random small obstacles. This tested the capability of the algorithm to slide through regular small obstacles.

After these tests, we studied the effect of the change in the initial size of the cell to the performance metrics of the algorithm. We used the 4th test case to study these metrics. The various results are given in the next sub-sections.

Free Path

The first test we applied was on a plane map. There was no obstacle on the way from source to destination. We saw that the robot was easily able to travel from the source to destination in a simple line. The map given was of dimensions 1000X1000 grids, which was initially broken down into chunks of 50X50 grids. The value of α was kept as 5 and β was kept as 10. The path traced by the robot at various generations is given in Figure 9. Here, we observe that the same path was traced at every iteration.

The condition of the grid at start, 3rd generation and 6th (last generation) are given in Figure 10a, b, and c, respectively (only the cells are shown, the grayness is not shown in figure). The total cost, fitness, and grayness of the final path at various generations are given by Figure 11. Note that for this case no major change is seen in the costs as the total path at every generation was found to be the same. The cells that are passed from one generation to other are shown in Figure 12a, b, and c. These were recorded at three consecutive runs of the algorithm. It can be seen form Figure 12 that the algorithm first doubted the fitness of the path and tried to explore distant points

Figure 19. The path traced in scalability test at (a) 1st generation, (b) 3rd generation, (c) higher generations

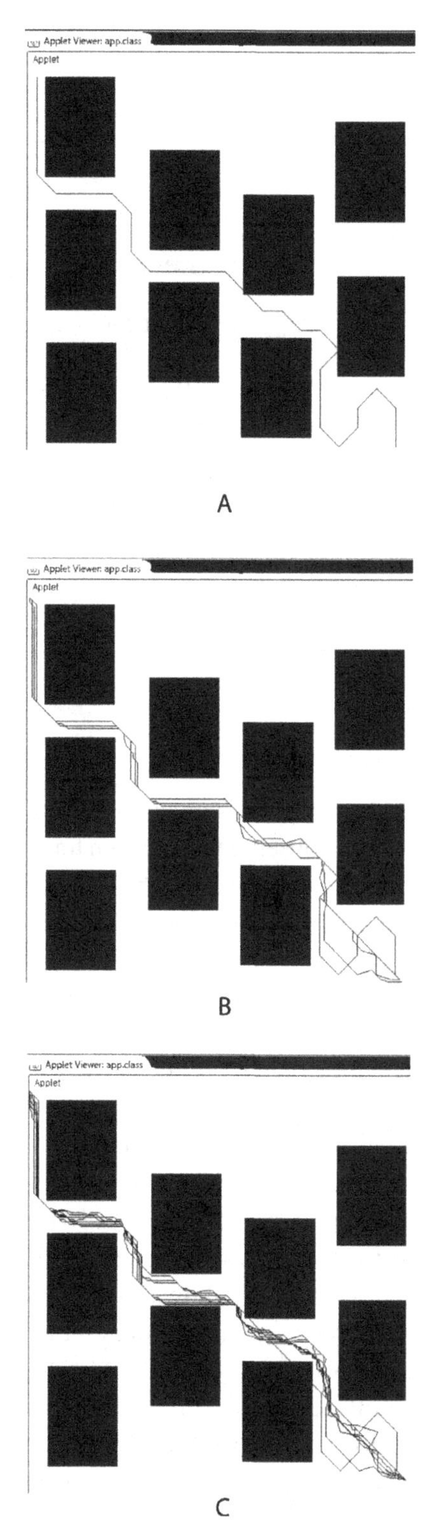

Figure 20. The condition of the grid at (a) generation 2, (b) generation 4, (c) higher generations

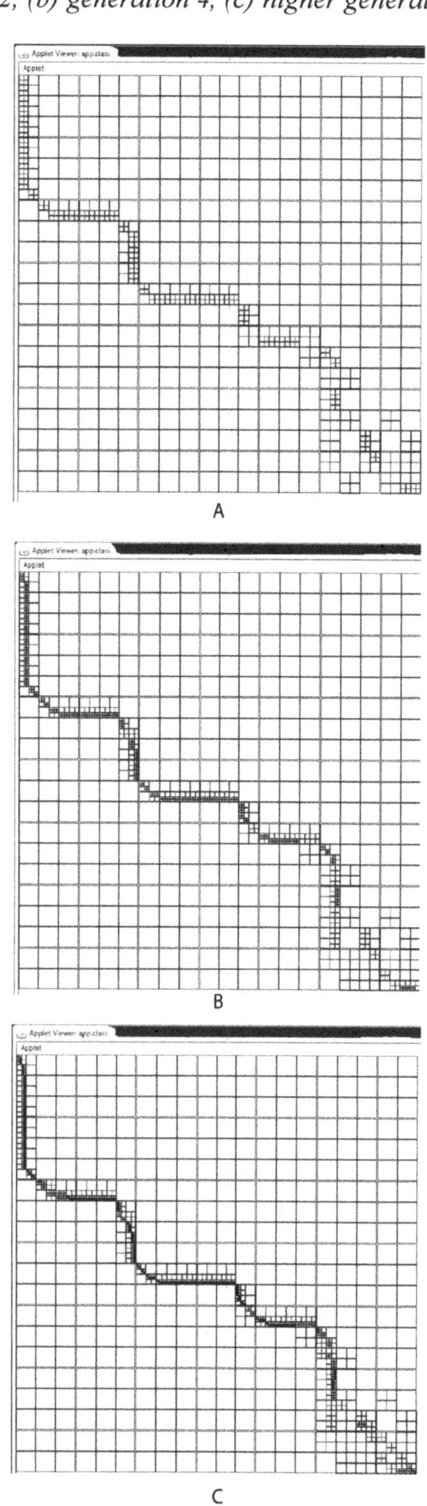

Figure 21. The various costs between generations

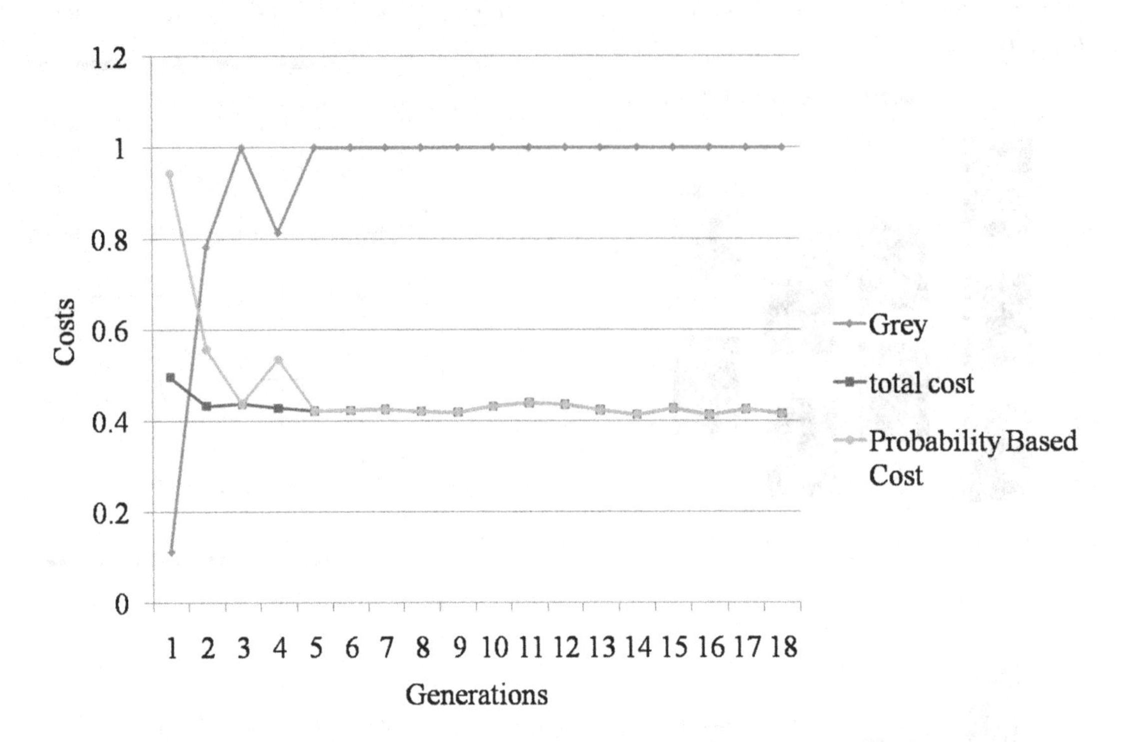

that could have acted as a backup in case the main path failed. As the algorithm continued, due to the very high probability (probability of 1), the effect of other nodes reduced and finally the main path was converted into the final path. The other nodes are still expanded and tried to pass in case the main path fails. Figure 13 shows the time required (in total) between the various generations.

Single Obstacle

The second test we applied was on a single obstacle placed on the path between the source and destination. The map is given in Figure 14. We saw that the robot was again easily able to travel from the source to destination in a simple line. The map given was of dimensions 1000X1000 grids, which was initially broken down into chunks of 50X50 grids. The value of α was kept as 5 and β was kept as 10. The path traced by the robot at various generations is given in Figure 14a, b, and c. Here, we observe that the path traced at the 1st iteration was later found to be with obstacle.

The algorithm in the next generation discovered another path. The condition of the grid at 2nd generation, 4th and higher generation are given in Figure 15a, b, and c, respectively (only the cells are shown, the grayness is not shown in figure). The total cost, fitness, and grayness of the final path at various generations is given by Figure 16. Note that for this case no major change is seen in the costs as the algorithm very easily found the correct path. The cells that are passed from one generation to other are shown in Figure 17a, b, and c. It can be seen form Figure 16 that the algorithm always generated multiple paths. Even though the 1st generated path failed, the information given to the 2nd generation carried points that helped it come up with another path that was quite different. Later this path became predominant as

Figure 22. The passing of nodes between (a) 1st to 2nd generation, (b) 3rd to 4th generation, (c) between higher generations

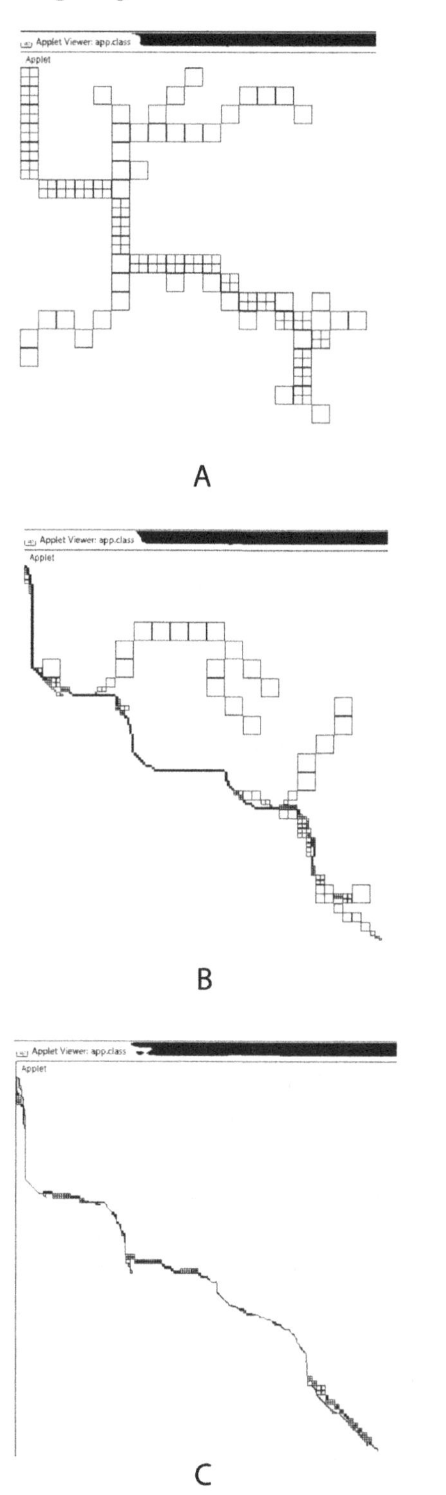

generations increased and hence gave the final path. Figure 18 shows the time required (in total) between the various generations.

Scalability Test

The next test we applied to test the scalability of the algorithm. Here multiple obstacles were placed in the path of the robot. We saw that the robot was easily able to travel from the source to destination in a simple line. The map given was of dimensions 1000X1000 grids, which was initially broken down into chunks of 50X50 grids. The value of α was kept as 5 and β was kept as 10. The path traced by the robot at various generations is given in Figure 19a, b, and c. The condition of the grid are given in Figure 20a, b, and c, respectively (only the cells are shown, the grayness is not shown in figure). The total cost, fitness, and grayness of the final path at various generations is given by Figure 21. The cells that are passed from one generation to other are shown in Figure 22a, b, and c. Figure 23 shows the time required (in total) between the various generations.

Scatter Test

The last test we applied to the algorithm was the scatter test. Here the obstacles were scattered all over the board. The map has been shown in Figure 24. We saw that the robot was easily able to travel from the source to destination in a simple line. The map given was of dimensions 1000X1000 grids, which was initially broken down into chunks of 50X50 grids. The value of α was kept as 5 and β was kept as 10. The path traced by the robot at various generations is given in Figure 24a, b, and c. The condition of the grid are given in Figure 25a, b, and c, respectively (only the cells are shown, the grayness is not shown in figure). The total cost, fitness, and grayness of the final path at various generations are given by Figure 26. The change in cost is clearly visible across the generations. The cells that are passed from

Figure 23. Time required by algorithm v/s generations

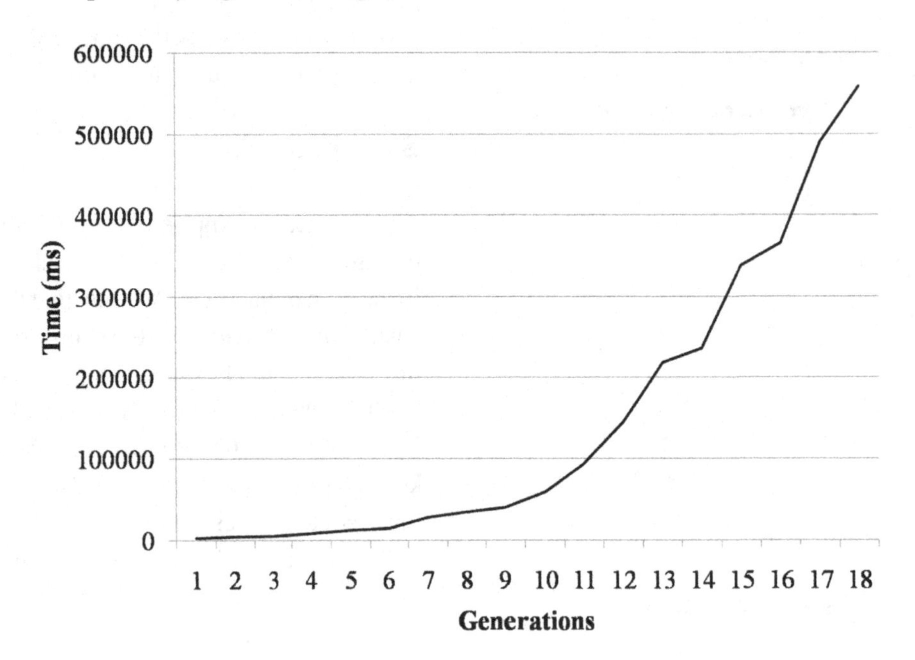

one generation to other are shown in Figure 27a, b, and c. Figure 28 shows the time required (in total) between the various generations.

Effect of Change in Initial Grid Size

We stated that we start with map of size M x N grids, which is initially broken down into chunks of smaller sizes that make up the initial probabilistic map. In this section, we analyze the effect of change in this parameter of the initial grid size (henceforth mentioned as size). The grid in the subsequent runs was decomposed to give finer grids. We tried to see the performance of the algorithm by running the algorithm for various sizes. It is clear that if the grid size becomes unity, the algorithm would get generalized to the conventional MNHS. If, however, the size becomes the size of the graph, the algorithm would generate highly un-probabilistic paths and would take large generations (not necessarily large time) to get high paths with high degree of certainty.

We applied all the simulations using the third test case as described above. The graph size was 1000X1000. We studied the algorithm in two heads. The first head consisted of the higher order graph sizes. Here we used the sizes 500, 250, 125, and 100 and compared them with each other. The other head consisted of the sizes 50, 25, 10, and 5. The various metrics of the algorithm were compared between various sizes in these two heads.

We first plotted the probability (or grayness) at various times for the paths generated by these algorithms. The results are given in Figure 29. The results clearly show that as the time increases, all the paths reach the grayness of 1. However, the higher sizes usually take more time to reach the grayness of 1 as compared to the smaller sizes. This is because the uncertainty is very high as the sizes are high. The decomposition in larger sizes takes time till a point comes where the certainty is 1. We also study the total cost of the algorithm at various algorithmic runs. The results are shown in Figure 30.

The various sizes at higher times converge to almost the same range. Please note that 30a and b are plotted against different data scales. No major conclusion can be drawn from this. Based

Figure 24. The path traced in scatter test at (a) 1st generation, (b) 3rd generation, (c) higher generations

Figure 25. The condition of the grid at (a) generation 2, (b) generation 4, (c) higher generation

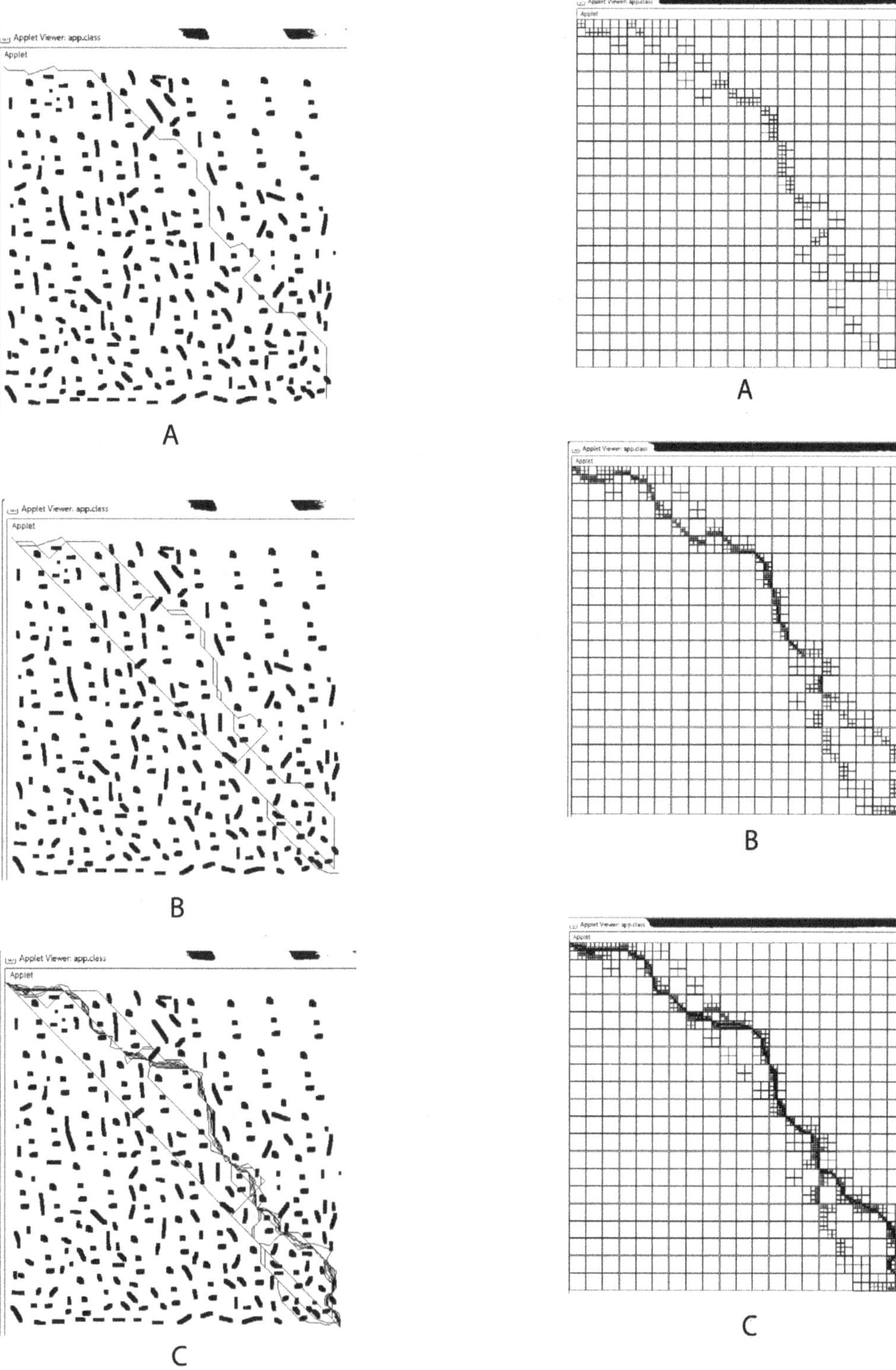

Figure 26. The various costs between generations

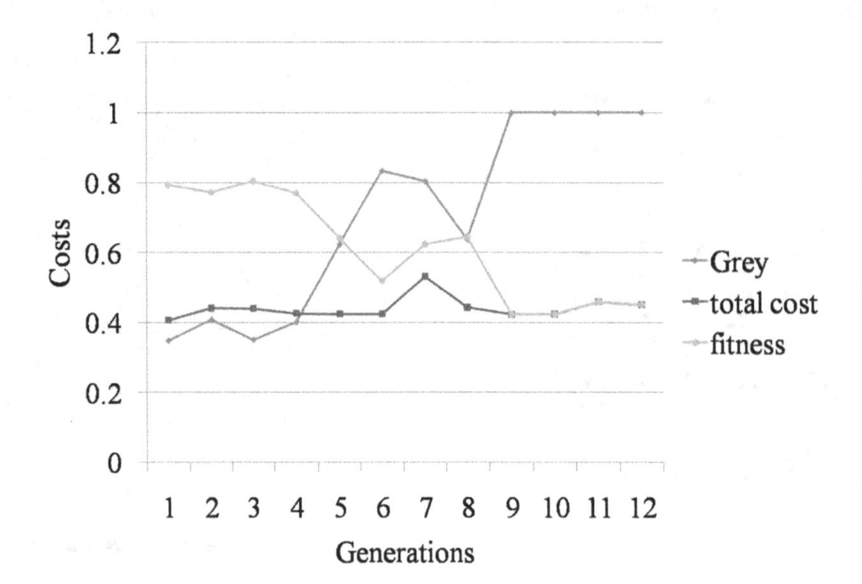

on the readings and analysis of both the presented studies, we may easily analyze the behavior of the probability-based fitness of the algorithm at various sizes. The results at various times are given in Figure 31. The various generations that the algorithm completes in a specific amount of time is studied and shown in Figure 32. It can be seen that the number of generations increase for smaller sizes. By analogy we see that the MNHS becomes A* algorithm when α is taken to be 1 it becomes a breadth first search when α is infinity (Shukla & Kala, 2008). The effect of changing α for various graphs can be hence easily visualized.

HYBRID MNHS AND EVOLUTIONARY ALGORITHMS

The second approach to be discussed in this chapter is fusion of MNHS with Evolutionary Algorithms (EA). We take this task as an optimization problem of the EA. The MNHS serves inside the EA, returning a fitness value, as we shall see later. This section and subsequent sections covering the entire algorithm are reprinted with few modifications from (Kala, et al., 2009a) © 2009, IEEE. Reprinted with permission from Kala, Shukla, & Tiwari, (2009a).

The primary work of EA is time optimization of the entire algorithm. It is known that the A* algorithm may be very time consuming when practically applied to complex maps. The runtime of the A* algorithm depends a lot upon its search space. This is the number of nodes given to the algorithm. The MNHS being a modified version of A* follows similar trends where the total number of points in the map or the number of grids affect the complexity of the algorithm. The EA controls the algorithm optimality by limiting the number of points in the graph that the MNHS has to consider while optimizing the total path length. There are three major parts of this algorithm. Each of which is discussed in the subsequent sub-sections.

Individual Representation

One of the foremost tasks in this algorithm is a good individual representation strategy. Here, we represent an individual by a collection of points (P_i) on the robotic map. They may be scattered

Figure 27. The passing of nodes between (a) 1st to 2nd generation, (b) 3rd to 4th generation, (c) between higher generations

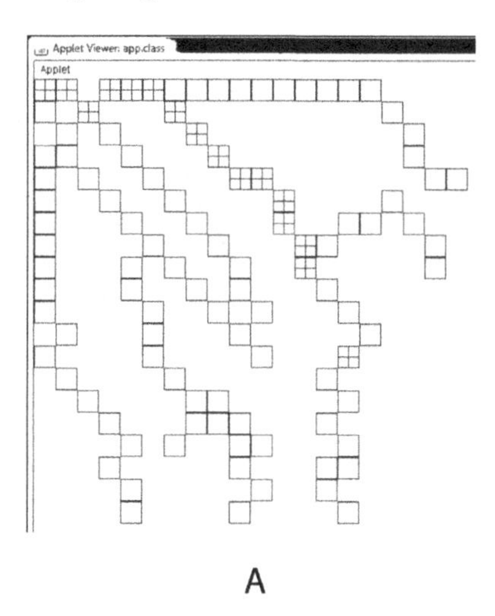

A

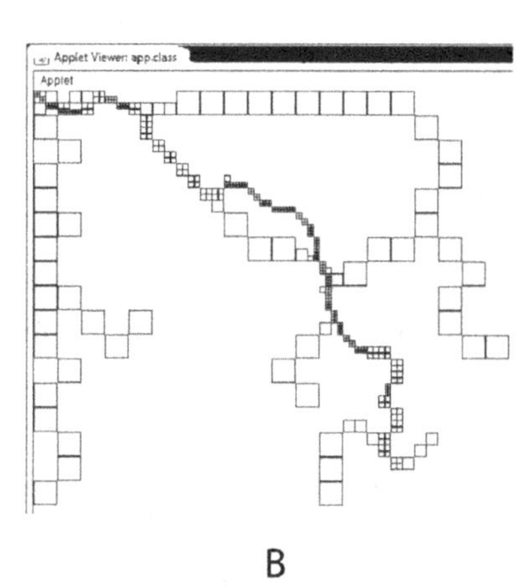

B

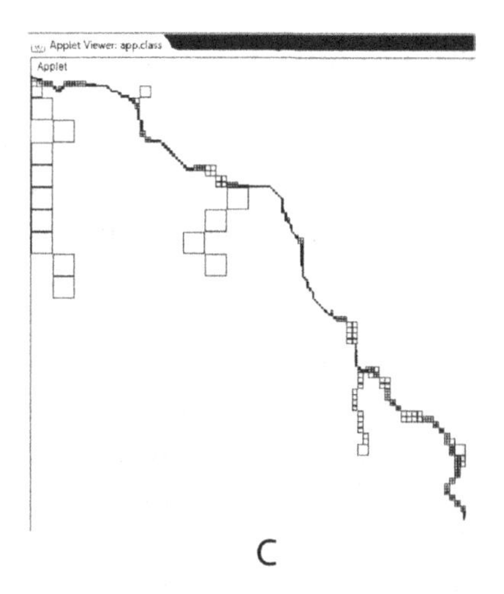

C

throughout the robotic map at various locations. We place a restriction here that the individual size is fixed to β (which is different from term β used in previous approach). This means that the individual can have a maximum of β points in its collection. The complete individual hence becomes <P_0, P_1, P_2, P_3, ... $P_β$, $P_{β+1}$>. Here P_0 is the source and $P_{β+1}$ is the goal. This collection is given to the MNHS to work over the optimal path out of this collection of points.

Each point is a collection of x and y coordinates and may be denoted by (x_i, y_i). The x-axis that we take for this problem is the straight line joining the source and the goal. The y-axis is perpendicular to the x-axis as given in Figure 33.

Let us suppose that the map is represented in the coordinate system X'-Y' given in Figure 33. Now any point needs to lie within the range of (0',0') to (m',n') so as to lie within the map. Here ' represents the use of X'-Y' coordinate system. This range needs to be converted into equivalent range in the X-Y coordinate system of the individual to generate valid points in the robotic map. This is done by a rotation of angle 'a' in the clockwise direction, where 'a' is the angle between the two coordinate systems given in Figure 33.

Another important characteristic of the individual representation is that the various points in the map are always sorted along the X-axis. Here the source is the first point and goal is the last point. Sorting the points shows a good strategy from the perspective of genetic operators. As a result of sorting, it is more likely that two points to be found in same genes in two different individuals would be near each other physically in the map. Hence, crossover is more realistic.

The entire length of the chromosome is hence fixed to a maximum value of 2β. If β is set to 10 and each P_i denotes the point (x_i,y_i) then the genetic individual of this path would be represented as <x_1 y_1 x_2 y_2 x_3 y_3 x_4 y_4 x_5 y_5 x_6 y_6 x_7 y_7 x_8 y_8 x_9 y_9 x_{10} y_{10}>.

Figure 28. Time required by algorithm v/s generations

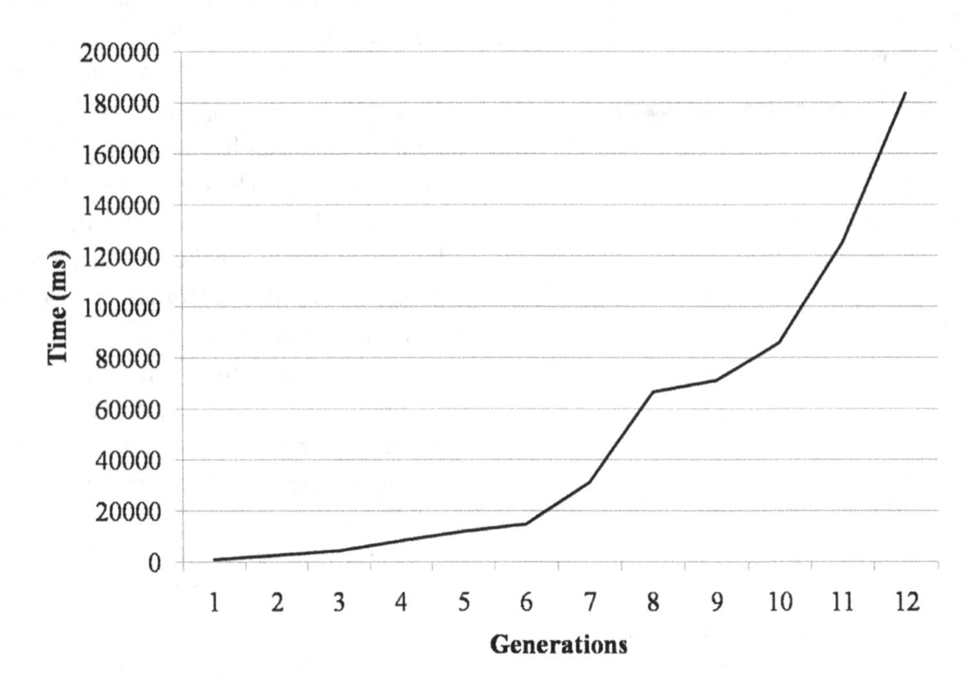

Conversion to a Graph

So far, we have a collection of β+2 points in the map. However, the MNHS works on a graph. Hence, we need to convert this collection of points into a graph so that the MNHS can work over the optimal path over this pool of points.

The graph that we give to the MNHS has the collection of β+2 points as the vertices. An edge exists between any vertex i to any vertex j if the robot can travel from i to j without any collision with obstacles. The weight of the edge is taken to be the physical distance between the vertices i and j. One of the major problems comes in the determination of feasibility of the path. To determine the feasibility of the path between any two vertices, we travel between the two vertices and check for the presence of obstacles. If an obstacle exists, the path is considered as infeasible and the traversal stops.

The adjacency matrix representation of graph has been used. This is a matrix whose every element a_{ij} denotes the existence of an edge between the vertex i and j. The element stores the weight of the weight if an edge exists, or else it stores infinity. Sine this is an undirected graph, hence $a_{ij} = a_{ji}$. In addition, the diagonal elements are all kept as infinity.

This graph is given to the MNHS in the fitness function of the EA. The first node is specified as the source and the last one is specified as the goal.

Evolutionary Operators

The EA uses various operators to carry out the task of optimization. It uses a rank-based fitness method and a uniform stochastic selection technique. The two major operators used are crossover and mutation. Both these operators have been adapted as per the problem. Crossover is point-based where the new individual gets half the points in the form of (x,y) from the first parent and the other half from the second parent. Scattered crossover is used for this purpose. Similarly, mutation is point-based where the mutation operator physically moves points represented by individuals

Figure 29. The grayness at various algorithmic runs v/s time for (a) set of higher initial grid sizes, (b) for set of lower initial grid sizes

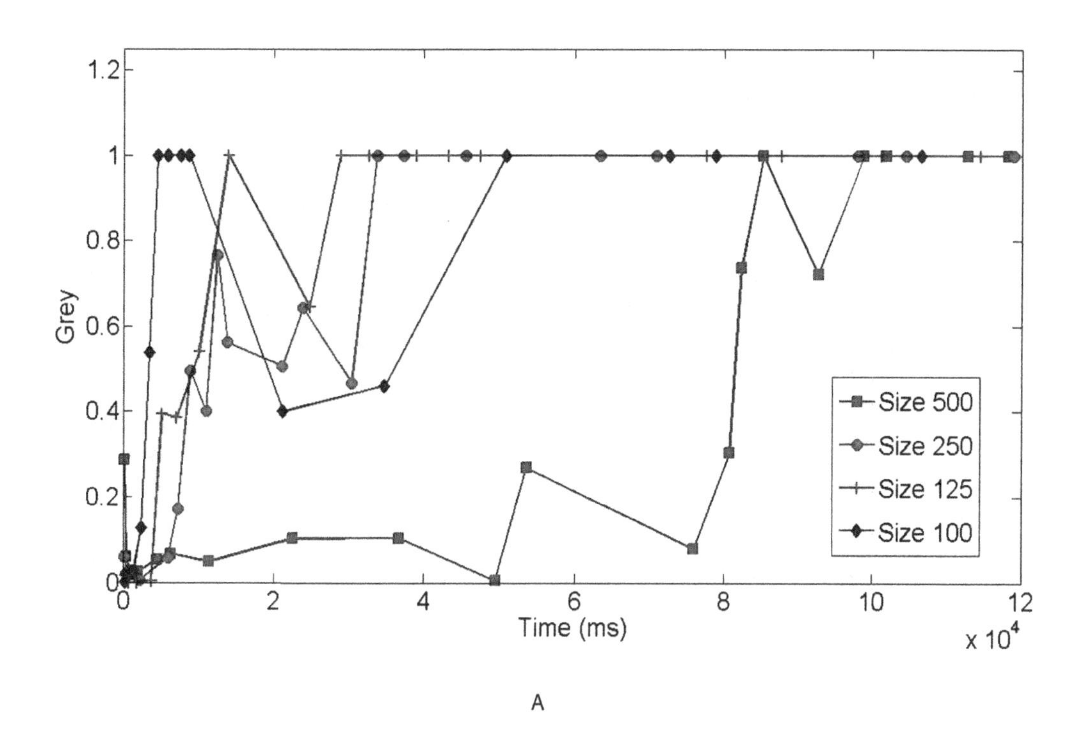

A

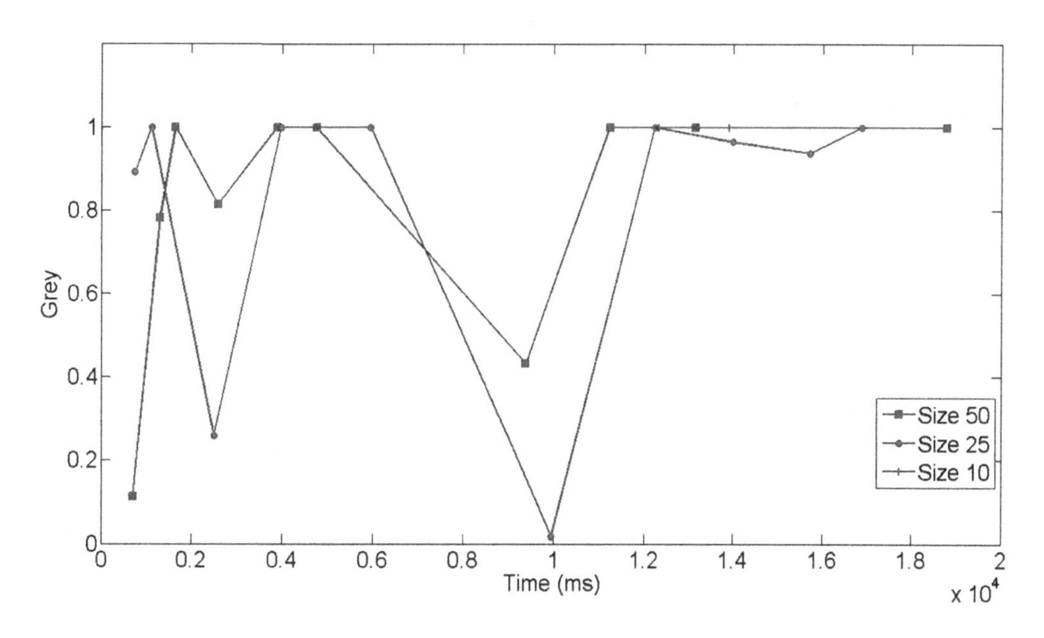

B

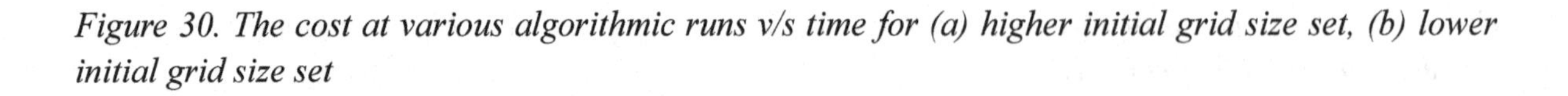

Figure 30. The cost at various algorithmic runs v/s time for (a) higher initial grid size set, (b) lower initial grid size set

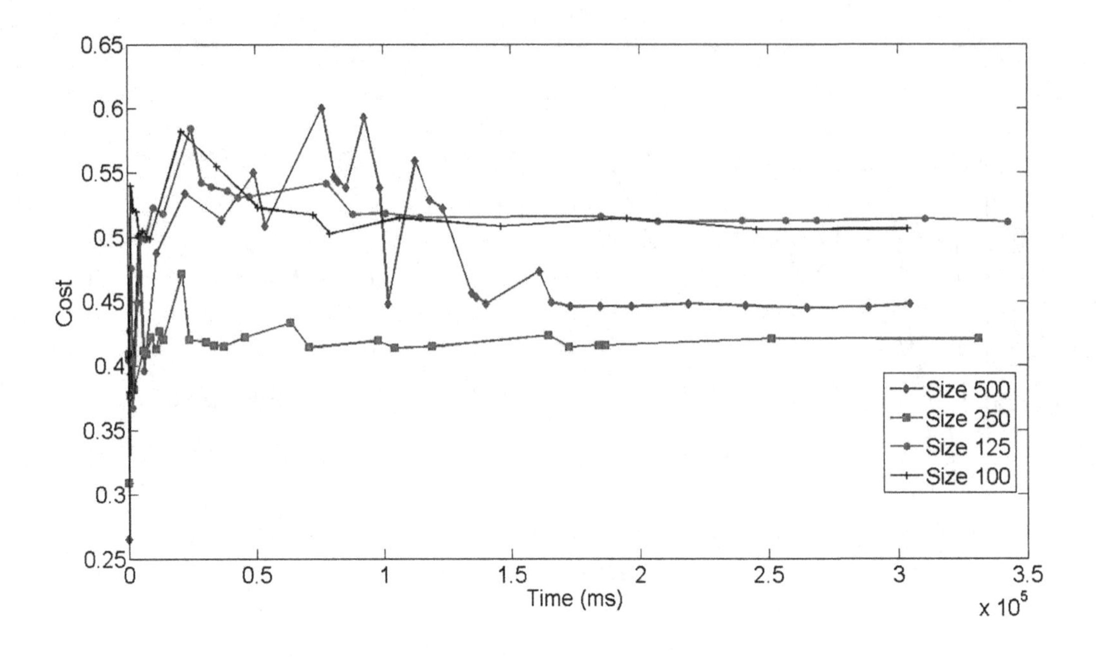

A

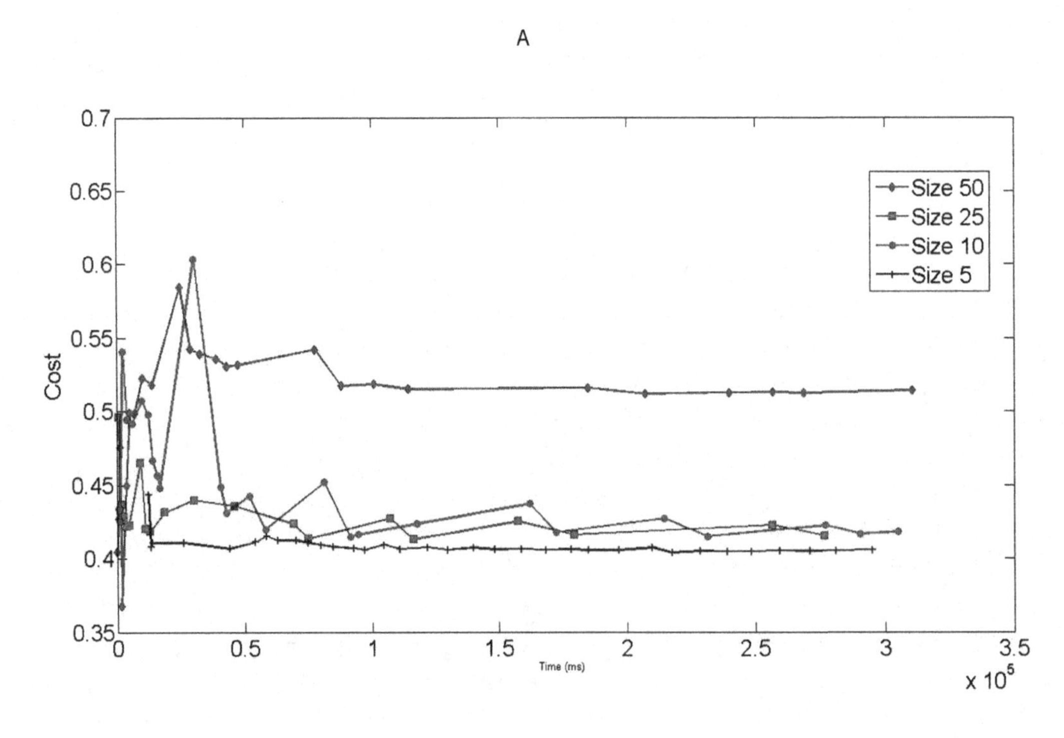

B

Figure 31. The probability based fitness at various algorithmic runs v/s time for (a) higher initial grid size set, (b) lower initial grid size set

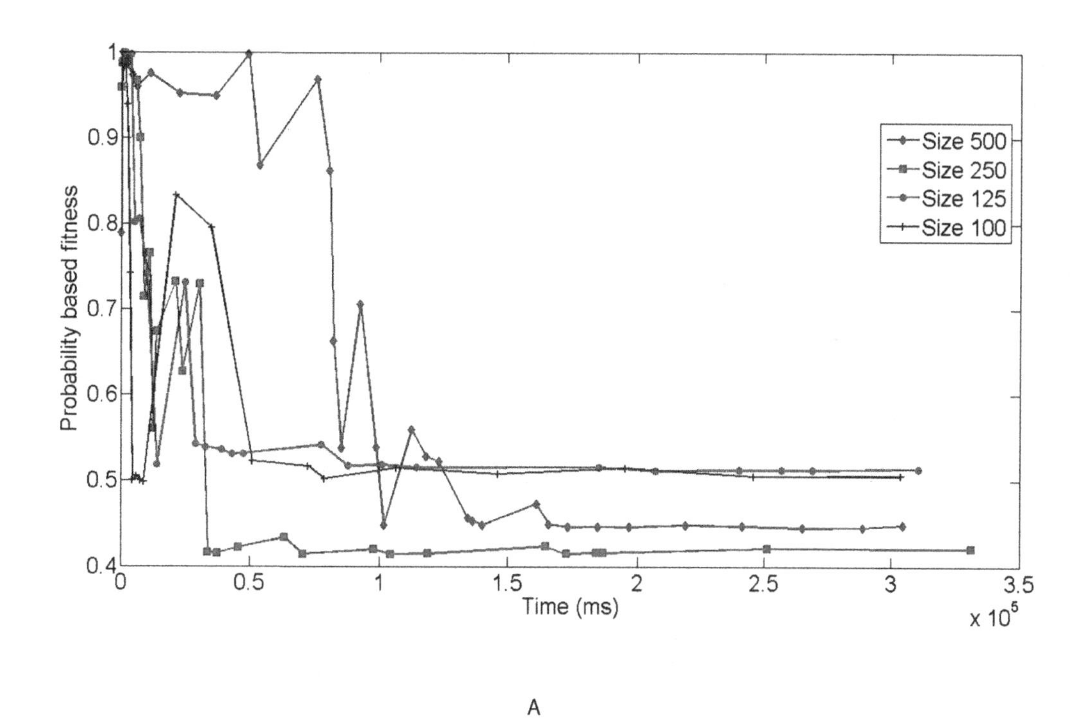

A

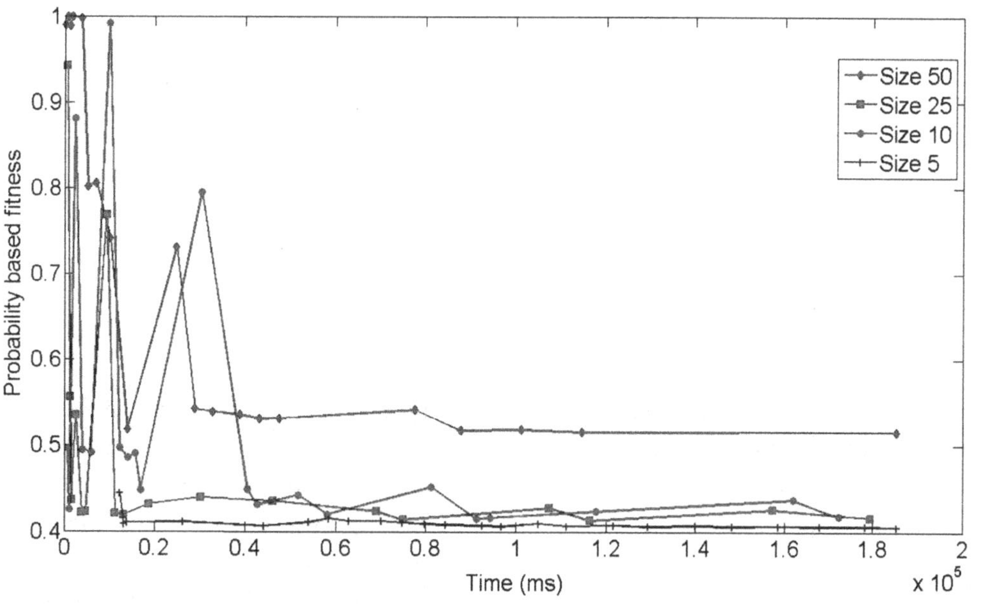

Size 50
Size 25
Size 10
Size 5
Probability based fitness
Time (ms)
x 10^5

B

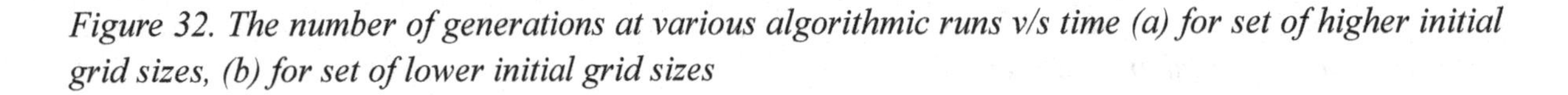

Figure 32. The number of generations at various algorithmic runs v/s time (a) for set of higher initial grid sizes, (b) for set of lower initial grid sizes

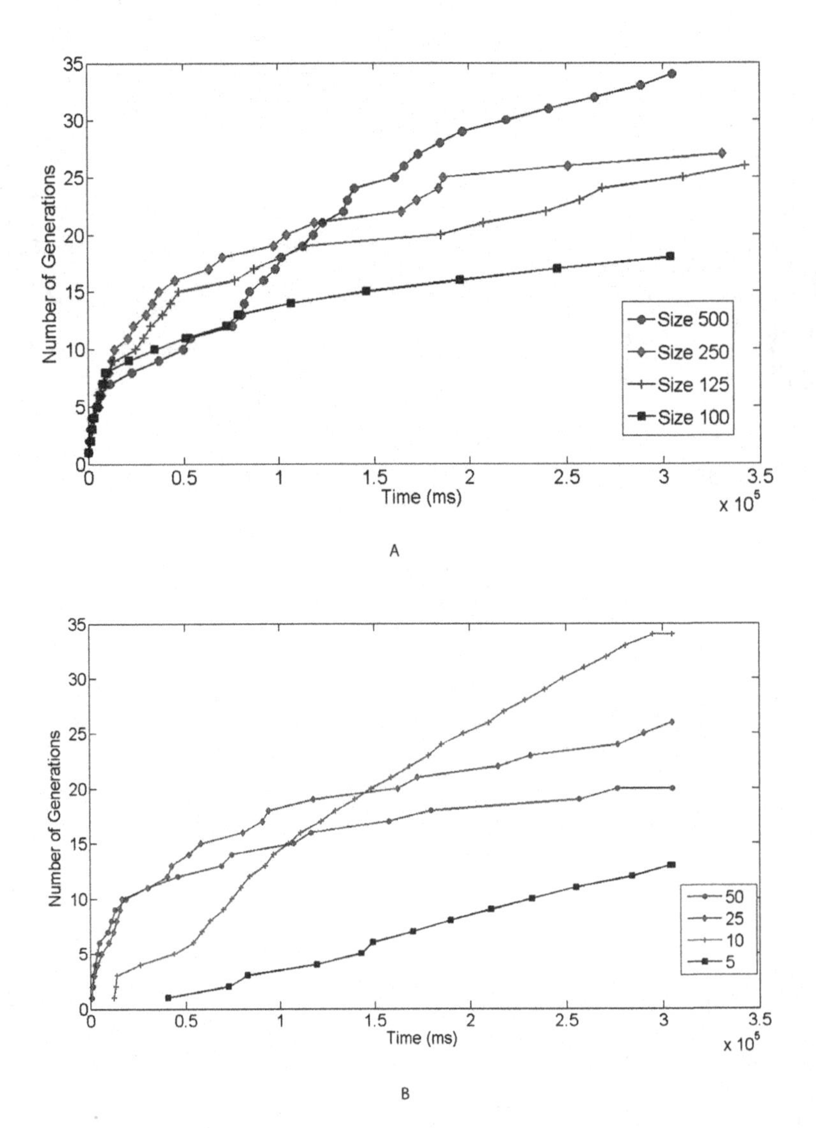

on the map where the magnitude of movement depends upon the mutation rate. Elite is another operator used to pass on the best few individuals from one generation to the other.

RELATION BETWEEN EA AND MNHS

We have already stated that the EA ensures results are available within small computation times and the MNHS carries forward the task of path optimality. In this section, we explicitly state this relation that also speaks about the algorithm optimality as compared to the other simple and hybrid approaches.

The MNHS algorithm fails to work in the presence of a large number of nodes due to the large time complexity. Since it is not an iterative algorithm, we cannot break its run to get the path. This makes it impossible to use MNHS or other

Figure 33. The coordinate system for individual representation and robotic map

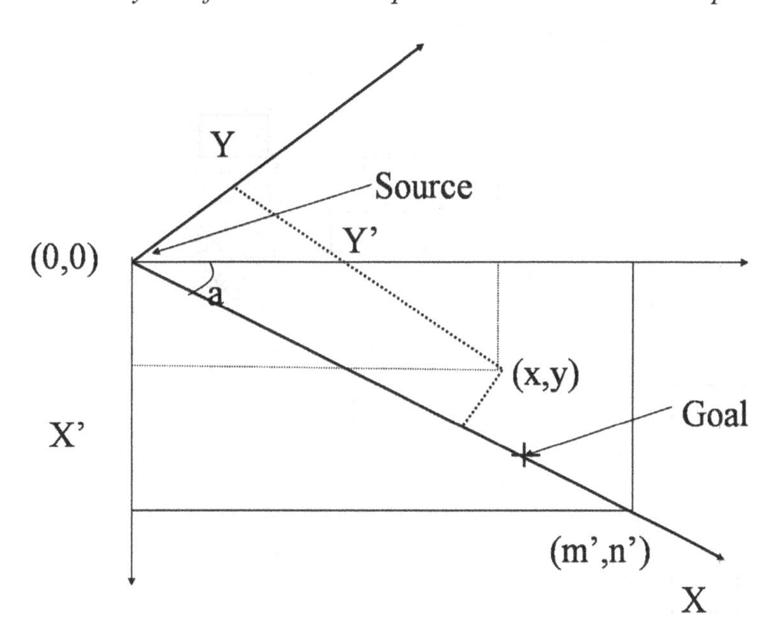

heuristic algorithms in most real life scenarios. As a result, we need an algorithm that can reduce the dimensionality and select the best nodes for the MNHS to perform in a finite time. The MNHS can easily work over a reduced set of points to give the optimal paths.

The EA is an iterative manner of solving problems. One of the main disadvantages of the algorithm is that it only tries to generate paths based on the fitness of the paths of the previous generations. This makes the algorithm make slow convergence in path optimality. The solutions come early but are not optimal. Hence, this algorithm needs the assistance of some heuristics that enable it to form good paths and figure out good and bad points that make up a path.

Hence, the combination of both these algorithms solves the twin problem of path optimality and computational time. It may be easily seen from the algorithm that the mutual contribution of the two algorithms is controlled by the factor β. If β is very small, the resultant algorithm would be dominated by EA. The MNHS would have very little choice between the nodes selection. On the other hand, if β is very large, the algorithm would be primarily MNHS in nature. The placement of

Table 2. Comparative advantages and disadvantages of graph search algorithms and evolutionary algorithms

S. No.	Graph Search		Evolutionary Algorithms	
	Advantages	Disadvantages	Advantages	Disadvantages
1.	Complete	High computation time, work only in small resolution maps	Small computation time to generate initial (sub-optimal) solutions, can work in high resolution maps	Non-complete especially in narrow corridors and complex paths
2.	Optimal	Non-iterative	Iterative	Non-optimal in complex paths

Figure 34. (a) The path generated by MNHS for no obstacle, (b) the run for single obstacle in case 2, (c) the run for complex obstacles in case 3, (d) the second run for complex obstacles in case 3

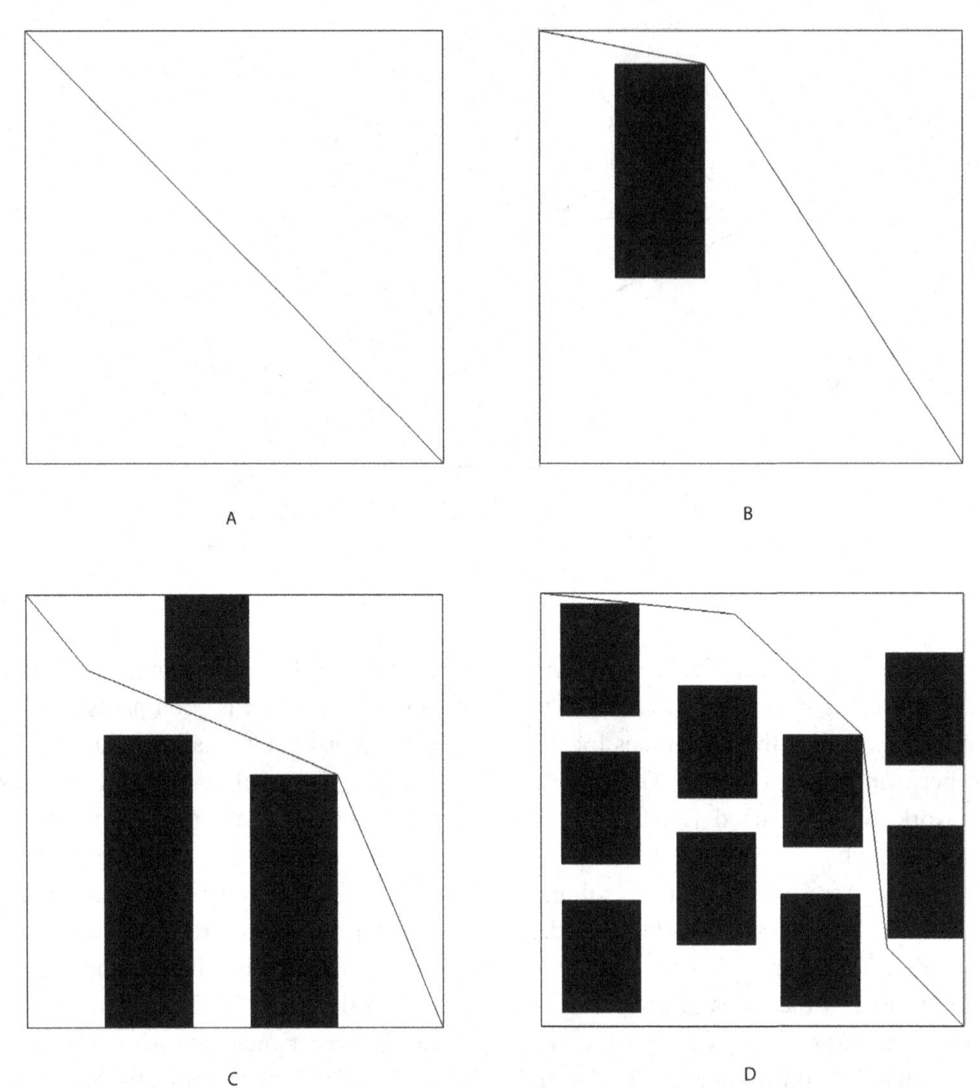

nodes would lose importance as compared to the path formulation between them. The relations are summarized in Table 2.

RESULTS

The problem of path planning deals with the determination of a path, which navigates the robot in such a way that no collision occurs. In order to solve the problem, we assume that the input is already available in form of a map. Here, we assume that the map is available in form of grid of size M x N. Each of the cells of this grid contains 0 or 1. A 0 in such a grid signifies that the region has an obstacle present. Similarly, a 1 signifies that the region is traversable and may be used for the purpose of travelling. The obstacles may span across multiple cells. The black regions here signify the presence of obstacles.

It is further assumed that the grid given as input is of considerable size. If the grid exceeds a certain threshold of size, it would become computationally impossible for the algorithm to find a result. Hence, we restrict the size of the map according to the computational capability and time constraints in whatever real life specific problem is being considered. The algorithm would generate as its output a path that can be used by the robot for the navigation purposes. The path may be traversed using any robotic controller. This is for the execution of the steps given by the planning algorithm.

We applied various tests to the algorithm in order to ensure that the algorithm behaves well in each and every condition. In all the cases, the map was of size 1000X1000, and the robot was supposed to move from the top left corner to the bottom right corner. All simulations were made on a 256 MB RAM and 585 MHz processor. All runs took less than 50 seconds with a very high convergence of the path length. The parameters were:

- $\alpha = 2$
- Number of individuals = 125
- Number of generations = 100
- Elite count = 2
- Crossover rate = 0.78
- Mutation rate = 0.06
- $\beta = 5$

Initially, we did not place any obstacle in the path from the source to destination. We observed that the algorithm traced the path from the source to the destination following a straight-line path. The result is shown in Figure 34a. The second case we considered was of a single obstacle in the path from the source to destination. The robot easily avoided the obstacle and marched towards the goal position. The result is shown in Figure 34b. The last case we presented before the algorithm was to test its ability to handle complex inputs. Various complex obstacles were placed in the path of the robot from the source to destination. The robot again easily avoided the obstacle and marched towards the goal position. The results are shown in Figure 34c and 34d.

CONCLUSION

The chapter focused upon the hybridization of algorithms for the problem of motion planning of mobile robots. For the same two approaches were presented. One of the algorithms for hybridization was fixed in both the approaches to be MNHS. The other algorithm was varied to provide diversity. In the first approach, MNHS was applied in a multi-layer hierarchical manner, wherein it was hybridized with itself. In the second approach, MNHS was fused with EA, wherein the EA selected optimal set of points and MNHS returned optimal path using those points as a graph.

First, we used a method to solve the problem of path planning in static environment using a hierarchical approach involving MNHS and probability-based fitness. We tested the algorithm for various test cases. In all the test cases, we observed that the algorithm was able to find the correct solution. The time required, number of generations, etc. depended upon the problem being considered. We again tested the power of the algorithm to react to variable number of obstacles. This is very practical in nature. We saw that again the algorithm was able to find the correct path in a few generations. Then the process of decomposition went on to further break down the path and give finer paths.

Earlier, the algorithm had given paths that were prone to collisions. However, the algorithm soon found out the path that had no obstacles. Any practical scenario is a uniform mixture of all the 4 tested cases that we have considered. Since the algorithm could solve each of the 4 cases, we can be assured that this algorithm would give a timely response in any situation. Another important aspect associated with the algorithm is that it can be terminated at any point of time. We saw that the improvement in path was rapid at the start and then the improvement was slower. Hence, even if we terminate the algorithm after a few iterations, we would still get good solutions that would most probably be collision free and of shortest possible length. Hence, using this algorithm we were able to make modified A* algorithm that was computationally much less expensive and solved the problems presented. The algorithm hence works better than the classical A* algorithm.

In the second approach, we made the use of MNHS and EA to solve the problem of robotic path planning. We saw that we were able to solve the problem in almost all given scenarios well in time. The MNHS proved to be a great algorithm for the purpose of optimality of path, which is a very important factor for the algorithm. At the same time, the optimality of the total time of execution was provided by the use of EA. The EA gave limited points in the entire map to the MNHS. Hence, using the two algorithms, we were able to optimize both the time and the total path length. This proved to be a very important algorithm that could solve much of the problems present in the earlier techniques of robotic path planning. A beautiful relation exists between the A* algorithm as well as the MNHS. Suppose we want better paths. A very natural choice would be to make the MNHS dominant in the entire algorithm. Similarly, say we want results early. This can be achieved by making the EA more dominant in the algorithm. Hence, a controlling measure must exist between the two algorithms. This controlling measure is provided by the use of the EA size constant β. Suppose the value of β is very large. Now the MNHS would be dominant and vice versa.

In both algorithms, the static nature of the map is a major limitation. The algorithms take time to generate result, and it is assumed that the map cannot change during the same duration. Further, neither of the work guarantees completeness. If it is very likely that the case path is very complex in nature or involves narrow corridors, the solution cannot be determined. However, the capability of both algorithms is still better than all the base algorithms taking the considered modeling scenario.

REFERENCES

Alvarez, A., Caiti, A., & Onken, R. (2004). Evolutionary path planning for autonomous underwater vehicles in a variable ocean. *IEEE Journal of Oceanic Engineering*, *29*(2), 418–429. doi:10.1109/JOE.2004.827837

Camilo, O., Collins, E. G. Jr, Selekwa, M. F., & Dunlap, D. D. (2008). The virtual wall approach to limit cycle avoidance for unmanned ground vehicles. *IEEE Transaction on Robotics and Autonomous Systems*, *56*(8), 645–657.

Carpin, S., & Pagello, E. (2009). An experimental study of distributed robot coordination. *ACM Robotics and Autonomous Systems*, *57*(2), 129–133. doi:10.1016/j.robot.2008.07.001

Chen, L. H., & Chiang, C. H. (2003). New approach to intelligent control systems with self-exploring process. *IEEE Transactions on Systems, Man, and Cybernetics. Part B, Cybernetics*, *33*(1), 56–66. doi:10.1109/TSMCB.2003.808192

Cortes, J., Jaillet, L., & Simeon, T. (2008). Disassembly path planning for complex articulated objects. *IEEE Transactions on Robotics*, *24*(2), 475–481. doi:10.1109/TRO.2008.915464

Goel, A. K., Ail, K. S., Donnellan, M. W., Gomez de Silva Garza, A., & Callantine, T. J. (1994). Multistrategy adaptive path planning. *IEEE Expert*, *9*(6), 57–65. doi:10.1109/64.363273

Hazon, N., & Kaminka, G. (2008). On redundancy, efficiency, and robustness in coverage for multiple robots. *IEEE Transactions on Robotics and Automation*, *56*(12), 1102–1114.

Hui, N. B., & Pratihar, D. K. (2009). A comparative study on some navigation schemes of a real robot tackling moving obstacles. *Robotics and Computer-integrated Manufacturing*, *25*(4–5), 810–828. doi:10.1016/j.rcim.2008.12.003

Hwang, J. Y., Kim, J. S., Lim, S. S., & Park, K. H. (2003). A fast path planning by path graph optimization. *IEEE Transactions on Systems, Man, and Cybernetics. Part A, Systems and Humans*, *33*(1), 121–128. doi:10.1109/TSMCA.2003.812599

Jan, G. E., Chang, K. Y., & Parberry, I. (2008). Optimal path planning for mobile robot navigation. *IEEE Transactions on Mechatronics*, *13*(4), 451–460. doi:10.1109/TMECH.2008.2000822

Jolly, K. G., Kumar, R. S., & Vijayakumar, R. (2009). A Bezier curve based path planning in a multi-agent robot soccer system without violating the acceleration limits. *Robotics and Autonomous Systems*, *57*(1), 23–33. doi:10.1016/j.robot.2008.03.009

Juidette, H., & Youlal, H. (2000). Fuzzy dynamic path planning using genetic algorithms. *IEEE Electronics Letters*, *36*(4), 374–376. doi:10.1049/el:20000314

Kala, R., Shukla, A., & Tiwari, R. (2009a). Fusion of evolutionary algorithms and multi-neuron heuristic search for robotic path planning. In *Proceedings of the IEEE Conference on Nature & Biologically Inspired Computing*, (pp. 684-689). IEEE Press.

Kala, R., Shukla, A., & Tiwari, R. (2011). Robotic path planning in static environment using hierarchical multi-neuron heuristic search and probability based fitness. *Neurocomputing*, *74*(14-15), 2314–2335. doi:10.1016/j.neucom.2011.03.006

Kala, R., Shukla, A., Tiwari, R., Rungta, S., & Janghel, R. R. (2009b). Mobile robot navigation control in moving obstacle environment using genetic algorithm, artificial neural networks and A* algorithm. In *Proceedings of the IEEE World Congress on Computer Science and Information Engineering*, (vol. 4), (pp. 705-713). IEEE Press.

Kambhampati, S., & Davis, L. (1986). Multiresolution path planning for mobile robots. *IEEE Journal on Robotics and Automation*, *2*, 135–145. doi:10.1109/JRA.1986.1087051

Lai, X. C., Ge, S. S., & Al Mamun, A. (2007). Hierarchical incremental path planning and situation-dependent optimized dynamic motion planning considering accelerations. *IEEE Transactions on Systems, Man, and Cybernetics. Part B, Cybernetics*, *37*(6), 1541–1554. doi:10.1109/TSMCB.2007.906577

Lin, H., Xiao, J., & Michalewicz, Z. (1994). Evolutionary algorithm for path planning in mobile robot environment. In *Proceedings of the First IEEE Conference on Evolutionary Computation*, (pp. 211-216). IEEE Press.

O'Hara, K. J., Walker, D. B., & Balch, T. R. (2008). Physical path planning using a pervasive embedded network. *IEEE Transactions on Robotics*, *24*(3), 741–746. doi:10.1109/TRO.2008.919303

Peasgood, M., Clark, C. M., & McPhee, J. (2008). A complete and scalable strategy for coordinating multiple robots within roadmaps. *IEEE Transactions on Robotics*, *24*(2), 283–292. doi:10.1109/TRO.2008.918056

Pozna, C., Troester, F., Precup, R. E., Tar, J. K., & Preitl, S. (2009). On the design of an obstacle avoiding trajectory: Method and simulation. *Mathematics and Computers in Simulation*, *79*(7), 2211–2226. doi:10.1016/j.matcom.2008.12.015

Pradhan, S. K., Parhi, D., & Panda, A. K. (2009). Fuzzy logic techniques for navigation of several mobile robots. *Applied Soft Computing*, *9*(1), 290–304. doi:10.1016/j.asoc.2008.04.008

Shibata, T., Fukuda, T., & Tanie, K. (1993). Fuzzy critic for robotic motion planning by genetic algorithm in hierarchical intelligent control. In *Proceedings of 1993 International Joint Conference on Neural Networks*, (pp. 77-773). IEEE.

Shukla, A., & Kala, R. (2008). Multi neuron heuristic search. *International Journal of Computer Science and Network Security*, *8*(6), 344–350.

Shukla, A., Tiwari, R., & Kala, R. (2008). *Mobile robot navigation control in moving obstacle environment using A* algorithm. Intelligent Systems Engineering Systems through Artificial Neural Networks* (*Vol. 18*, pp. 113–120). ASME Publications.

Tsai, C., Lee, J., & Chuang, J. (2001). Path planning of 3-D objects using a new workspace model. *IEEE Transactions on Systems, Man and Cybernetics. Part C, Applications and Reviews*, *31*(3), 405–410. doi:10.1109/5326.971669

Urdiales, C., Bantlera, A., Arrebola, F., & Sandoval, F. (1998). Multi-level path planning algorithm for autonomous robots. *IEEE Electronic Letters, 34*(2), 223-224.

Xiao, J., Michalewicz, Z., Zhang, L., & Trojanowski, K. (1997). Adaptive evolutionary planner/navigator for mobile robots. *IEEE Transactions on Evolutionary Computation*, *1*(1), 18–28. doi:10.1109/4235.585889

Zhu, D., & Latombe, J. (1991). New heuristic algorithms for efficient hierarchical path planning. *IEEE Transactions on Robotics and Automation*, 7(1), 9–20. doi:10.1109/70.68066

KEY TERMS AND DEFINITIONS

Fusion of Graph Search and Evolutionary Algorithm: Planning algorithm where the iterative nature of the evolutionary algorithm and its ability to work with high resolution maps fuses with the optimality and completeness of the graph-search algorithm. The evolutionary algorithm attempts to generate a number of points in the map that are used as inputs to the graph-search algorithm, which gives the best path to the goal (which is used as a fitness measure of evolutionary individuals).

Hierarchical Graph Search: A planning algorithm that controls the resolution in a multi-resolution map representation, so as to return the best path in the smallest time. Planning is done in multiple iterations in multiple map resolutions, with the solution of one iteration carried forward as a roadmap for developing the solution of the next iteration for which resolution of the map is increased.

Hierarchical Planning: Task of breaking up the planning process into two or more levels, where each level may be planned using an algorithm which suits the level. Higher-level plans may be vague, which may be detailed by lower-level plans.

Hybrid Planning: Using two or more algorithms for the task of planning, such that the advantages of the various algorithms are improved and the limitations are reduced. Hybridization must complement the disadvantages of one algorithm by the advantages of the other algorithm.

Multi-Resolution Map Representation: Simultaneously representing the map in a number of resolutions from high to low. Parts of map may be in high resolution, while other parts may be in lower resolution.

Chapter 8
Hybrid Evolutionary Methods

ABSTRACT

The limitations of single algorithm approaches lead to an attempt to hybridize or fuse multiple algorithms in the hope of removing the underlying limitations. In this chapter, the authors study the evolutionary algorithms for problem solving and try to use them in a unique manner so as to get a better performance. In the first approach, they use an evolutionary algorithm for solving the problem of motion planning in a static environment. An additional factor called momentum is introduced that controls the granularity with which a robotic path is traversed to compute its fitness. By varying the momentum, the map may be treated finer or coarser. The path evolves along the generations, with each generation adding to the maximum possible complexity of the path. Along with complexity (number of turns), the authors optimize the total path length as well as the minimum distance from the obstacle in the robotic path. The requirement of evolutionary parameter individuals as well as the maximum complexity is less at the start and more at the later stages of the algorithm. Momentum is made to decrease as the algorithm proceeds. This makes the exploration vague at the start and detailed at the later stages. As an extension to the same work, in the second approach of the chapter, the authors show the manner in which a hybrid algorithm may be used in place of simple genetic algorithm for solving the problem with momentum. A Hybrid Genetic Algorithm Particle Swarm Optimization (HGAPSO) algorithm, which is a hybrid of a genetic algorithm and particle swarm optimization, is used in the same modeling scenario. In the third and last approach, the authors present a hierarchical evolutionary algorithm that operates in two hierarchies. The coarser hierarchy finds the path in a static environment consisting of the entire robotic map. The resolution of the map is reduced for computational speed. The finer hierarchy takes a section of the map and computes the path for both static and dynamic environments. Both these hierarchies carry optimization as the robot travels in the map. The static environment path gets more and more optimized

DOI: 10.4018/978-1-4666-2074-2.ch008

along with generations. Hence, an extra setup cost is not required like other evolutionary approaches. The finer hierarchy makes the robot easily escape from the moving obstacle, almost following the path shown by the coarser hierarchy. This hierarchy extrapolates the movements of the various objects by assuming them to be moving with same speed and direction.

INTRODUCTION

The attempt to eliminate the limitations of the individual algorithms leads us to the use of hybrid techniques planning. The attempt is to study the limitations of an algorithm of choice, and to later attempt to find a manner or an algorithm that can remove the limitations and enhance the advantages. We have been studying this notion of hybridization of the algorithms to solve the problem of motion planning for mobile robots. The base algorithm that we study in this chapter is evolutionary algorithm. The aim is to make some changes in the conventional problem solving in evolutionary robotics, which was discussed in detail in chapters 4 and 5.

Evolution has always been seen as the future of robotics. Increasing efforts are being made to enable evolutionary algorithms specify complete design of robotics from hardware and/or algorithmic point of view. The ability of evolutionary algorithms to evolve complicated structures is the sole reason behind the same. A number of component designs of robots are current being produced by evolutionary algorithms. Evolutionary algorithms are increasingly used in industry for on-board circuit designs. The control algorithms used in modern robots are further worked over by evolutionary algorithms. These algorithms enable robots do complex tasks as per requirement. All these works can be done off-line, and hence, computational time is not a constraint.

One of the major limitations of the evolutionary algorithm is its time consuming nature. Especially in maps having very large resolution, these algorithms struggle to give decent results within time. This makes these algorithms inefficient. Most robotic problems require the solutions to be real time in nature. For these reasons, it is a requirement for the planning algorithm to work in real time. Hence, in this chapter, we attempt to make the algorithm as time efficient as possible. The various advantages and disadvantages of evolutionary algorithms, that forms the basis of this chapter, are discussed in Table 1.

The chapter introduces three approaches for enhancing the conventional evolutionary algorithms. In the first approach, we introduce the concept of momentum in evolutionary algorithms, which controls the granularity of the checking of path feasibility. The factor is made adaptive and controlled for enhancing the performance of the evolutionary approach. In the second part of the chapter, we take the same system of evolutionary algorithm with momentum, and replace the evolutionary algorithm by a more powerful algorithm called Hybrid Genetic Algorithm Particle Swarm Optimization (HGAPSO). This algorithm itself represents a hybrid of genetic algorithm and

Table 1. Advantages and disadvantages of evolutionary algorithms

S. No.	Advantages	Disadvantages
1.	Small computation time to generate initial (sub-optimal) solutions	High overall computational time for high resolution maps
2.	Iterative	Possibility of immature convergence
4.	Probabilistically optimal (not too complex paths)	Non-complete in narrow corridors and complex paths
5.	Probabilistically complete (not too complex paths)	Non-optimal in complex paths

particle swarm optimization. In the third part of the chapter, we hybridize evolutionary algorithms with themselves. In other words, the algorithm is implemented in a hierarchical manner. The resultant algorithm is able to give solutions in real time.

The first method discussed deals with use of momentum. Here we discuss our work reported in Kala et al. (2011). Discussions in the introduction, conclusions and sections covering the entire algorithm are reprinted with few modifications from Kala, Shukla, and Tiwari, (2011). Reprinted by permission of Taylor and Francis Ltd. (http://www.tandfonline.com). The evolutionary algorithms make excessive calls to fitness function to evaluate the individuals. It is hence necessary to limit the complexity of the fitness function at the same time not compromising with the fitness algorithm. The application of conventional genetic algorithm (Gerke, 1995; Han, et al., 1997; Sedhigi, 2004; Toogood, et al., 1995; Woong-Gie, 1997; Yanrong, 2004) might involve the fitness algorithm traversing the represented path to ensure a collision free navigation. Ideally, the navigation needs to be cell by cell which would make the algorithm very slow. For this, we use the concept of momentum. A higher momentum means a higher speed of navigation where the algorithm traverses by skipping some cells. Initially the solutions are premature and may be checked for feasibility by higher momentums. The momentum gradually reduces as the solutions mature. This is the time when we would like to ensure their feasibility.

Complexity in the problem of robotic path planning may be indicated by various parameters depending upon their interpretation. In solutions generated by A* algorithm, the map dimensions indicate the problem complexity. For behavioral solutions the number of behavioral rules or weights, and for evolutionary systems the genomic length is an indicative of the complexity. These more or less convey the same meaning for the same problem. In this problem, we assume the complexity denotes the maximum number of turns in the computed robotic path. The solution of the robotic path problem may be very simple or may be very complex. Lesser turns naturally mean a greater scope for the robot to traverse the path with greater speeds for a better drive at less time. In addition, this may denote a straighter and shorter path. The constraints include the feasibility of the path and the avoidance of obstacles in the path.

Using the suggested approaches, we intend to solve 3 major problems that exist in most algorithms. The first is the problem of path complexity control. It may easily be seen that the complexity of the path can never be predicted as it depends upon the complexity of the map, which largely varies with map. The second is the optimization of the time complexity of fitness function by using a cautious investment of time. This may be seen as a similar approach to the probability-based map representation (as used in chapter 7) or a radix tree map representation (Kambhampati & Davis, 1986) with the role of probability being played by momentum. The third is the need-based genetic individual deployment that we maintain by varying the number of individuals in the genetic search. The number of individuals used in the optimization is varied in the implementation. Only the required number of individuals is deployed that goes a long way in optimization of time.

In the second approach, we go further with hybridization and use a hybrid optimization algorithm HGAPSO in place of evolutionary algorithm for evolution of robotic path. Here, we discuss our work reported in Kala et al. (2012). Discussions in introduction, conclusions, and sections covering the entire algorithm are reprinted with few modifications from Kala, Shukla, and Tiwari, (2012) with permission from Inderscience Publishers.

In the third approach, we break the problem of path planning into two related sub-problems, namely coarser path planning and finer path planning. The finer path planning gets inputs of reasonably simple size. Nevertheless, it is expected to give precise outputs in real time scenarios. On the other hand, the coarser path planning may take time to optimize the complete path. The path may be vaguely built as further optimizations would be carried by the finer path-planning module. To incorporate sudden and dynamic obstacles, both these techniques need to work hand-in-hand. The coarser planning has a role to play in case of some sudden blockage where global path needs to be changed in sufficiently less time, the finer planning technique not only tunes the path, but also helps in escaping from regular obstacles. An only coarser or finer planning would make the algorithm computationally very expensive, and would not allow dynamic or sudden obstacles. Here, we discuss our work reported in Kala et al. (2010). Discussions in the introduction, conclusions, and sections covering the entire algorithm are reprinted with few modifications from Kala, Shukla, and Tiwari (2010). Reprinted by permission of Taylor and Francis Ltd. (http://www.tandfonline.com).

We briefly discuss some of the existing literature in use of genetic and some hierarchical approaches. Readers may skip and go to next section without any loss of continuity. Because of the underlying complexity of the problem, a significant use of evolutionary techniques to solve the problem is prevalent. A variety of mechanisms are applied to customize the conventional genetic algorithm to give an enhanced performance in solving the problem. Kala et al. (2009) represented a graph path as a genetic individual. The genetic operators of mutation and crossover were designed to work over these paths. Other customizations include the use of restrictions in genetic operations to ensure feasibility of the solution generated at every step (Alvarez, et al., 2004). Another representation technique uses a set of sparse points from source to destination with the line-joining source and destination as one of the axis (Toogood, et al., 1995; Han, et al., 1997). Many times the corners of the obstacles may be numbered and the individual is a sequence of numbers that the robot visits (Sadati & Taheri, 2002). Tu et al. (2003) give the concept of variable length representation of chromosome. The robot visits these points in a sequential manner. The use of Bezier curve is also done that caters to the non-holonomic constraints in the path generated by the evolutionary algorithm (Jolly, et al., 2009).

Many times the map may be of a very large dimensionality, and hence, it may be very difficult to work with such a graph. For such cases, the problem of planning is usually solved in a hierarchical manner, which is known as Multi-Resolution Path planning. This may represent the map in multiple resolutions using a variety of representation techniques (Kambhampati & Davis, 1986; Urdiales, et al., 1998). This further makes hierarchical path planning algorithms possible (Wang, et al., 2002). Other related works using evolutionary computation include (Doitsidis, et al., 2009). Here the authors make use of optimization power of the evolutionary algorithms to evolve a fuzzy-based robotic controller. The genetic algorithm here tries to evolve the optimal set of rules. The algorithm of Dittrich et al. (1998) is another novel work that presents the application of Genetic Programming for the problem with the fitness being measured in a combination of simulator and physical robot.

EVOLUTIONARY ALGORITHM WITH MOMENTUM

The first approach dealt in the chapter deals with the use of Evolutionary Algorithm (EA) with momentum. We first formulate a general evolutionary framework for solving the problem, and then touch upon the concept of momentum. Taking it as a general evolutionary problem, we need a technique for individual representation, a fitness evaluation mechanism, and genetic operators. Each of these is discussed. This section and subsequent sections covering the entire algorithm are reprinted with few modifications from Kala, Shukla, and Tiwari (2011). Reprinted by permission of Taylor and Francis Ltd. (http://www.tandfonline.com).

Individual Representation

One of the foremost tasks in this algorithm is a good individual representation. Here, we represent an individual by a series of points (P_i) on the robotic map. The complete individual hence becomes <P_0, P_1, P_2, P_3, ... P_n, P_{n+1}>. Here P_0 is the source and P_{n+1} is the goal. Each point is a collection of x and y coordinates and may be denoted by (x_i, y_i). The x-axis that we take for this problem is the straight line joining the source and the goal. The y-axis is perpendicular to the x-axis as given in Figure 1.

Let us suppose that the map is represented in the coordinate system X'-Y' given in Figure 1. Now any point needs to lie within the range of (0',0') to (m',n') so as to lie within the map. Here ' represents the use of X'-Y' coordinate system. This range needs to be converted into equivalent range in the X-Y coordinate system of the individual to generate valid points in the robotic map. This is done by a rotation by an angle α in the clockwise direction, where α is the angle between the two coordinate systems given in Figure 1.

Complexity of a path or a solution is defined as the total number of points in the path or solution. For the path represented by <P_0, P_1, P_2, P_3, ... P_c, P_{c+1}>, the complexity is c. The complexity for any path may lie between 0 (straight line from source to goal) to a maximum permissible value C_{max}.

Another important characteristic of the individual representation is that the various points in the map are always sorted along the X-axis. The final path is the path traversed by touching the various points in a straight-line one after the other. Here, the source is the first point and goal is the last point. Hence, we assume that in the final path, the robot cannot move such that the projection of its motion in X' axis is backwards. Sorting enables to get fit individuals quickly.

The entire length of the chromosome is fixed to a maximum value of $2C_{max}$. The unoccupied positions in any solutions are filled with Inf (Infinity). A sample path is given in Figure 2. If C_{max} is set to 10 and each P_i denotes the point (x_i,y_i) then the genetic individual of this path would be represented as <x_1 y_1 x_2 y_2 x_3 y_3 x_4 y_4 Inf Inf Inf Inf Inf Inf Inf Inf Inf Inf Inf Inf Inf Inf Inf Inf>.

Figure 1. The coordinate system for individual representation and robotic map

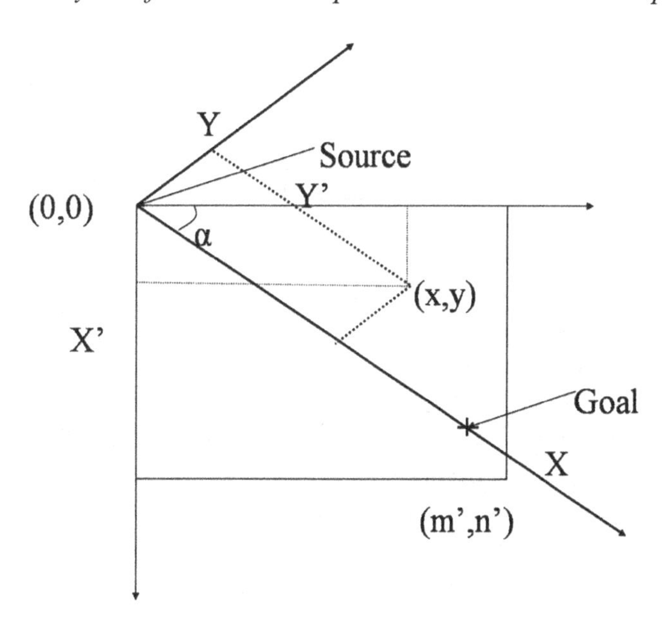

Figure 2. A sample path of the problem

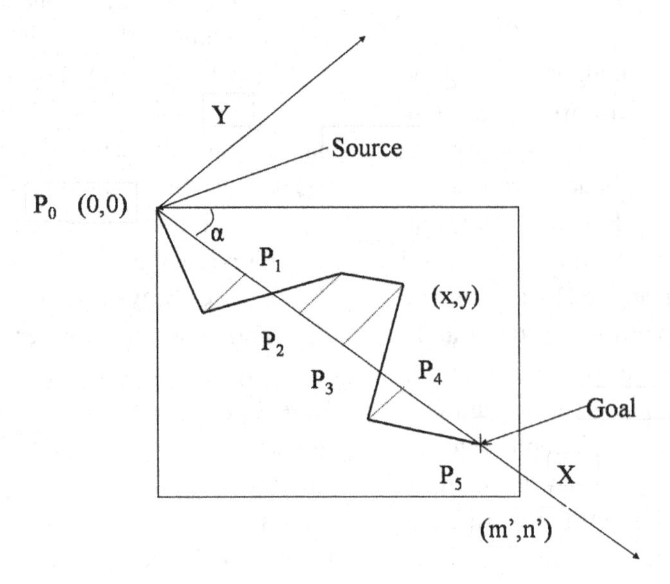

Fitness Function

The next important factor in the algorithm is the fitness function. The fitness function is used to judge the quality of any path. Here, we try to work upon three major factors that all contribute to the final fitness of the path. These are the total length of the path (l), the distance from the closest obstacle (d), and the path complexity (c). The objective is to minimize each one of l, d, and c. Using the principles of Multi-Objective Optimizations, we frame the equation of the total fitness by weighted factors given by Equation 1.

$$Fit(P) = \alpha\, l + \beta\, d + \gamma\, c \tag{1}$$

Here, P is any path or individual of the form of consecutive points P_i denoted by $\prod P_i$, α, β and γ are objective weights with the constraint that $\alpha + \beta + \gamma = 1$, l is the total path length, d is the distance from closest obstacle, c is the path complexity.

Total path length l is a measure of the total length of the path from the source to the destination. Consider the path shown in Figure 2. Here, we traverse the path consecutively between the points represented by the individual. The first point is taken as the source and the last as the destination. Hence, we always start from the source and end at the destination. This path may be measured by the physical distances between the points P_i and P_{i+1} given by Equation 2. It is natural that the lower values of this factor are desirable.

$$T = \sum_{i=0}^{c} \left\| P_{i+1} - P_i \right\| \tag{2}$$

Here, T is the path length, P_i is the point represented by the individual, P_0 is the source, P_{c+1} is the goal, || x || denotes the Euclidian norm. This length is normalized to lie between 0 and 1 by Equation 3:

$$l = \frac{T}{(m+n)} \tag{3}$$

Here l is the total path length (normalized), T is the path length (un-normalized), m+n is the maximum length possible for any path (normalizing factor), m and n are the length and breadth of the map, respectively.

Distance from closest obstacle returns the smallest distance between any point in the path of the robot and the obstacle. The motivation for this factor comes from the physical movement of the robot. It is known that if the robot path goes through a region that is very close to an obstacle, it may have to slow down in order to avoid a collision. This is especially important if the obstacle is close to some turn of the robot. Another way to look at the same problem is through non-holonomic constraints. The algorithm we build might result in very sharp turns, but it may be impossible for the robot to make such turns due to physical limitations or the non-holonomic constraints. For this, the robotic controllers or sometimes even the path planning algorithms artificially smooth the path. This requires a comfortable distance from the obstacle. The control algorithms may themselves be erroneous and hence taking a safety margin is desirable.

In this algorithm, we assume the factor distance from closest obstacle (d) measures the ease of the movement of the robot that decays in a Gaussian manner as the distance increase. The relation between the minimum physical distance of the obstacle and robot (D) and d is given by Equation 4:

$$d = e^{-\frac{D^2}{2(aM)^2}} \tag{4}$$

Here, d is the distance from closest obstacle (normalized), D is the physical distance from closest obstacle (un-normalized), a is the decay constant governing shape of the Gaussian curve, M is the radius constant controlling the effective region of effectiveness of this function. After M units of distance, the function returns almost a zero value.

It may be easily seen that effectively the Equation 4) works only for distances (D) in the range of 0 to M. After this, the value becomes almost zero. We apply Equation 4) only to D that is less than M units of length. Practically M represents the distance, which is comfortable enough for the robot to easily make its way out.

In order to avoid repetition (and save time) we make a lookup table that stores the values of D for all combinations of points in the graph initially at start. This is done using the principles of Dynamic Programming whose recurrence relation and initial condition is given by Equation 5. This process is known as the fuzzification of the graph.

$$D(P,k) = \begin{cases} 0 & (5-i) \\ Inf & (5-ii) \\ \min\{D(P,k-1), D(\delta(P),k-1)+1\} & (5-iii) \end{cases} \tag{5}$$

(5-i) if k = 0 and P is an obstacle
(5-ii) if k = 0 and P is not an obstacle
(5-iii) for all other cases

Here P is any point, k is the Dynamic Programming iterator that lies between 0 and M, δ(P) is the neighborhood function that returns all points in the neighborhood of P (the 8 surrounding points).

It may happen that paths represented by individuals happen to have obstacle in between. This may be easily found out by a path traversal from source to goal. These are in-feasible solutions of the EA and are assigned a fitness value of Inf (Infinity).

The factor path complexity refers to the number of turns that a robot is allowed to make in its entire path. It is a measure of the straightness of the path that is a very desirable property of a good path in the field of robotics. If the path complexity is low, the path would be as straight as possible. On the other hand, a large path complexity enables a robot to follow a zig-zag path that is usually longer and more difficult to implement in the physical motion of the robot. Hence, we try to restrict or minimize the path complexity in this algorithm. The path complexity may also be taken as the used size of the individual as discussed. The path complexity c is normalized by dividing with the maximum allowable complexity C_{max} to make it lie in the interval of 0 to 1.

Many times a higher complexity path may drive the robot closer to the straight-line path from source to destination and hence may be significantly smaller as compared to the lower complexity path. An analogy of this may be derived from the fact that many times the shortest path from source to destination is filled with many obstacles that require a large number of turns (with low speed) than another path which may have a very long route but only a couple of turns. Hence, complexity (c) and length (l) do not necessarily mean the same thing. It may again be seen that the preference between complexity and length is not easy. In the same analogy, one may prefer to go with the more complex path since its length is small and we may be able to reach the destination early at lower speeds and lesser comfortable journey. One may also prefer a longer length path with lesser complexity as it may lead to goal early due to possibility of higher speeds and more comfortable journey at the cost of length. Choosing between a low traffic/longer route or a high traffic/shorter route between New Delhi to Noida is always a difficult choice. This necessitates the possibility to keep multiple diverse paths in possibility while the algorithm runs. We implement the same in this algorithm by diversity preservation (Badran & Rockett, 2007; Laumanns, et al., 2001).

Another analogy may be seen between the evolution of the path in this algorithm and the evolution of Artificial Neural Networks (ANNs). We know that adding neurons in ANN results in larger complexity that needs to be controlled. The larger number of neurons results in an over-fitting of the trained curve between the data points (Shukla, et al., 2010). This is accounted for in most of the modern evolutionary ANN designs that evolve an ANN. This is one of the motivations behind the algorithm.

MOMENTUM

Momentum refers to the speed of traversal while working with the optimization of the path in EA. With this concept, we try to optimize the total time of fitness function, which would adversely affect the final computation time as well as the results.

Recall our discussion of previous section where we had to traverse the entire path, grid by grid, from source to goal in order to calculate the distance from obstacle (d) as well the feasibility of the path. In a practical sense if we traverse the path in such a manner for every call of the fitness value, the algorithm speed would decrease by a large amount. We need a mechanism to optimize this time. This is done by the concept of momentum. Here, we only check for the presence of obstacle and calculate the factor d between any point P_i and P_{i+1} for the set of points Q given by Equation 6:

$$Q = \left\{ P_i, mc \frac{P_{i+1} - P_i}{\left\| P_{i+1} - P_i \right\|} t, P_{i+1} \right\} \tag{6}$$

Here P_i and P_{i+1} are the consecutive points in the robot path, $\frac{P_{i+1} - P_i}{\left\| P_{i+1} - P_i \right\|}$ is the unit vector in direction of P_{i+1} - P_i, mc is momentum, t is the traversal step (t=1, 2, 3…).

It may be observed that if the momentum is large, lesser number of points is checked, and correspondingly the execution time is less. However, at large, two types of conditions are possible. The first condition that is rare to happen is that we completely step over an obstacle. Here, the algorithm assumes that the path is feasible. However, in reality, there would be an obstacle in the path and there would be no where out to avoid this obstacle. We call this condition as the false state. The other case that can happen is that there might be obstacle on the path being considered, but a feasible path may lie very close. This is a

Figure 3. The concept of momentum

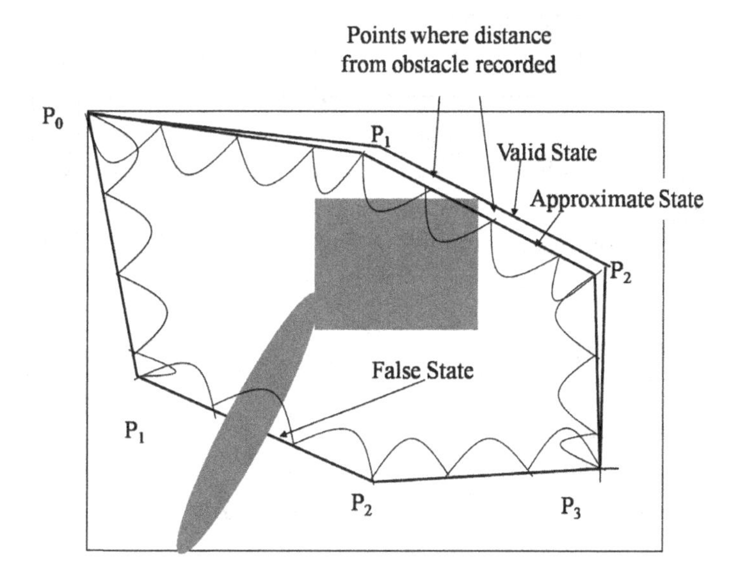

very frequent phenomenon to happen. We call this condition as the approximate state. Both these states are given in Figure 3. The curved lines denote the traversal with momentum.

The ill effects of large momentum do not only apply to the path feasibility, rather they also apply to the distance from obstacle (d). Consider the approximate state in Figure 3. Here even if we adjust P_2 and P_3 to make path feasible it may lie very close to the obstacle that might not be recorded at large momentum. This case is given in Figure 3 as the valid state. The two points report a sufficient distance of separation from obstacle, which is not true. This condition is referred to as the valid state.

Hence large momentum increases speeds but wrongly records path feasibility as well as distance from obstacle (d). The smaller values of momentum on the other hand result in high computational time but give correct feasibility measure. Hence a constant momentum based technique would not work. We make the momentum variable that changes in a Gaussian manner along with the generations. Initially the solution pool has all random solutions and even a large momentum would be able to work and give initial idea of the path feasibility as well as quality. As the generations go on, we keep decreasing the momentum to get the real values of the feasibility and parameters. The Gaussian decay of momentum is given by Equation 7.

$$mc = mc_{\min} + (mc_{\max} - mc_{\min})e^{\frac{-g^2}{2(bG)^2}} \quad (7)$$

Here mc_{max} is the maximum possible momentum, mc_{min} is the least possible momentum, g is the generation number of EA, b is the decay constant, G is the radius constant or the maximum number of generations possible.

Another important factor here is of diversity preservation in regard to false state and approximate state discussed above and presented in Figure 3. Suppose that the best-known solution of the EA after a number of generations is found to be infeasible in regard to approximate state. To overcome such a condition it is necessary for the EA to always retain sufficient number of weakly mutated individuals corresponding to the best few kinds of solutions. Now suppose that the infeasibility was due to false state. In such a case the solution is maintaining a population diversity which has been inbuilt in the EA (Badan & Rockett, 2007; Laumanns, 2001). Diversity preservation prohibits all individuals representing more or less similar paths, which may later turn to be infeasible as a result of false state. We discuss more about application of diversity preservation in next section.

EVOLUTIONARY ALGORITHM

The base of the path-planning algorithm is EA that tries to generate paths of increasing complexity. The EA tries to solve the whole problem of path planning as an optimization problem where it tries to optimize the location of various points so as to give the best path as per the set criterion. In this case, the EA is also responsible to figure out the number of points that must be present in the solution or the final path. As discussed this complexity is variable and keeps on increasing as we proceed with the algorithm. In this section, we present the various genetic operators that contribute to the working of the EA. We have used a total of 7 genetic operators. These are Selection, Crossover, Soft Mutation, Hard Mutation, Elite, Insert, and Repair. We discuss each one of these one by one.

Selection

Using this genetic operator, we select the individuals for the purpose of participation in reproduction or crossover. Rank based fitness scaling with stochastic uniform selection scheme has been used. A point that we have been insisting is the diversity preservation. Here a set of points qualify for crossover only when they are separated by a distance less than a threshold distance η in the input space. The separation (S) is measured as the average separation between corresponding points existing in the path.

Say that two solutions A and B have x and y number of points in their paths. Further suppose $x < y$. We calculate the x closest set of points (a,b) such that a lies in A and b lies in B and neither of a or b are repeated. The average distance between a and b for all points gives the separation S. This is given by Equation 8. This is easy to implement as A and B are already sorted.

$$S = \frac{\sum_{a \in A, b \in B, b \text{ is least mapping of } a} \|a - b\|}{\min(x, y)} \quad (8)$$

Crossover

Crossover mixes two individuals for the generation of a next generation individual. Here we use a scattered crossover technique for crossover between the individuals. Suppose the two parents are A and B that have x and y number of points existing. We first make a pool of points R that carries all points from A and B. The points common to A and B are taken only once. The points in R are sorted according to the X' axis values. Now we distribute the points in R to the new children such that each of the 2 generated children get (x+y)/2 points and each of the point in R belongs to either of the two children.

Soft Mutation

This mutation moves the solutions or individuals by little amounts. In other words, we follow a mutation technique with very small mutation value. It may be noted that the various individuals formed as a result of this mutation are added to the solution pool without affecting the parents. This is a solution to the approximate state. Besides as a general evolutionary principle, this helps in the exploration of the search space or the optimization space.

Hard Mutation

This mutation tries to add more diversity and possibility of the path by exploring new areas. In other words, we follow a mutation technique with high mutation value. No change is made to the number of points or the path complexity in both these mutations.

Elite

In this genetic operator, we simply pass the individuals with high fitness value to the next generation. This operator plays a big role in the preservation of the good solutions in the population pool at all generations.

Insert

Using this operator, we add new individuals to the population pool. This operation helps in exploring newer areas in the search space that might have been left out. Since the complexity is on a constant rise, it is important to add the higher complexity individuals in the solution pool. They would participate and interact with other individuals to increase their complexity as well, if higher complexity is desired and may provide a good solution. As per the Darwinian principles, these individuals survive and influence the solution if they result in high fitness value.

In order to use this operation we select the individuals from the existing population pool that have the best fitness. We add random points to it until the number of points become equal to the maximum allowable number of points or complexity c_{max}. The entire individual is then sorted as per our set criterion. It may be noted that c_{max} increases with generations to mark an increase in complexity. We discuss the details in next section.

Repair

This is the last genetic operator. At any generation, we are likely to have a large number of infeasible solutions or individuals. The reasons for this may be the un-optimized stage of the individual, the individual has a lesser complexity as per requirement, detection of an obstacle due to decrease in momentum, etc. We make a single attempt to repair such individuals by replacing them with newer individuals of the maximum permissible complexity c_{max}. This does not guarantee the feasibility of the newly generated individual, but solves two major problems. The first is that it drives the algorithm faster from stages where it is within a complexity that is less than the complexity needed to solve the problem. Every map has a least complexity below which it would not give any feasible solution. While the algorithm is into these stages, all solutions are passed from lower complexity to higher complexity without wasting time at the lower complexity regions. Depending only upon the insert operator to drive the algorithm to regions with sufficient complexity would have been at a very low pace.

The second problem that is solved by this operator is cleanliness. The infeasible solutions do not contribute at all to the EA. It is better to clean them by replacing them with random solutions of complexity high enough to represent feasible solution.

VARIABLE GENETIC PARAMETERS

We have already discussed the variable nature of the momentum in previous section. In this section, we discuss the variation or the variable nature of the number of individuals used by the EA as well as the maximum complexity c_{max}.

Variable Number of Individuals

The number of individuals plays a vital role in EA that help in the exploration of the search space. A very low number of individuals would result in a very little search space exploration but this would make more generations possible in the same computation time that may give enough time for crossover to converge or mutation to search nearby or even distant areas. Ideally, the numbers of individuals depend upon the search space.

We start the algorithm in a condition where only limited size complexity is allowed. It may be seen that this limits the search space to a very small extent. Hence, we require only a very small number of individuals. As the algorithm goes on, the maximum complexity is allowed. As a result, the search space also grows more and more. This requires more number of individuals for exploration. By this time, a few lower complexity paths are found, which reserve some individuals for their exploration. Hence, we model the algorithm such that the number of individuals is increased in a Gaussian manner along with generations. This is given by Equation 9.

$$I = I_{max} - (I_{max} - I_{min})e^{\frac{-g^2}{2(cG)^2}} \quad (9)$$

Here I_{max} is the maximum possible number of individuals, I_{min} is the least possible momentum, g is the generation number of EA, c is the decay constant, G is the radius constant or the maximum number of generations possible.

Variable Maximum Complexity

The natural world gives us a lot of inspiration to make effective systems (Holland, 1975). Life started millions of years back with uni-cellular organisms. Over the years, we have seen the evolution of many complex species, which better adapt to their environment. The beauty also lies in the fact that this adaptation is dynamic with the species adjusting themselves with every generation to the changing environment. A similar concept may be seen in human brain, which starts as a premature entity and later develops into complex forms as the child transforms into an adult. We try to imitate the same philosophy of life where complexity increases along with time and generations. We know from our understanding of natural (uni-cellular v/s multi-cellular organisms) as well as artificial systems (ANN or FIS) that best systems employ the use of minimum complexity needed to solve the objective.

The maximum complexity or the maximum allowable points also follow similar trends and discussions of the number of individuals. We know that in general a smaller complexity is better. More complexity may add to longer and non-straight paths. Hence, the complexity is made to rise in a Gaussian manner as given by Equation 10.

$$c_{max} = C\left(1 - e^{\frac{-g^2}{2(cG)^2}}\right) \quad (10)$$

Here C is the globally maximum possible complexity, g is the generation number of EA, d is the decay constant, G is the radius constant or the maximum number of generations possible

RESULTS

The testing of the algorithm was done on a simulation engine that was built by the authors themselves. We coded the algorithm using JAVA framework with Eclipse IDE. JAVA Applets was used for the depiction of the map and the solutions. The map was fed into the simulator as a JPEG image with the dimensions of the image as the dimensions of map. The white regions denoted the presence of accessible areas and the black denoted obstacles. Here, we present the results to four maps out of the various maps used for the testing purposes. All four of these maps are generated with some or the other inspiration to conditions that a robot might face in real life.

The first map denotes a simple map with a single obstacle in the path from source to destination. This may be a very common situation where a robot has a straight path with some obstacle in its way. The second and the third maps have a variety of obstacles and the robot needs to use some intelligent technique to figure out its way by avoiding all the obstacles. This tests the ability of the robot to calculate the shortest path. The fourth and the last map that we present is the path based map where robot tries to move itself on a non-straight and complex path from the source to the destination.

In all the cases, the maps were of size 1000 x 1000. The coordinate axis of the map had (0,0) point at the top left. This was the source specified for all cases. The goal was the bottom right point with the coordinates of (999,999). All the simulations were carried on a system with 3 GB RAM and 2.40 GHz 2.40 GHz Core 2 Duo Intel processor.

For all experiments the parameters were:

- α, β, and $\gamma = 0.33$ each (0.5, 0. 5 and 0.0 for fourth case)
- a, b, c, and d = 0.3
- Momentum = 10 to 1.5
- $\eta = 0.03$
- M = 10
- c_{max} = 0 to 5 (7.5 for third and fourth case)
- Number of individuals = 1 to 1000 (2000 for third and fourth case)
- Number of generations = 500 (2000 for third case and fourth case)
- Gaussian soft mutation = 0.06
- Gaussian hard mutation 0.25

At any generation, 68% individuals came from reproduction, 2% from eliticism, 15% and 5% from soft and hard mutation, and the remaining 10% from insert.

The maps and the path traced by the robot for each of these cases are given in Figure 4a, b, c, and d. It may be easily seen that in all the cases the robot was able to evolve a path from the source to the destination. The path served the set criterion in the fitness function. The graph shown in Figure 4a is an example of a simple problem where the solution was generated with a single intermediate step or a complexity of 1. It may be easily observed from the path traced that the robot kept a comfortable distance of separation from the obstacle and avoided going too close to it. A similar trend can be seen in the graph 4b, which represents a slightly more complex problem, even though the solution is just a level more in complexity.

Here, it is interesting to observe that the robot could have moved diagonally in the top section which might have resulted in shorter path but it preferred not to do but traverse straight. This is due to the heavy penalty of increase in complexity that we added. In the physical movement by a robotic controller the path would get smoothen up (we have left enough room for that). This would enable the robot to traverse the path smoothly at high speed. This would compensate the increase in length where the robot would have been forced to make two sharp turns.

The graph shown in Figure 4c marks a still more complex graph where the robot has taken a very steep turn in the top section. It can be seen that the conditions were highly chaotic and finding a way out was not very easy. Here a very special path is chosen by the robot as the optimal path. In the top section, it prefers to pass through a very complex structure by making two turns, which makes the complexity high. After this section, the work is relatively simple where the robot has a lot of straight path to traverse. This large amount of straight path compensates for the complexity added at start. Various other paths could have been possible of lower initial complexity but the straight path traversed gives this path the edge over other options. Figure 4d again represents an interesting map where the robot has to march in

Figure 4. Experimental results for EA (a) map 1, (b) map 2, (c) map 3, (d) map 4

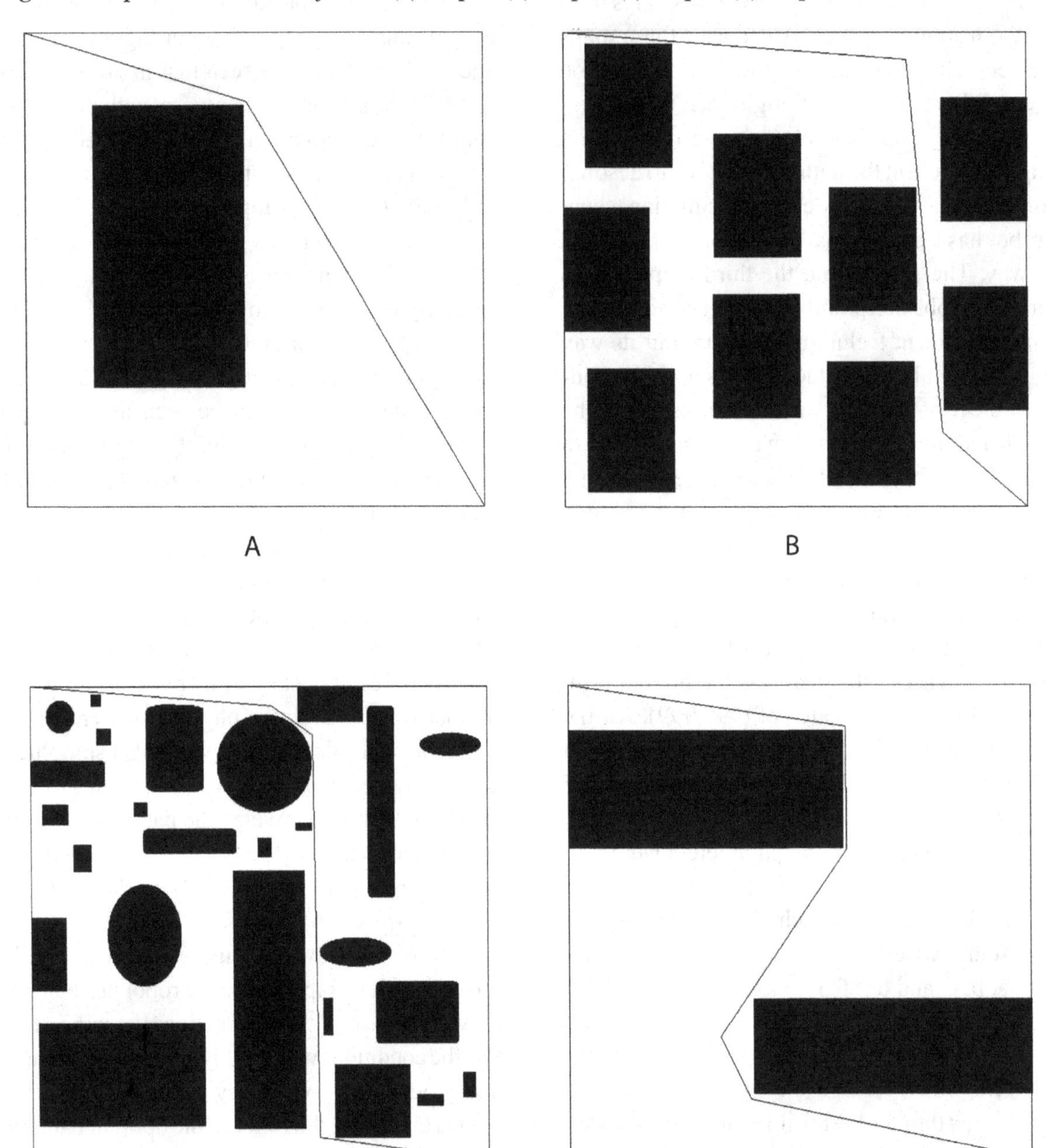

a predefined path. It may be easily seen that the robot happened to optimize the path length in this case. This is because the straight path penalty had been switched off by setting a weight age of zero.

We also study the behavior of the best fitness value in all the discussed paths and configurations. The plots of the best fitness against the generations are given in Figure 5a, b, c, and d. Here we do

Figure 5. Graph showing best fitness v/s generation for the EA for (a) map 1, (b) map 2, (c) map 3, (d) map 4

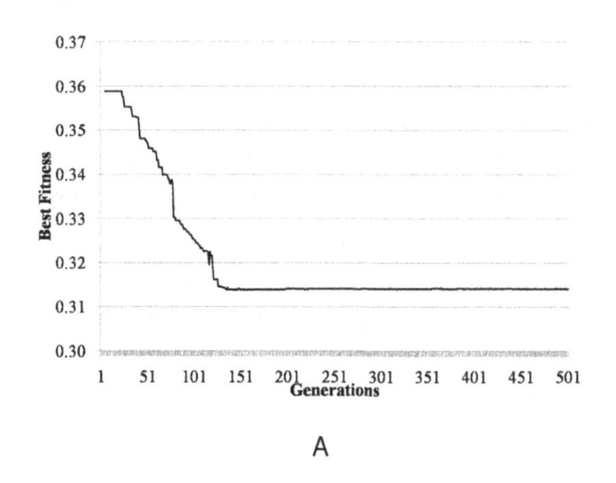

A

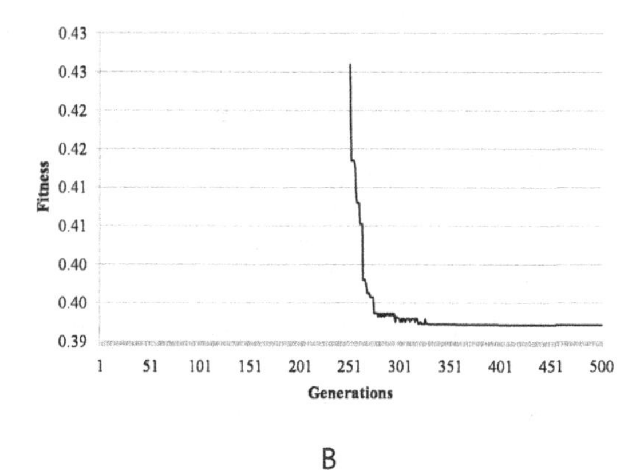

B

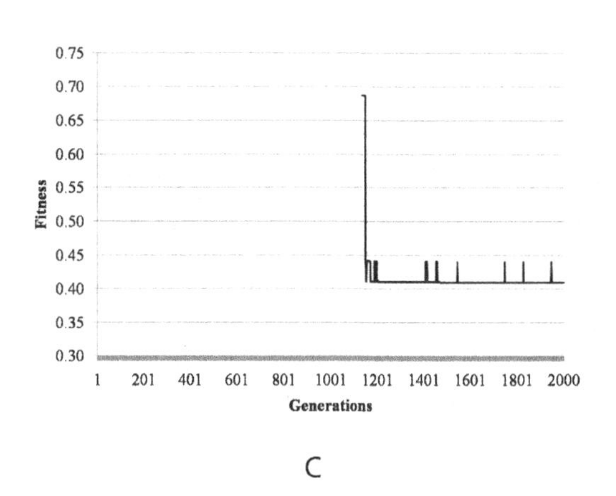

C

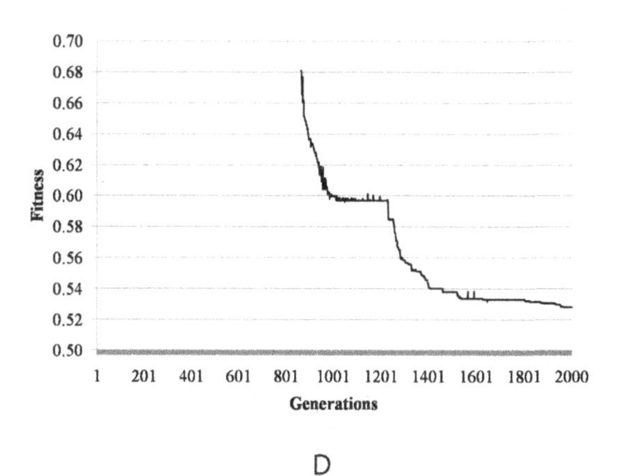

D

not plot the regions in the graph where the best fitness value was infinity. It may be noted that the fitness value would always be infinity until the time the algorithm acquired the minimum complexity needed to solve the problem. As the algorithm proceeds, the best value generally keeps improving.

In all 4 maps, it can be seen that the fitness values improves with generations after the needed complexity is met. The graph given in Figure 5a represents the algorithmic convergence where the decrease in value is very sharp at start and later on the fitness value converges to the most minimum value. A similar trend is observed in Figure 5b with a very sharp convergence. This may be due to the limited search space in the complexity in which the solution is found. Figure 5c represents a very interesting graph where the best fitness value very sharply decreases and then oscillates. The rise in fitness value represents the approximate state condition discussed above where the increase in momentum reveals infeasibility of a path. As similar paths are available in the close vicinity, the next higher fitness value is close enough. The soft mutation re-decreases the value of the fitness function.

Figure 5d is another very interesting graph. Here the fitness value decreases and converges and later re-decreases and re-converges. This is due to the fact that when the algorithm was at a complexity of 2, the algorithm optimized and converged. This continued until a complexity of 3 as well. When the algorithm entered a complexity of 4, the better paths were found. As a result, the value re-decreased and re-converged. The effect of approximate state condition can also be seen in this graph.

Another important factor that we would like to study is the total number of feasible solutions at every generation. This would give us an idea of how many individuals are actually contributing towards the search of the most optimal path or solution. The graph between the numbers of feasible solutions for all the 4 maps discussed is plotted in Figure 6a, b, c, and d.

The graph shown in Figure 6a represents a very expected Gaussian behavior that we had introduced in the total number of individuals. Figure 6b has no feasible solution until the minimal complexity is achieved. Afterwards, the expected Gaussian increase can be observed. However, we see a large oscillatory behavior here.

Figure 6. Graph showing number of feasible solutions v/s generation for the EA for (a) map 1, (b) map 2, (c) map 3, (d) map 4

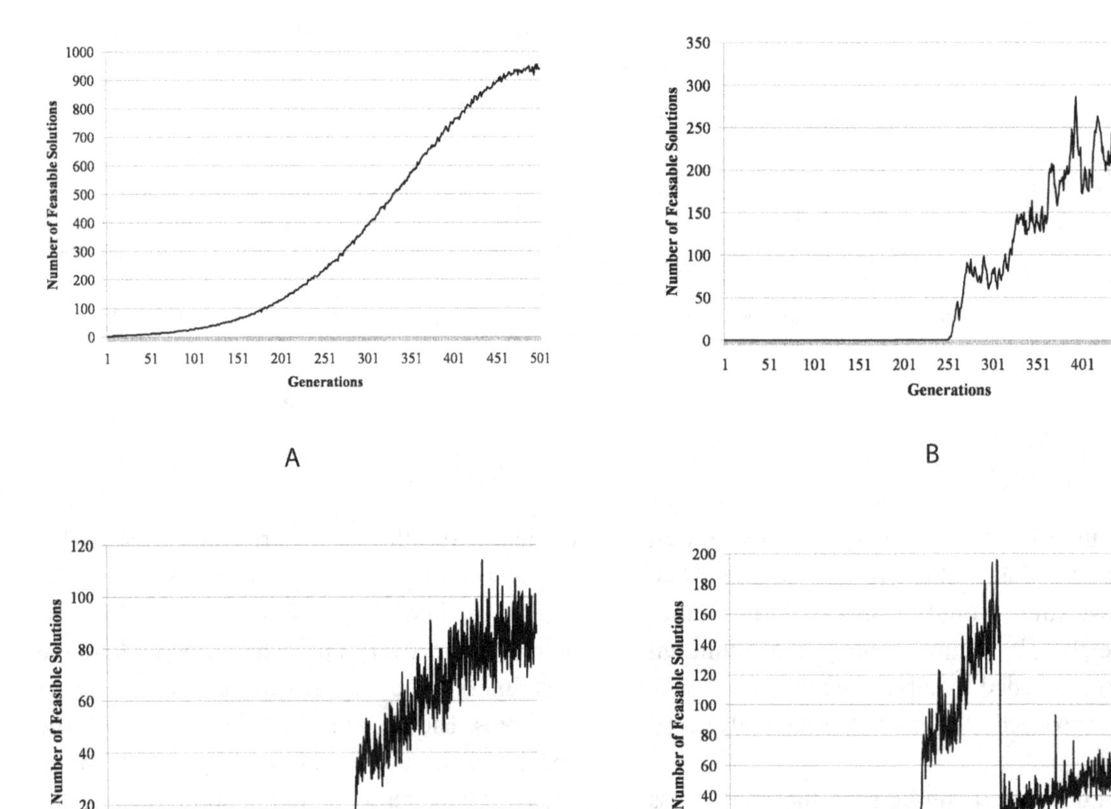

These oscillations depict stochastic nature of the algorithm, and hence, the general trend is of more value. Recall that the hard mutation and insert genetic operator were random in their functionality. The increase in randomness is again attributed due to the decreasing momentum values, which reveal the infeasibility of many individuals that were otherwise feasible in lower generations.

This trend is magnified when we look at Figure 6c. This figure reveals vital information that the total number of feasible solutions is very less as compared to the total number of solutions in the population pool. The reason for the same is the map on which the algorithm is applied. Even a small change in the coordinate values can make a good solution infeasible. The higher complexity or higher generation regions of the graph make the path more flexible, which adds to the possibility of infeasibility. This is the reason why the oscillation is of even larger magnitude. Similar discussion may be done in Figure 6d as well. Here the two levels denote the convergence of the algorithm at a complexity of 2 and 4 as discussed in the earlier paragraph. As the complexity 4 possibility arrives,

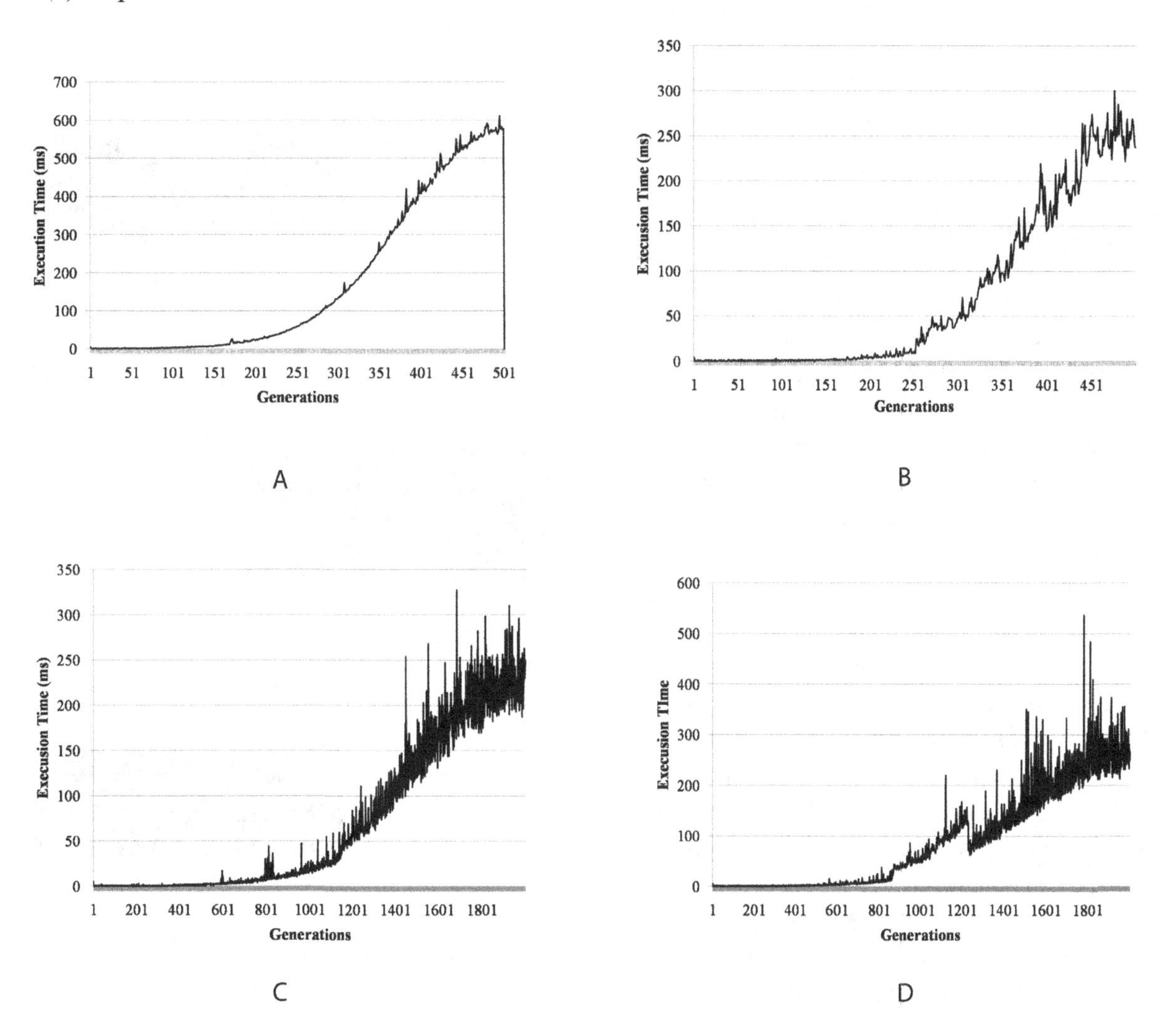

Figure 7. Graph showing time of execution v/s generation for the EA for (a) map 1, (b) map 2, (c) map 3, (d) map 4

many individuals migrate to a higher complexity region. In this migration, they happen to lose their feasibility due to the random nature of insert operator. This makes the number of feasible solutions very low. As the algorithm continues, they regain their feasibility by the repair operation.

The last parameter of study is the execution time for various generations. The execution time per generation is plotted against generations in Figure 7a, b, c, and d. As the generations increase, the rise of time may be attributed due to the increase in number of individuals, increase in number of feasible solutions as well as decrease in momentum. Generally, infeasible solutions take lesser time as compared to feasible solutions. This is because we stop the traversal as soon as we meet any obstacle on our way. The effort of rest of the journey is saved. The oscillatory nature may again be attributed to the variation in number of feasible solutions.

Tradeoff between the Parameters of Multi-Objective Optimizations

In the fitness function, we had introduced the parameters α, β, and γ for path length (l), distance from obstacle (d), and path complexity (c_{max}). During our discussion as well as by experimentation, we observed that it was very easy to work over with the distance from obstacle factor which could be easily managed in most of the non-crowded maps. Mostly this factor retained one of its lowest values during experimentation.

The other two factors were a good point of discussion while we were discussing the algorithm. We stated that sometimes we may wish to traverse through high complexity areas compromising with the complexity and speed but keeping the length optimal. Many times, we would like to compromise with the path length and keep the speed high with low path complexity. We experimentally verified the same where the choice was controlled by varying the multi-objective parameters. Consider the graph used in fourth case. We present two paths traced by the robot in Figure 8a and b. Figure 8a shows the graph with the values of α, β, and γ as 0.5, 0.5, and 0. Here, we can see that the shortest

Figure 8. (a) The path traced by the robot with preference to shortest path, (b) the path traced by the robot with preference to least complex path

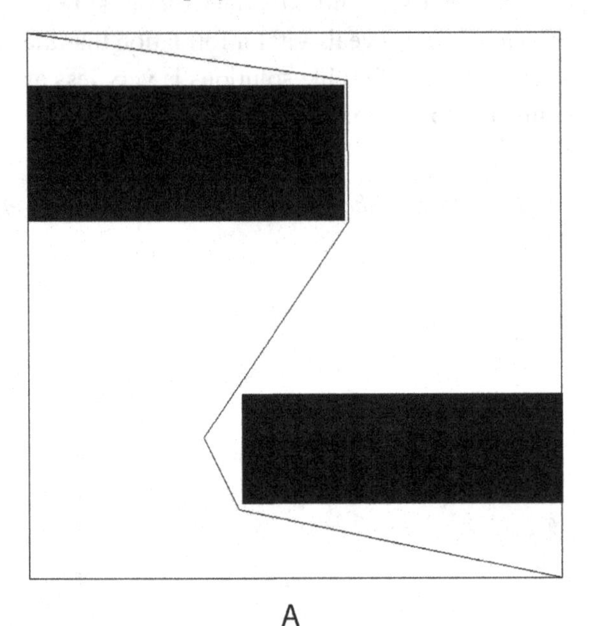

A

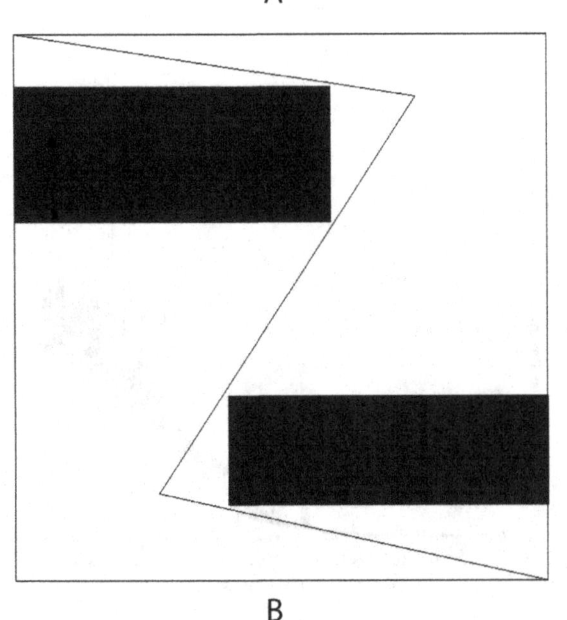

B

path was followed. Figure 8b shows the path for the values of α, β, and γ as 0.33, 0.33, and 0.33. Here, the straight path was preferred.

Figure 9. (a) The path traced by the robot with preference to shortest path, (b) the path traced by the robot with preference to least complex path

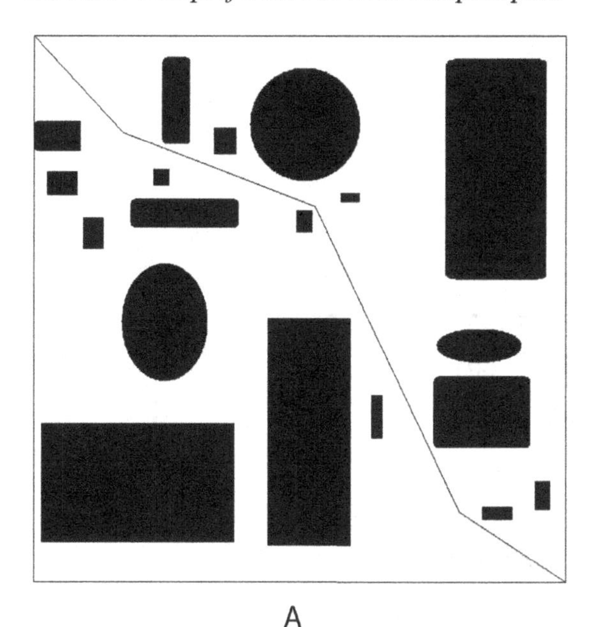

A

B

Another similar case was found with the same set of multi-objective optimization parameter values in map given in Figure 9a and b. In Figure 9a, a straight path was preferred while in Figure 9b, a smaller complexity path was preferred.

Effect of Momentum

The last set of discussion that needs experimental verification is the effect of momentum. We discussed that increased momentum would lead to shorter computational time at the same time may result in reporting many infeasible solutions as feasible. We also stated that large momentum would wrongly quote the value of the parameter distance to obstacle (d). We study the effect of momentum on each of these domains by applying variable momentum over a set of randomly generated solutions. A total of 1000 paths of a complexity of 3 were generated for the map 2. The momentum was varied from 1 to 1000 and all the parameters were measured. The relation of momentum with time of execution is given in Figure 10. Figure 11 gives the relation between the momentum and the predicting capability of the fitness function regarding the feasibility of the solution. Figure 12 shows the relation between the momentum and mean deviation of the physical minimum distance from robot to obstacle. As per our earlier discussion, all distances greater than 10 are taken to be zero.

An interesting observation is in Figure 12 that reports oscillatory behavior. Here it may be seen that many times larger jumps may lead the algorithm closer to the obstacle. Hence, it cannot be generalized that the factor d would be having large deviations as momentum increases.

HGAPSO WITH MOMENTUM

The evolutionary algorithm presented in the previous section was able to generate good results in finite time duration for the maps presented. The

Figure 10. Relation between time of execution and momentum

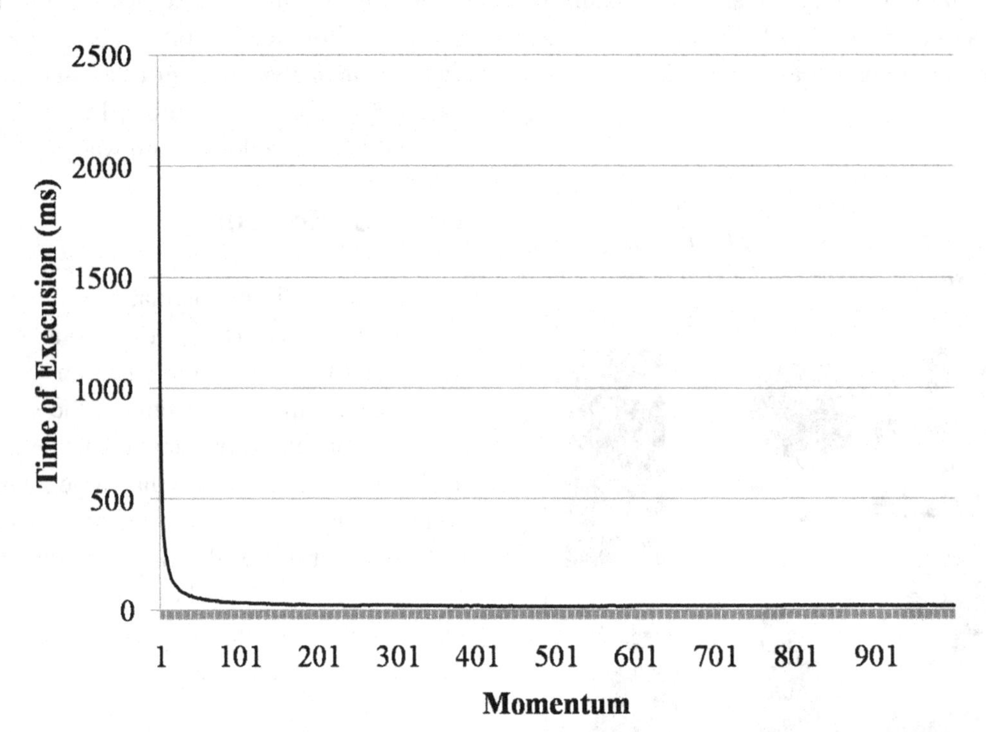

Figure 11. Relation between feasibility prediction and momentum

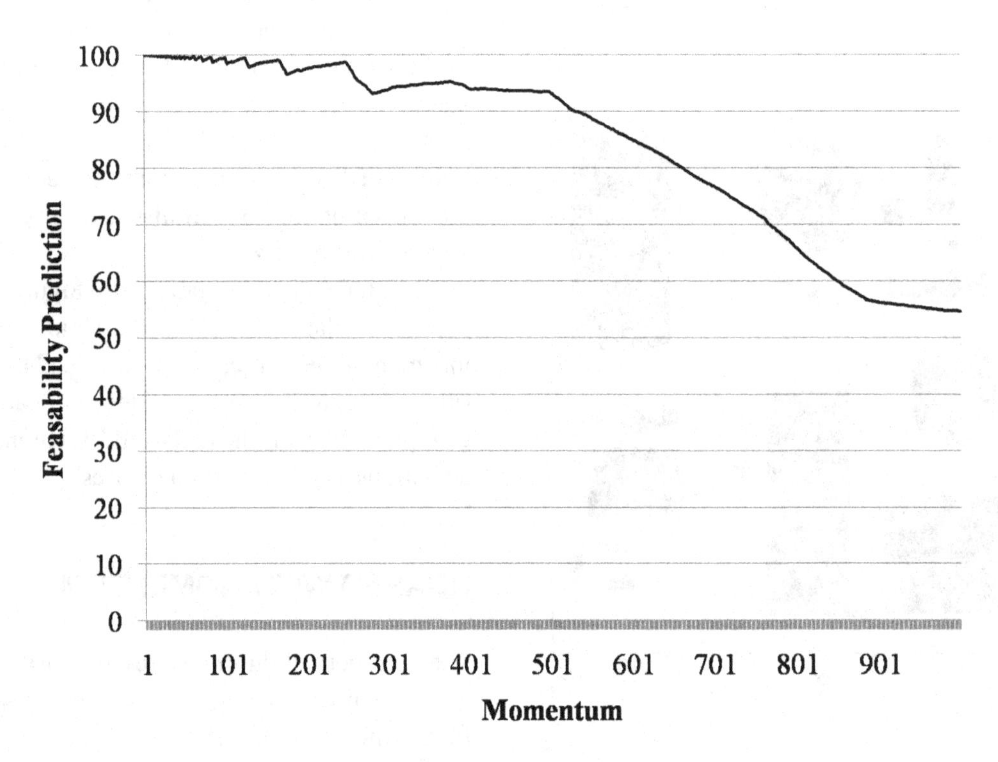

Figure 12. Relation between average minimum physical distance and momentum

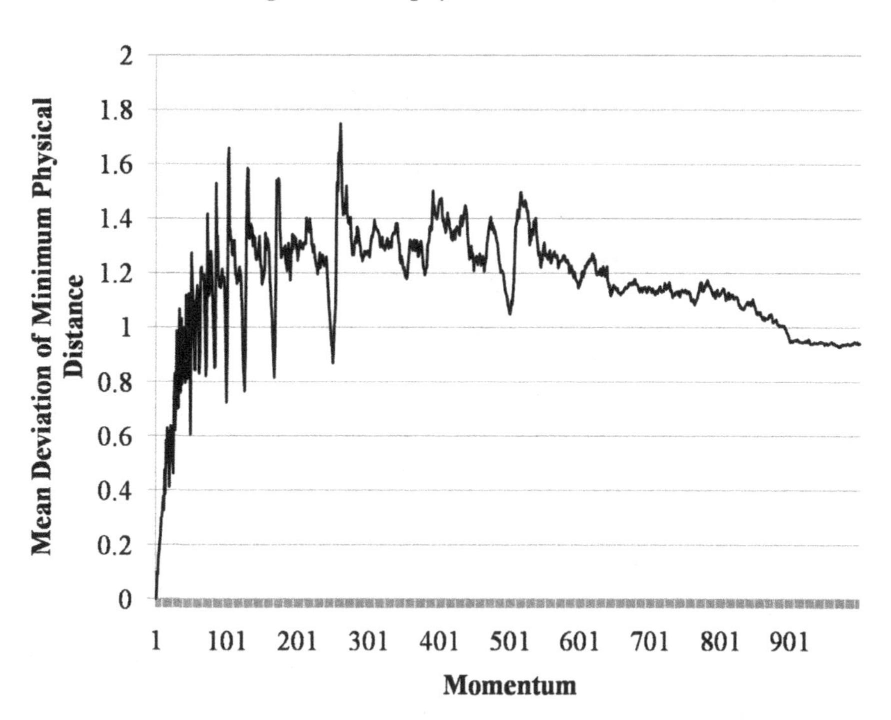

studied limitation of the work however was the possibility of the algorithm to return sub-optimal paths as many good individuals may be observed to be infeasible at higher generations of the algorithm. This is possible with various critical maps in which a blockage of path is only visible at a very fine resolution of the map. In this section, we further extend the algorithm by the use of Hybrid Genetic Algorithm Particle Swarm Optimization (HGAPSO) (Juang, 2004), which is a hybrid of GA and PSO. Diversity preservation is added to both the individual algorithms of GA and PSO separately. This further ensures the generation of paths with better characteristics as defined by the fitness function. For details of application of PSO please refer to chapter 5, which is a requirement for further discussion. This section and subsequent sections covering the entire algorithm are reprinted with few modifications from Kala, Shukla, and Tiwari (2012) with permission from Inderscience Publishers.

The basic algorithm used for the problem solving is HGAPSO. The feasible solutions out of the entire solution pool are divided into two halves. The fitter half participates in the evolutionary process. First PSO is applied. The generated individuals go directly into the next generation. The same individuals are also worked upon by GA that adds to the next generation population. The infeasible solutions are not allowed to participate. Other operators are applied for the production of rest of the solutions for next generations. The overall algorithm is given by Figure 13.

The Genetic Algorithm uses the operators of crossover and mutation for the generating the next generation individuals. The crossover operation of the algorithm assures that the diversity is maintained. This is done by selecting only those individuals that are separated by a distance smaller than threshold distance (η). Mutation operator used simply re-locates the points of the individual by moving them by small amounts within the map.

Figure 13. The hybrid genetic algorithm particle swarm optimization (HGAPSO) algorithm for solving path planning problem

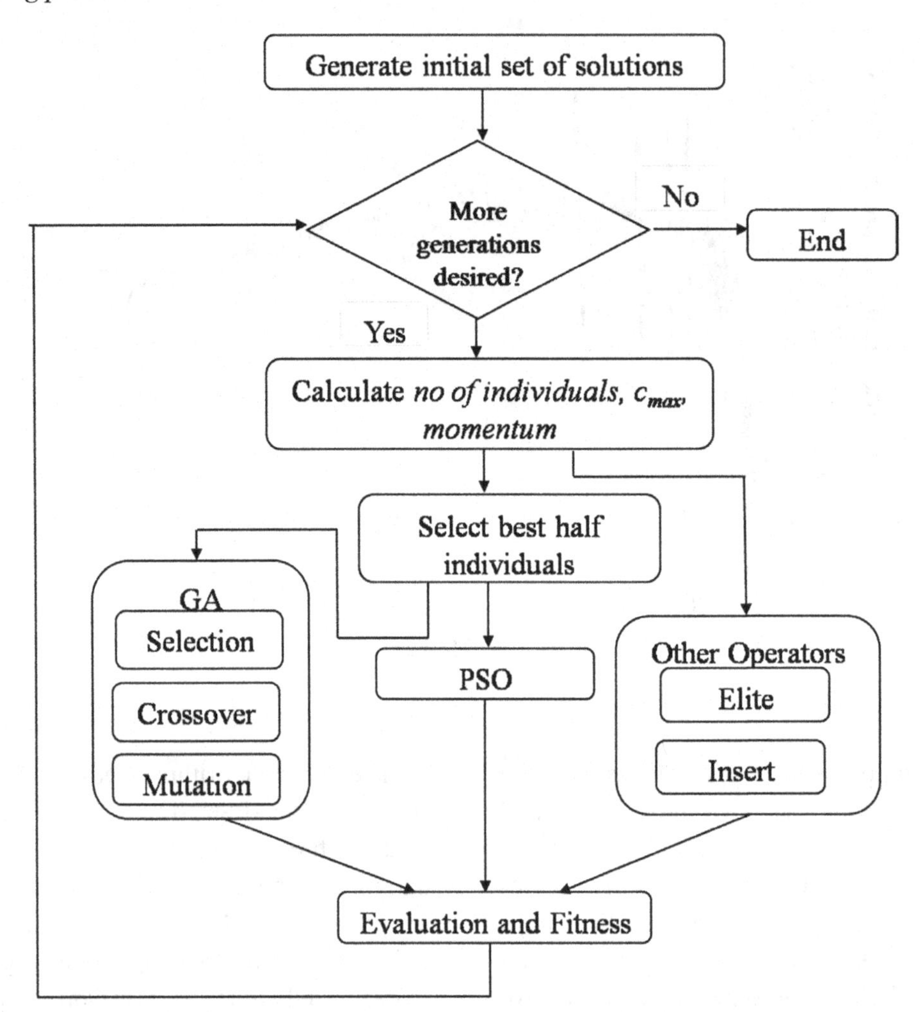

The PSO optimization technique generates solutions by moving a particle in the input space. The movement of the particle is based upon its past experience, best available solution, and the experience of the neighbors. Here also we only consider the non-null (infinity) positions in the individual representation while modifying the individual position or velocity. An instance of Fitness Distance Ratio-Based PSO (FDR-PSO) is used for diversity preservation (Peram, et al., 2003; Veeramachaneni, et al., 2003). In this algorithm, the modification of every particle has an added term that checks for the performance of the neighboring particles. Using this algorithm, the velocity v^i_d update for d^{th} dimension for any particle i with position x^i_d is done by Equation 11. The update on position is given by the conventional equation.

$$v^i_d(t+1)=w\ v^i_d(t)+c_1.rand.(p^i_d-x^i_d)+c_2.rand.(p^g_d-x^i_d)+c_3.rand.(p^n_d-x^i_d) \quad (11)$$

Here w is the inertial factor, c_1, c_2, and c_3 are PSO constants, rand is a random number, p^i_d is the best solution in the path of particle, p^g_d is the globally known best path, p^n_d is the position of best neighboring particle j that maximizes term given in Equation 12.

$$\frac{fitness(p_j) - fitness(x_i)}{\left\| p_d^j - x_d^i \right\|} \tag{12}$$

The other operators used in the algorithm include elite and insert, which have the same manner of implementation as discussed in the previous section.

RESULTS

The algorithm was tested over four benchmark maps that varied from each other. All the different kinds of maps used are given in Figure 14. In all the cases, the maps were of size 1000 x 1000. The coordinate axis of the map had (0,0) point at the top left. This was the source specified for all cases. The goal was the bottom right point with the coordinates of (999,999).

For all experiments the parameters were:

- α, β, and $\gamma = 0.33$
- a, b, c, and d = 0.3.
- c_1, c_2, and $c_3 = 1$.
- Momentum = 10 to 1.5.
- Mutation rate = 0.04
- Crossover rate = 0.6.
- Elite count = 5%
- $\eta = 0.03$.
- M = 10.
- $c_{max} = 0$ to 5.
- Number of individuals = 1 to 1000.
- Number of generations = 500

The maps and the path traced by the robot for each of these cases are given in Figure 14a-d. In all figures, the path denoted by straight line denotes the experimentally computed path as per discussed parameters. The path with a circular legend at corners shows the computed path with no complexity control (α, β =0.5, γ =0). Trends are similar to those observed while experimenting with evolutionary algorithm.

We further study the best fitness value in all the discussed paths and configurations. These for all the graphs are given in Figure 15. Closely observing any of the maps, we would be able to observe small oscillations in the fitness value while the algorithm proceeds. In other words, the best fitness value increases even after the application of elite genetic operator. This is attributed to the presence of approximate state.

We further plot the total number of feasible solutions at every generation. This would give us an idea of how many individuals actually contribute towards the search of the most optimal path or solution. This is shows in Figure 16. Here also oscillatory behavior may be seen which is mainly due to the occurrence of approximate state. In addition, the randomly generated individuals may be infeasible.

We also study the execution time requirement between the generations. The execution time per generation is plotted against generations in Figure 17. As the generations increase the rise of time may be attributed due to the increase in number of individuals, increase in number of feasible solutions as well as decrease in momentum. Generally, infeasible solutions take lesser time as compared to feasible solutions. This is because we stop the traversal as soon as we meet any obstacle on our way. The effort of rest of the journey is saved. The oscillatory nature may again be attributed due to the variation in number of feasible solutions.

We earlier stated that an increase in momentum might lead to an incorrect recording of the path feasibility as well as the distance to obstacle (d). We try to study the same concept here. A total of 10000 paths of a complexity of 3 were generated for the map 2. The momentum was varied from 1 to 1000 and all the parameters were measured.

Figure 14. Path traced by robot with and without complexity control objective function for (a) map 1, (b) map 2, (c) map 3, (d) map 4

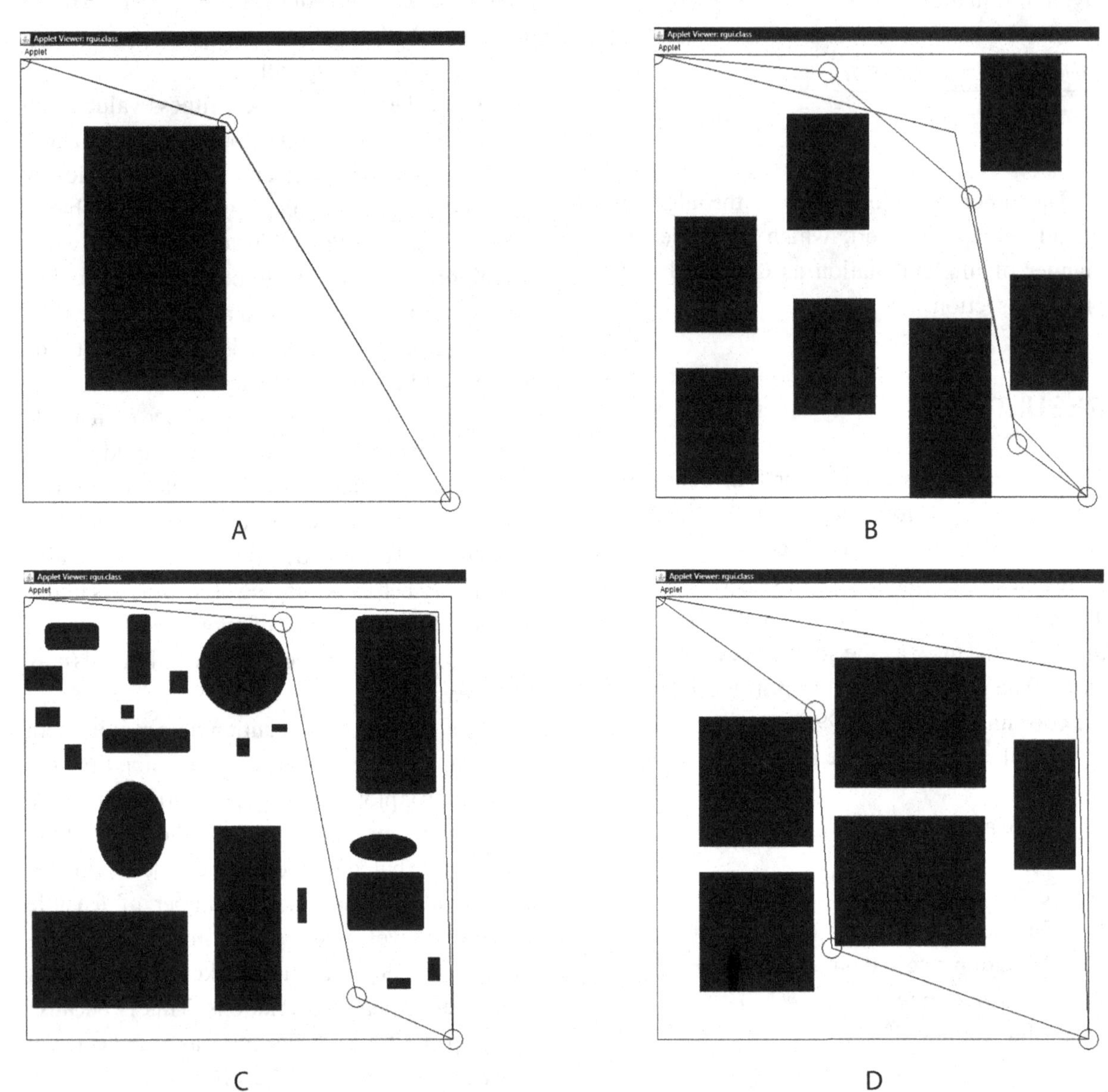

The relation of momentum with time of execution is given in Figure 18. Figure 19 gives the relation between the momentum and the predicting capability of the fitness function regarding the feasibility of the solution. Figure 20 shows the relation between the momentum and mean deviation (or error in measuring) of the minimum distance from robot to obstacle (d).

An interesting observation is in Figure 20 that reports oscillatory behavior. Here, it may be seen that many times larger jumps may lead the algorithm closer to the obstacle. Hence, it cannot be generalized that the factor d would be having large errors as momentum increases.

Figure 15. Graph of fitness v/s generations showing convergence in the four maps

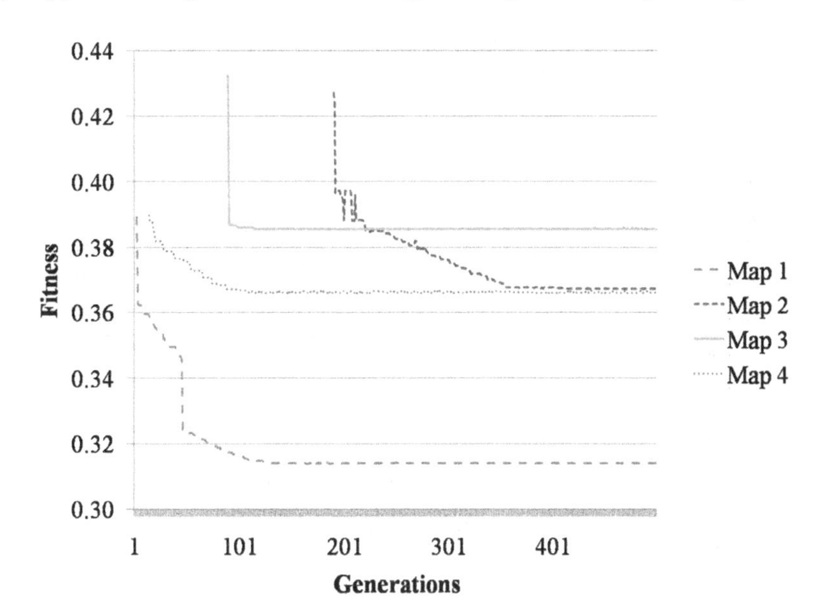

HIERARCHICAL EVOLUTIONARY ALGORITHM: HIERARCHY 1

The next approach in this chapter is planning for the robot using hierarchical evolutionary algorithm. Here there are two levels of planning: coarser level and finer level. The coarser level is discussed here, whereas the finer level would be discussed in next section. This section and the subsequent sections covering the entire algorithm are reprinted with few modifications from Kala, Shukla, and Tiwari (2010) reprinted by permission of Taylor and Francis Ltd. (http://www.tandfonline.com).

Figure 16. Graph of total number of feasible solutions in population pool v/s generations for four maps

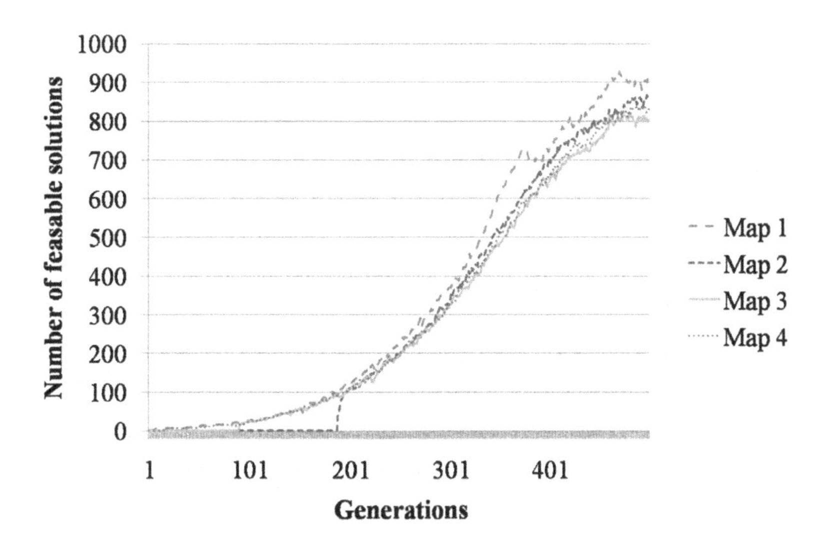

Figure 17. Graph of time of execution per generation v/s generations for four maps

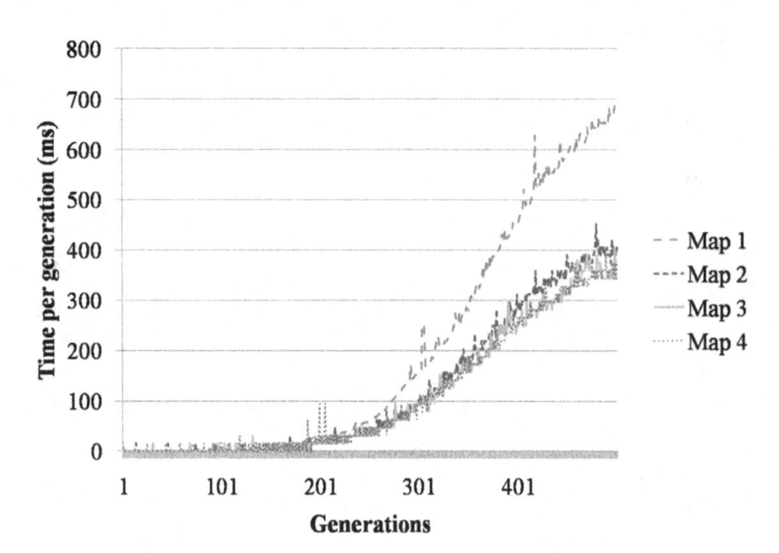

Using hierarchical evolutionary algorithm for generation of a robotic path makes a strategy where two different evolutionary algorithms work at different resolutions. Such a technique is largely used in hierarchical evolutionary optimization or memetic algorithms. These algorithms use two or more optimization algorithm for the problem considered. One algorithm is given the task of working over placement of individuals in the entire evolutionary space, while the other may be a local search algorithm, which shifts the individuals towards nearest optima. The combination of a local search technique to an evolutionary algorithm works well in exploring the landscape, as well as quickly exploiting or getting into minima. This is similar to our algorithm wherein global or coarser

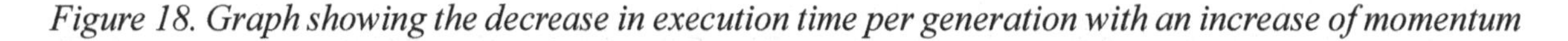

Figure 18. Graph showing the decrease in execution time per generation with an increase of momentum

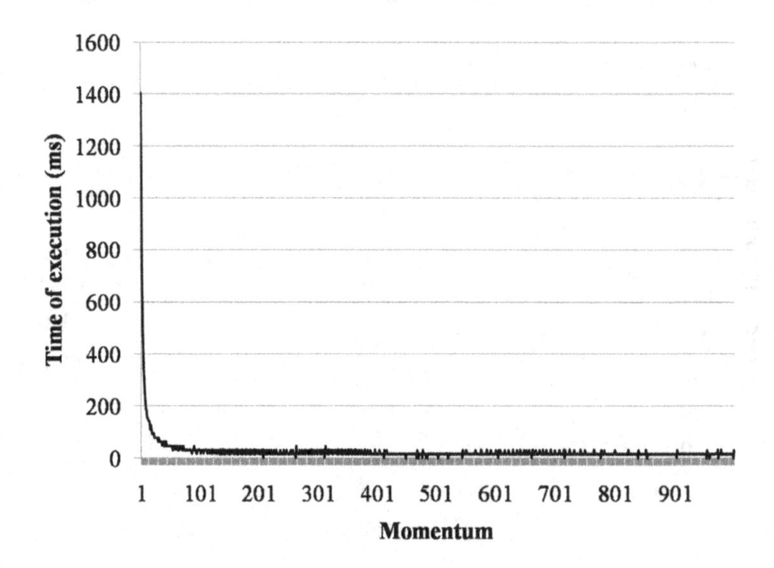

Figure 19. Graph showing the decrease in correctness of feasibility prediction with an increase of momentum

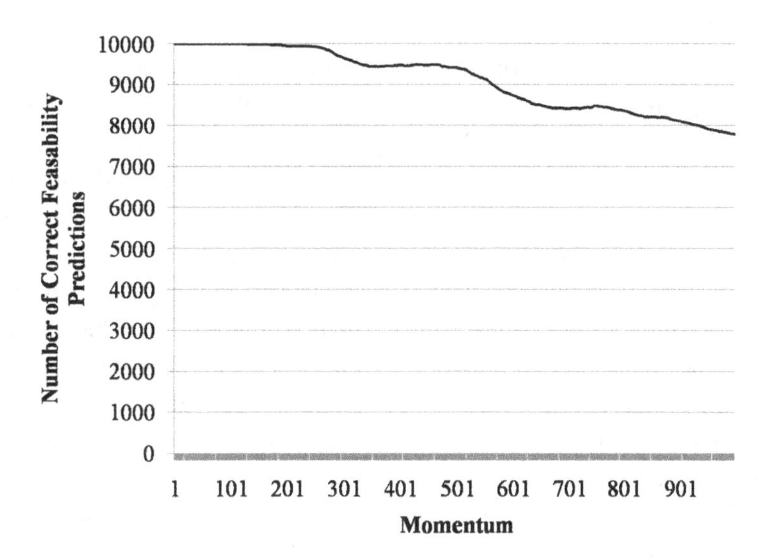

level constructs a rough path, and a finer level works for computing the precise path whose part path lies at the local optima. If the coarser path is optimal, it would lie at the global optima.

Another analogy may be drawn from classification. The task is to classify an input into one of the output classes. Hierarchical classifiers are widely used for classification of complex classes. Here at the global level, a classifier first gives a general idea of the classes to which the input may belong to. Based on the output of the global level classifier, a local level classifier is

Figure 20. Graph showing correctness in prediction of distance to closest obstacle (d) for various values of momentum

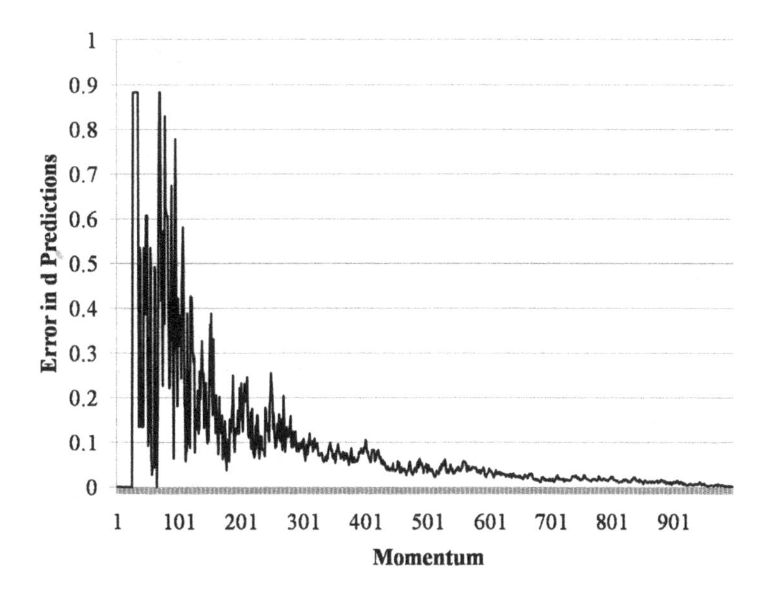

then invoked. This classifier outputs the precise class to which the input belongs. The global level classifier sees the classificatory space at a coarser level and is mainly responsible for selecting the local classifier. The local classifier has a view of only a part of the classificatory space, but it has a finer view of data.

The first planning is done at a coarser level. Here the map consists of only static obstacles. This requires a classification of all the obstacles into static obstacles and dynamic obstacles. In practice, this task may be easily carried out by scanning the environment in few successive times. This planning considers only static obstacles. The dynamic or moving obstacles are neglected. Then a higher resolution map of this environment is made. The higher resolution map may be too difficult for the evolutionary algorithm to solve. Hence, we first reduce the map resolution to make it computationally feasible for the evolutionary algorithm to evolve the robotic path. Each of the steps is discussed in details in the coming sections.

Resolution Reduction

The original map is taken as a grid of size M x N that is given to the algorithm to solve. Here M and N are usually reasonably large numbers. This makes computing the optimal path with evolutionary approach very difficult due to the vast nature of the evolutionary search space. Each cell c_{ij} in this map denotes 1 or 0 depending upon the presence or absence of obstacle. Any cell c_{ij} in the map may be given by (13):

$$c_{ij} = \begin{cases} 0 & \text{if no obstacle exists at location (i,j) of map} \\ 1 & \text{if an obstacle exists at location (i,j) of map} \end{cases} \quad (13)$$

Here $0 \leq i \leq M,\ 0 \leq j \leq N$

The map resolution is reduced by a factor of α. This means that the resultant map has dimensions ceil(x/ α) x ceil(y/α), where ceil(x) is the greatest integer greater than x function. In other words, a block of size α x α of the original high-resolution map makes up a unit cell of the reduced resolution map. Consider α to have a value of 5. The original and the reduced resolution map are given in Figure 21a.

The value of any cell of this reduced resolution map is aggregated value of all the cells in the block of the higher resolution map. This aggregation produces values between 0 and 1 corresponding to the cell of the lower resolution. The resultant map is a gray map as given in Figure 21b. Here the aggregated value denotes the shade of gray with 1 denoting complete black and 0 denoting complete white. Let d_{kl} be any aggregated cell of the lower resolution map. This is given by Equation 14:

$$d_{kl} = \sum_i \sum_j c_{ij} \quad (14)$$

For working, we need to convert this map into a binary map. All cells above a threshold value are assumed to be 1 and all others are assumed to be 0. This gives us a black and white lower resolution map shown in Figure 21c. Any cell e_{kl} of this map is given by Equation 15:

$$e_{kj} = \begin{cases} 1 & d_{kl} > Th \\ 0 & d_{kl} < Th \end{cases} \quad (15)$$

Th represents the threshold that may be set to any convenient value.

Evolutionary Algorithm

The task of the evolutionary approach is to work over the coarser map and find a feasible and optimal path using which the robot may be able

Figure 21. (a) Original map, (b) gray map, (c) binary low resolution map

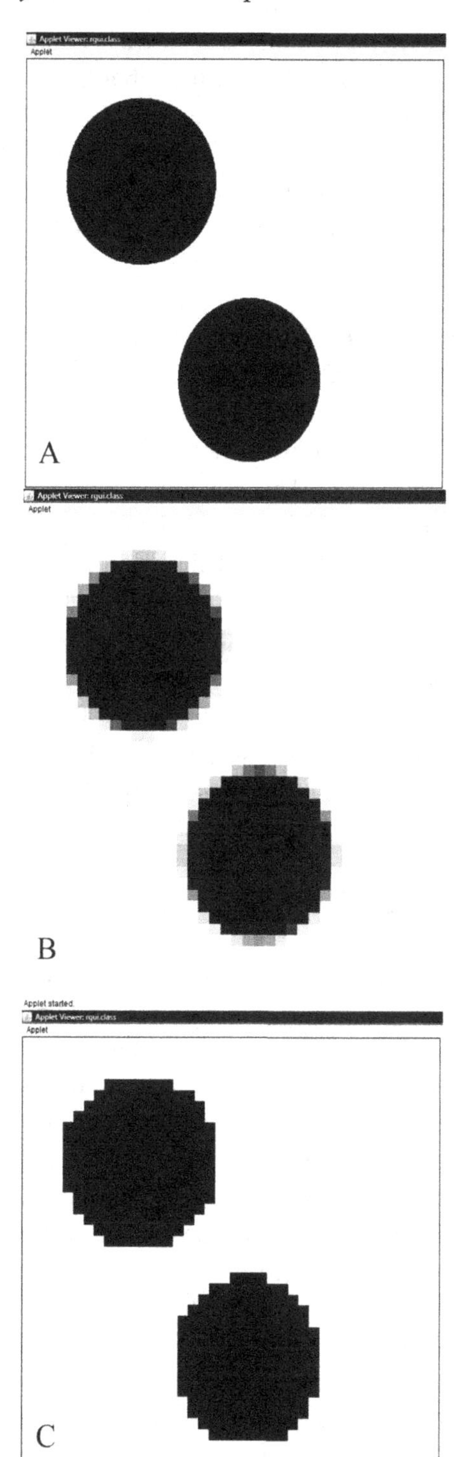

to reach the destination from the source. Here we use the evolutionary operators to construct a higher generation population from a lower level population. The fitness of the population keeps improving along with time and hence the latter solutions are optimal.

The design of evolutionary algorithm is similar to one presented in earlier section. The individual of the EA is a collection of points of the form $<P_1, P_2, 3, \ldots P_n>$. The first point is the source and the last point is always the destination. Further each point P_i is a combination of x and y coordinates and may be represented by (x_i, y_i). The number of points n is variable.

There are a total of 2n points in the EA individual representation. We have a variable size genetic individual. The fitness of any individual is the combination of two factors. These are total path length and complexity. The maximum complexity and number of individuals are made to increase with time. The EA uses a total of 7 operators for the generation of the higher generation from the lower. These are Selection, Crossover, Soft Mutation, Hard Mutation, Elite, Insert, and Repair.

Iterating with Time

The unique feature of this evolutionary technique is that the algorithm runs while the robot is moving. The algorithm keeps trying to find more and more better paths as the robot continues walking. The present position of the robot is the source. Hence, the source of the EA is variable in nature that changes along the evolutionary process. Whenever the robot makes any move, all the individuals of this EA are updated. Since the source was not present in the EA individual representation, it may not be needed to update the source.

We check if the projection of the line joining the present position of the robot or the source and the first point of the individual representation is positive or negative on the line joining the original source and the goal. In case the slope is negative, the first point is deleted. This signifies the robot has crossed the first point in the course of its journey. Let the path of the coarser hierarchy at any point of time be as given in Figure 22a. Further let the robot be at any location A shown in Figure 22b. To find whether the robot has crossed the point P (or Q) represented in the coarser path, we find the slope based on angle θ_1 (or θ_2). If this is positive, the robot is yet to cross the point, else the robot has crossed the point. This is shown in Figure 22b. It may be easily seen that the robot has crossed point Q, but is yet to cross point P.

This complete process is then repeated with the new point that emerges as the first point. This process is repeated for all the individuals.

Figure 22. (a) Path computed by coarser evolutionary algorithm, (b) deletion of crossed points

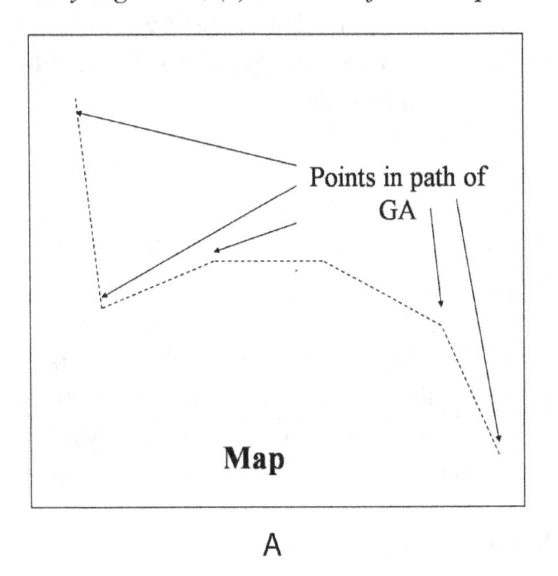

A

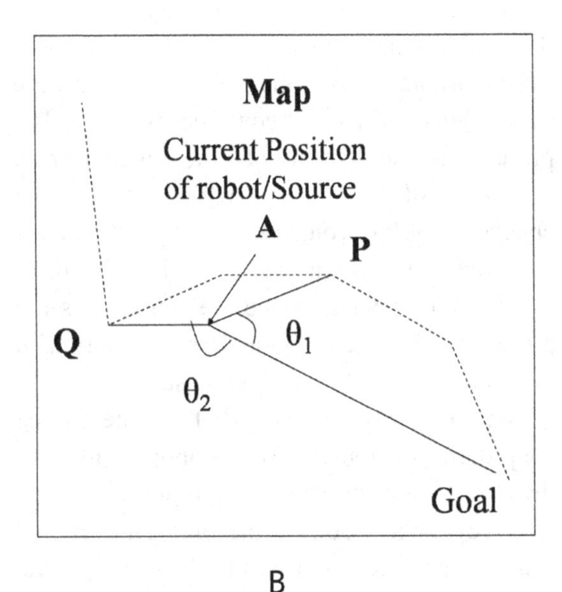

B

HIERARCHICAL EVOLUTIONARY ALGORITHM: HIERARCHY 2

The other level at which the planning is done is at the finer level. This contains the actual high-resolution map. This level of planner is not given the complete map. It is rather given a small part of the map around the area across the present location of the robot. Hence, a part of the entire map is cut for this optimization. The evolutionary algorithm is used to decide every move of the robot. The robot keeps moving as guided by the EA till the goal is reached. The various stages of the algorithm are discussed in the following sub-sections.

Map Segmentation

The first task is to segment the map. Here, we only give a small section of the entire map to the planner for figuring out the optimal path. The section of extracted map depends upon the current position of the robot. The extracted map is a high-resolution map built around the robot. The current position of the robot becomes the source and the first position outside the map in the path returned by the coarser EA becomes the goal. While segmenting we have to cut a square of size $\gamma \times \gamma$. The placement of this square is such that the robot is always at a distance of 0.25γ from two sides of the square and $0.75\ \gamma$ from the other two sides of the square. The direction of the square is towards the goal. The segmentation of map is shown in Figure 23a-d for the various possibili-

Figure 23. Map segmentation (a) case 1, (b) case 2, (c) case 3, (d) case 4

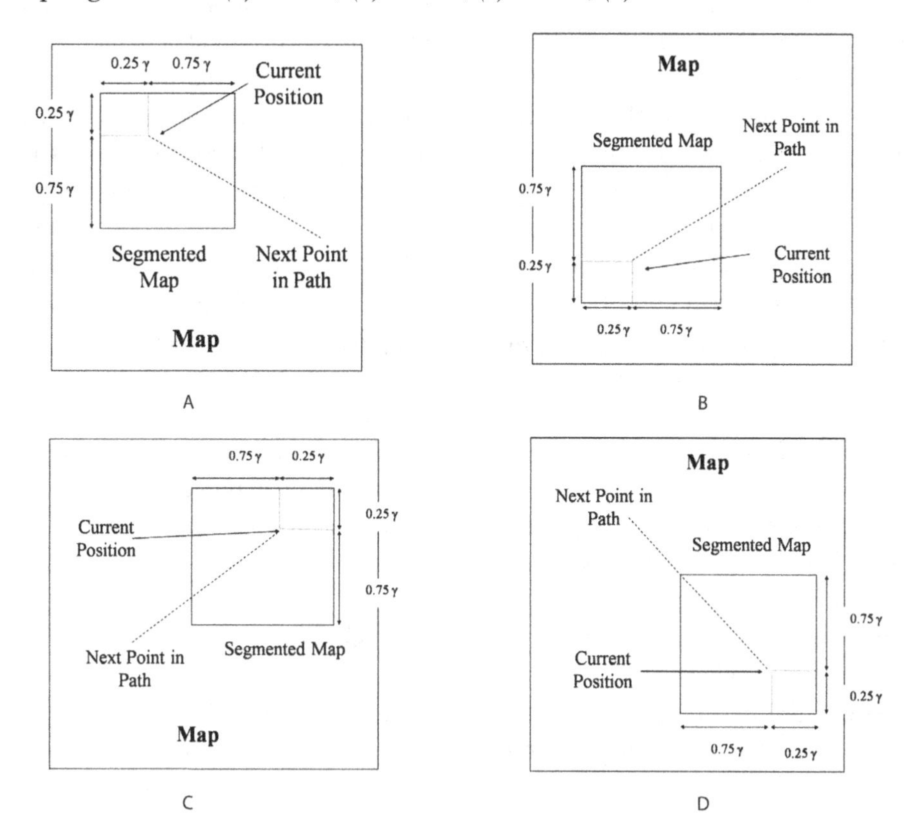

ties. The selection of source and goal position is showed in the same figure.

Handling Dynamic Obstacles

A key aspect of the algorithm is that it needs to handle dynamically moving obstacles. We assume that the robot knows the position and the velocities of all the obstacles in the surroundings. This may be easily scanned by the robot. The planner, while planning the next move, assumes that these obstacles continue moving with the same velocity. This may naturally not be true in practice, but the robot only needs to plan a unit move extrapolating the present observations of position and velocity. This enables robot, not to reach a position, which may be occupied by some other robot at a later stage of time.

Consider that an obstacle is moving at a velocity v whose projections at the x- and y-axis is v_x and v_y. Let the current position of the obstacle be (x, y). Hence the position at any time t may be given by (16):

$$(x',y') = (x + v_x t, y + v_y t) \tag{16}$$

Consider the robot moves with a constant speed of r. The robot must plan its motion such that at time t it would not be located at any location such that it collides with obstacle located at (x',y').

Evolutionary Algorithm

The framework of the evolutionary algorithm used in this approach is similar to the Evolutionary Algorithm (EA) of the coarser planner. The individual of the EA is a collection of points of the same form <P_1, P_2, P_3, ... P_n>. The first point is the source, which is always the current robot position and the last point is always the destination. The number of points n is variable. The fitness of any individual is the combination of two factors, total path length, and number of turns or path complexity. The maximum number of points and the number of individuals increase with time and generations. The same 7 operators of are used. These are Selection, Crossover, Soft Mutation, Hard Mutation, Elite, Insert, and Repair.

Passage of Individuals

The finer EA may get reasonably less time for the computations. It is natural that it cannot carry out the complete working of the EA within this time. Hence, every EA gets half individuals from the previous EA run. The other half individuals are generated as per the conventional procedure. Again recall that the source and the goal are not represented in the EA individual. Hence, the individuals in most cases do not require a conversion for being used in the next EA. In addition, fit individuals of one EA run would be fit enough for the other EA run, since there might not be a substantial change in the map.

While passing the points we have to ensure that the robot has not passed some point represented in the individual. We check if the projection of the line joining the present position of the robot or the source and the first point of the individual representation is positive or negative on the line joining the original source and the goal. In case the slope is negative, the first point is deleted. This signifies the robot has crossed the first point in the course of its journey as shown in Figure 22. This complete process is then repeated with the new point that emerges as the first point. This process is repeated for all the individuals.

The complete algorithm may be summarized by Figure 24. The algorithm is an evolutionary approach, where a series of individuals representing solution are taken. As per our previous discussion, there are two different populations maintained for the coarser level and the finer level. The robot speed is fixed and within allowable time, we need to optimize both these populations. Both finer and coarser optimizations are by separate EAs. The role of finer EA is to optimize the current path, which would play a major role in deciding the immediate move of the robot. The coarser EA tries to optimize the global strategy being followed by the robot. In place of performing this step after a series of move of the robot, we prefer to perform this step at every robotic move to keep constant track of the changing environment. The algorithm stops once the robot reaches the final goal.

RESULTS

A number of executions were done with a variety of maps. These had both static and dynamic obstacles. In all these cases, we saw that the robot was easily able to reach the goal position, starting from the specified goal position. The robot path was always optimal in nature. Further, there was no visible collision of the robot in anywhere during its path.

The first experiment was done with a variety of static obstacles. The complete setup of the algorithm was executed. The map was of size 1000 x 1000. The parameters were:

- $\alpha = 20$
- $\gamma = 200$
- Mutation rates for master = 0.06 and 0.25
- Maximum number of individuals (master) = 1000
- Maximum number of individuals (slave) = 100
- Number of generations (slave) = 100 per step
- Multi-objective parameters (master and slave) = 0.75 and 0.25

Figure 24. The complete hierarchical evolutionary planning algorithm

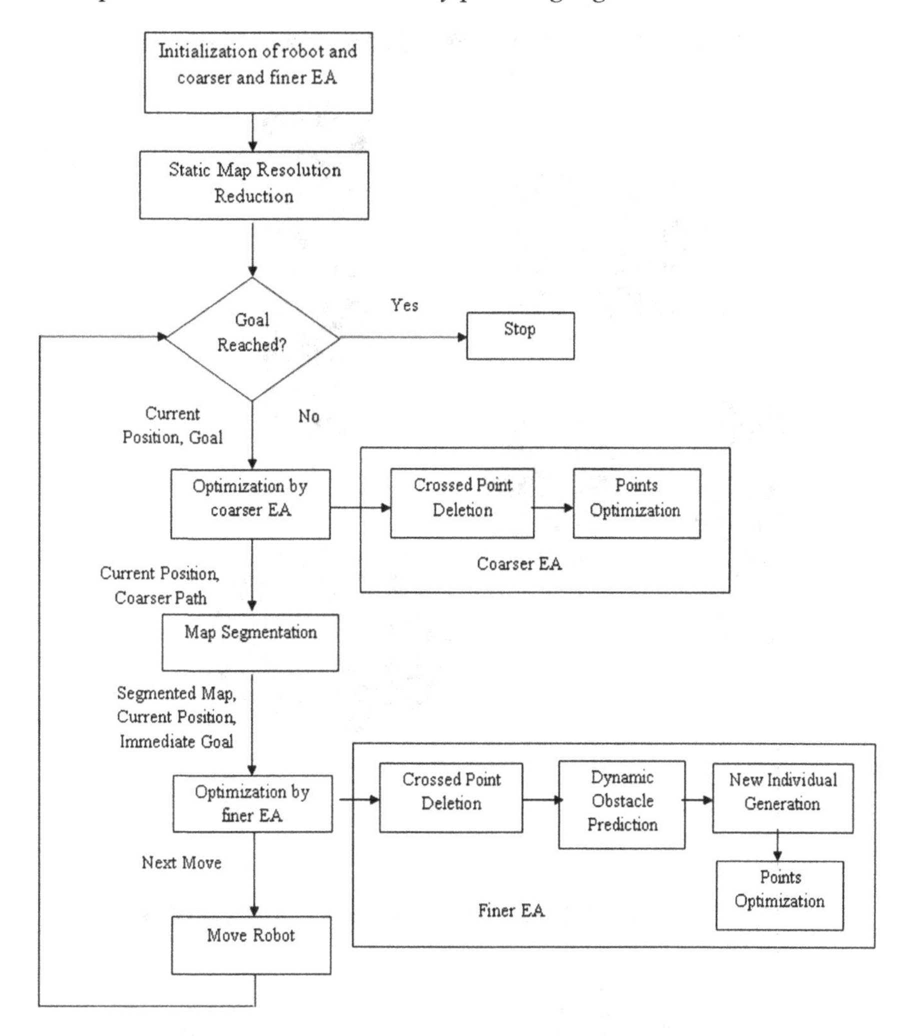

At any generation, 68% individuals came from crossover, 15% from soft mutation, 5% from hard mutation, 2% from elite, and 10% from the operator new. The basic methodology of setting the parameters was that the algorithm execution time for every robotic step must be close to the speed of an average robot. The path traced by the robot is shown in Figure 25. It can be easily seen that the robot was able to steer its way out of all the complex obstacles and reach the goal in an optimal path. This depicts the robot capability to solve complex maps.

The second experiment was done using two static obstacles and two dynamic obstacles. The two static obstacles were circular in structure. The two dynamic obstacles were rectangular in structure and marched towards the goal as the robot made its move. The same set of parameters was used as stated in the previous experiment. The path traced by the robot is given in Figure 26. The two solid rectangles denote the final positions of the dynamic obstacles and the empty rectangles show their initial positions. It can be easily seen that the robot could adjust its movements in such a manner so as to avoid collision with both these obstacles. In the entire path, the robot did not have to wait for path clearance by the obstacles. It made moves such that it could easily cross the obstacles.

Figure 25. The execution of the algorithm in complex map

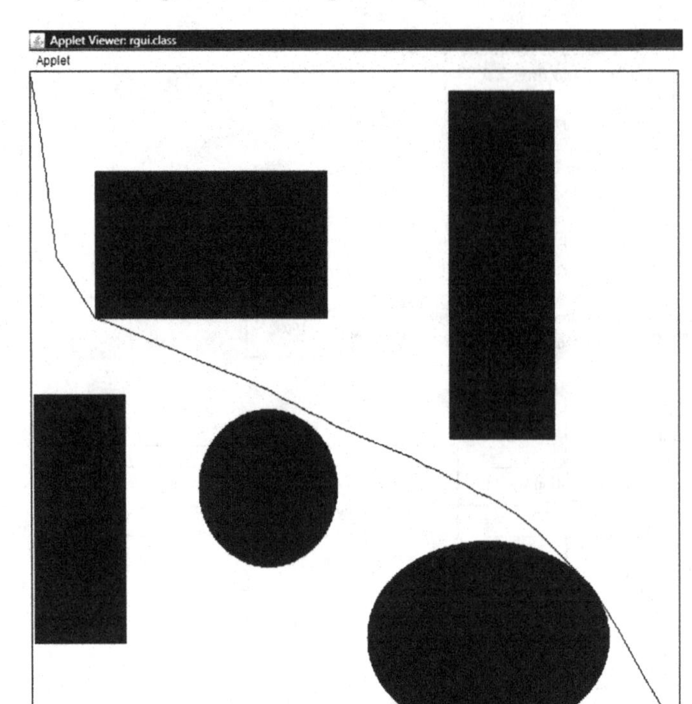

Figure 26. The execution of the algorithm with moving obstacles

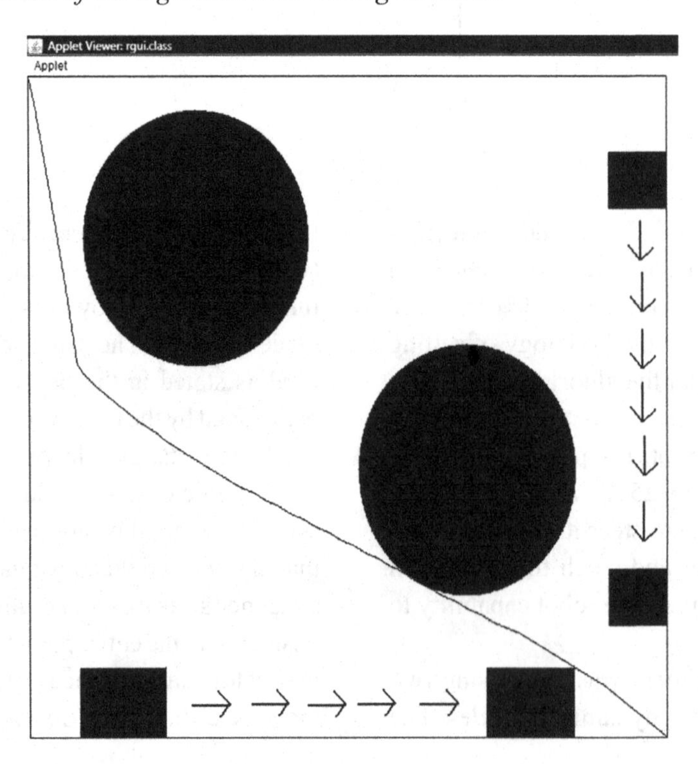

This execution clearly denotes the capability of the robot to escape from static and dynamic obstacles and reach the goal in an optimal manner.

Since we have used an evolutionary approach, one of the major factors to consider in this approach is algorithms capability to respond to sudden occurrence of obstacles. A robot moving might suddenly see some obstacle in front of it and would be expected to make its way out of it and reach the goal in an optimal manner. For this, we give a simple map to the robot and allow it to move. However, as soon as the robot is somewhere mid-way in its journey, we suddenly place an obstacle on its way. The robot is still able to steer out its way and reach the destination. The complete path traced by the robot is given in Figure 27. This clearly shows that the robot is able to react to any sudden change in environment.

In these experiments, we have demonstrated the ability of the robot to solve complex maps, escape from static and dynamic obstacles, and to react to the sudden emergence of obstacles. Any real life situation would primarily involve these conditions in different forms. Hence, we may assume that the robot would be able to solve a variety of situations that it encounters in real life.

An important characteristic of the algorithm is the role of parameters α and γ in its execution. The parameter α denotes the distribution between the coarser and finer hierarchies. A very large value of α would result in a very big map size of the finer hierarchy. This would necessitate the need of too much of computation time for the finer hierarchy to compute the optimal path. This would make the algorithm equivalent to a single EA. The entire path may be optimal, but the real time nature would be

Figure 27. The execution of the algorithm with sudden obstacle emergence

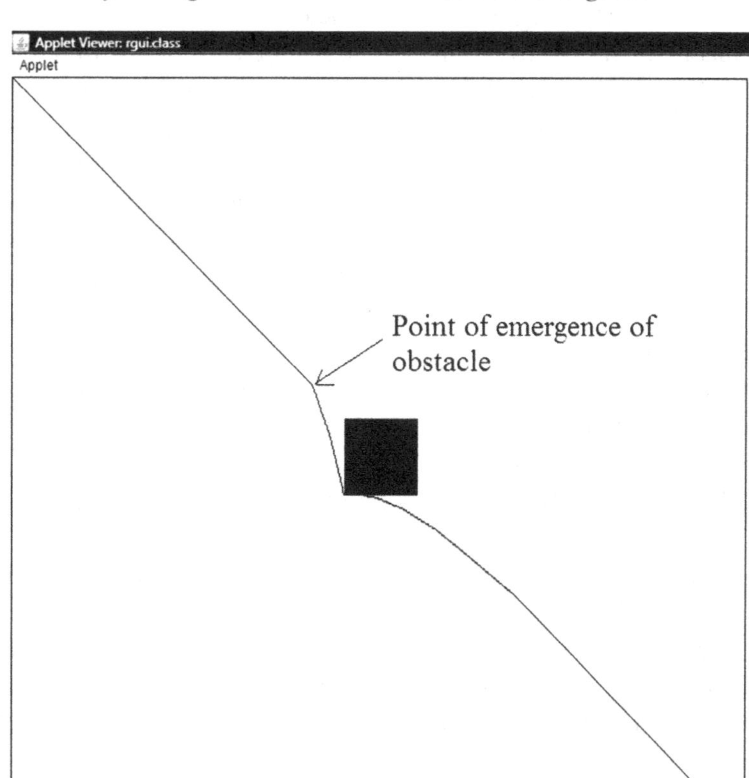

gone. A very small value of this parameter would result in excessive size for the coarser hierarchy. The coarser hierarchy algorithm may now need a lot of computation time to give a guiding path for the finer hierarchy. This path would be next to the optimal path that might not require further optimizations by the finer hierarchy.

It may be easily observed that this behavior is also like the use of a single evolutionary approach. Accordingly, the two parameters need to be set. The parameters also have a dependence on the map. A reasonably simple map with few simple obstacles may be degraded to a good extent without much loss of information. A large value of α is workable. This, however, would not be the case with a complex maps having too many complex obstacles in various parts. Keeping α large in such a case might show two unconnected regions of the map as connected, which would be wrong guidance to the finer hierarchy. The other factor γ denotes the vision of the finer hierarchy. A very large value might give a reasonably large part of the map to the finer hierarchy to work over. This would result in a very large computational time, but in return would lead to better results in terms of path optimality.

CONCLUSION

This chapter focused upon the techniques of using evolutionary algorithm as a base for the motion planning of mobile robots. Hybridization was seen as the key for improving the performance of these algorithms. The chapter presented three approaches.

In the first approach, the motivation was to evolve the optimal path considering the fact that the complexity of the path is completely unknown that depends largely upon the map. The algorithm optimized the fitness function by making use of a variable momentum-based approach that decayed in a Gaussian manner. A similar approach was used for the number of individuals for the EA that increased in a Gaussian manner along with the number of generations. The fitness function used in the problem optimized the total path length (l), distance of the path from closest obstacle (d), and the total path complexity (c_{max}). The individual contribution of these could be controlled by the use of multi-objective optimization parameters of the algorithm.

We presented four kinds of maps with all varying complexities. We saw that the robot was easily able to solve all of the four maps and return a solution. The solution to all these cases could be visually seen to be highly optimized in terms of the set optimization parameters. The results conveyed a lot of additional information as far as the problem is considered. We observed that the multi-objective optimization parameters could be easily varied to obtain a tradeoff between the path length and complexity. We showed that the algorithm was flexible to optimize both of them to varying degrees. The final evolved path completely changed when we made a transition in the preference of one from the other. The actual choice depends a lot on the physical constraints of the robot. A speed restricted robot may chose to go for path optimization and a car like robot built to operate at higher speeds might want to optimize the total path complexity to be able to run at high speeds. Another interesting behavior was the momentum that showed us signs of expected trends as well as randomness when measured on the parameters of time, feasibility prediction, accuracy, and distance to obstacle prediction. While the trends were attributed to the reasons illustrated in theory, the randomness stated a typical behavior of the parameter.

Now one may argue that in order to replace a single parameter we happened to add too many parameters in the conventional evolutionary path-planning algorithm. This should increase the problem in place of decreasing. Here, we state two major points. The first is the role of parameter. Path complexity was a major parameter that we replaced with a set of passive parameters. While experimentation we observed that the added parameters depended little on the robotic and map constraints and could be assumed to have global constant values. The other major point what differs from the natural counterparts is that the robotic processing capability and the physical constraints are unknown and influence the parameters. A child knows his limitations and has a lifetime to train according to those limitations. People driving cars have initial problems driving buses or trucks. That is precisely what we are doing when we set parameter. We are trying to optimize the algorithm according to the unknown constraints and demands.

In the second approach, the evolutionary algorithm was extended to the use of HGAPSO. The combined effects of GA and PSO enabled the generation of path that was optimal based on the set criterion. The algorithm was simulated over four types of maps with all varying complexities. We saw that the robot was easily able to solve all of the four maps and return a solution.

In the last approach, the entire algorithm was broken into two stages of coarser and finer path planning. The coarser level used an evolutionary technique for path planning. This level worked only on a static map, which had a reduced resolution. The path generated by this level of path planning was used for guiding the path planning at the other level, which was the finer level. This level of path planning was done on part of the actual high-resolution map. The obstacles in this part were mobile which made the environment dynamic. For the planning purposes, this level extrapolated the motion of the various dynamic obstacles. It assumed that the obstacles keep moving with the same speed and in the same direction. In this manner, the robot is able to escape from the likely collisions with the dynamic obstacles.

The algorithm was tested against three types of scenarios. Results were found to be optimal and collision free. The algorithm made a variety of parameters for the generation of the path. In all the executions, these parameters were kept as constant and generated fair enough paths whose optimality was visible. This makes it appear that the various parameters are passive in nature whose values may be fixed to some constant values. In reality, all these parameters have a high degree of relevance to the map being used and the optimal values of the parameters would depend upon the map being used. This showcases the problem of optimal parameter setting of these parameters. The problem is similar to the problem of parameter setting in the problems of machine learning, pattern recognition, or other neural and evolutionary approaches, where the parameters depend upon the data set, fitness landscape, etc. The optimal parameters can only be computed after analyzing the map and possibly running the algorithm a number of times.

REFERENCES

Alvarez, A., Caiti, A., & Onken, R. (2004). Evolutionary path planning for autonomous underwater vehicles in a variable ocean. *IEEE Journal of Oceanic Engineering*, *29*, 418–429. doi:10.1109/JOE.2004.827837

Badran, K. M. S., & Rockett, P. I. (2007). The roles of diversity preservation and mutation in 700 preventing population collapse in multiobjective genetic programming. In *Proceedings of the 9th Annual Conference on Genetic and Evolutionary Computation*, (pp. 1551–1558). IEEE.

Dittrich, P., Burgel, A., & Banzhaf, W. (2006). Learning to move a robot with random morphology. *Lecture Notes in Computer Science*, *1468*, 165–178. doi:10.1007/3-540-64957-3_71

Doitsidis, L., Tsourveloudis, N. C., & Piperidis, S. (2009). Evolution of fuzzy controllers for robotic vehicles: The role of fitness function selection. *Journal of Intelligent & Robotic Systems*, *56*, 469–484. doi:10.1007/s10846-009-9332-z

Gerke, M. (1999). Genetic path planning for mobile robots. In *Proceedings of the American Control Conference,* (Vol. 4), (pp. 2424–2429). IEEE.

Han, W., Baek, S., & Kuc, T. (1997). GA based on-line path planning of mobile robots playing soccer games. In *Proceedings of the 40th Midwest Symposium on Circuits and Systems*, (Vol. 1), (pp. 522–525). IEEE.

Holland, J. H. (1975). *Adaptation in natural and artificial systems*. Ann Arbor, MI: University of Michigan Press.

Jolly, K. G., Sreerama, K. R., & Vijayakumar, R. (2009). A Bezier curve based path planning in a multi-agent robot soccer system without violating the acceleration limits. *Robotics and Autonomous Systems*, *57*, 23–33. doi:10.1016/j.robot.2008.03.009

Juang, C. F. (2004). A hybrid of genetic algorithm and particle swarm optimization for recurrent network design. *IEEE Transactions on Systems, Man, and Cybernetics. Part B, Cybernetics*, *34*(2), 997–1008. doi:10.1109/TSMCB.2003.818557

Kala, R., Shukla, A., & Tiwari, R. (2010). Dynamic environment robot path planning using hierarchical evolutionary algorithms. *Cybernetics and Systems*, *41*(6), 435–454. doi:10.1080/01969722.2010.500800

Kala, R., Shukla, A., & Tiwari, R. (2011). Robotic path planning using evolutionary momentum-based exploration. *Journal of Experimental & Theoretical Artificial Intelligence*, *23*(4), 469–495. doi:10.1080/0952813X.2010.490963

Kala, R., Shukla, A., & Tiwari, R. (2012). Robotic path planning using hybrid genetic algorithm particle swarm optimization. *International Journal of Information and Communication Technology*.

Kala, R., Shukla, A., Tiwari, R., Rungta, S., & Janghel, R. R. (2009). Mobile robot navigation control in moving obstacle environment using genetic algorithm, artificial neural networks and A* algorithm. In *Proceedings of the IEEE World Congress on Computer Science and Information Engineering,* (pp. 705–713). IEEE Press.

Kambhampati, S., & Davis, L. (1986). Multiresolution path planning for mobile robots. *IEEE Journal on Robotics and Automation*, *2*, 135–145. doi:10.1109/JRA.1986.1087051

Laumanns, M., Thiele, L., Deb, K., & Zitzler, E. (2001). *On the convergence and diversitypreservation properties of multi-objective evolutionary algorithms*. TIK Report No. 108. Zurich, Switzerland: Swiss Federal Institute ofTechnology (ETH).

Peram, T., Veeramachaneni, K., & Mohan, C. K. (2003). Fitness-distance-ratio based particle swarm optimization. In *Proceedings of the 2003 IEEE Swarm Intelligence Symposium*, (pp. 174-181). IEEE Press.

Sadati, N., & Taheri, J. (2002). Genetic algorithm in robot path planning problem in crisp and fuzzified environments. In *Proceedings of the IEEE International Conference on Industrial Technology,* (Vol. 1), (pp. 175–180). IEEE Press.

Sedighi, K. H., Ashenayi, K., Manikas, T. W., Wainwright, R. L., & Heng-Ming, T. (2004). Autonomous local path planning for a mobile robot using a genetic algorithm. In *Proceedings of the IEEE Conference on Evolutionary Computation,* (Vol. 2), (pp. 1338–1345). IEEE Press.

Shukla, A., Tiwari, R., & Kala, R. (2010). *Real life applications of soft computing*. Boca Raton, FL: CRC Press. doi:10.1201/EBK1439822876

Toogood, R., Hong, H., & Chi, W. (1995). Robot path planning using genetic algorithms. In *Proceedings of the IEEE Conference on Systems, Man and Cybernetics,* (Vol. 1), (pp. 489–494). IEEE Press.

Tu, J., & Yang, S. (2003). Genetic algorithm based path planning for a mobile robot. In *Proceedings of the IEEE International Conference on Robotics and Automation*, (pp. 1221-1226). IEEE Press.

Urdiales, C., Bandera, A., Arrebola, F., & Sandoval, F. (1998). Multi-level path planning algorithm for autonomous robots. *Electronics Letters*, *34*, 223–224. doi:10.1049/el:19980204

Veeramachaneni, K., Peram, T., Mohan, C., & Osadciw, L. A. (2003). Optimization using particle swarms with near neighbor interactions. In *Proceedings of Genetic and Evolutionary Computation Conference*, (pp. 110-121). IEEE.

Wang, C., Soh, Y. C., Wang, H., & Wang, H. (2002). A hierarchical genetic algorithm for path planning in a static environment with obstacles. In *Proceedings of IEEE Conference on Electrical and Computer Engineering,* (Vol. 3), (pp. 1652–1657). IEEE Press.

Woong-Gie, H., Seung-Min, B., & Tae-Yong, K. (1997). Genetic algorithm based path planning and dynamic obstacle avoidance of mobile robots. In *Proceedings of the IEEE Conference on Systems, Man, and Cybernetics,* (Vol. 3), (pp. 2747–2751). IEEE Press.

Yanrong, H., & Yang, S. X. (2004). A knowledge based genetic algorithm for path planning of a mobile robot. In *Proceedings of the IEEE Conference on Robotics 845 and Automation*, (Vol. 5), (pp. 4350–4355). IEEE Press.

KEY TERMS AND DEFINITIONS

Coarser and Finer Level Planning: Breaking up of task of planning into two levels. Coarser level planning gives a vague path and works on low-resolution maps. Finer level planning works on part of the map, which is given in high resolution, and attempts to generate a path as directed by the coarser-level planning. For real time nature, both levels work as the robot moves.

HGAPSO: Hybrid Genetic Algorithm Particle Swarm Optimization. A hybrid evolutionary algorithm of genetic algorithm and particle swarm optimization.

Momentum: A method of controlling granularity in computing robotic path cost. Momentum controls speed of traversal while judging feasibility and fitness objectives. High speeds mean lesser points in path are checked implying lesser computational time, which also implies high uncertainty regarding validity of measurements, and vice versa.

Parameter Adaptation: The optimal parameter values may change as the algorithm proceeds. Adaptation attempts to assign every parameter optimal value as per present situation.

Path Complexity: The total number of turns in map or the total size of the genetic individual. Lesser complexity or turns means the possibility to travel at high speeds covering most of journey travelling straight ahead, not having to slow down for many turns. Present turns may be made smoothly. It is possible that the shortest path by length may have a large number of turns, in which case the choice of path lies largely in the preference to travel faster and smoother with lesser turns, or slower with a shorter path.

Chapter 9
Hybrid Behavioral Methods

ABSTRACT

The chapter focuses upon the use of hybrid planning systems that are primarily behavioral in nature. Three approaches are discussed. In the first approach, the authors solve the problem of path planning using a combination of A algorithm and Fuzzy Inference System. The A* algorithm does the higher-level planning by working on a lower detail map. The algorithm finds the shortest path, while at the same time generating the result in a finite time. The A* algorithm is used on a probability-based map. The lower-level planning is done by the Fuzzy Inference System (FIS). The FIS works on the detailed graph where the occurrence of obstacles is precisely known. The FIS generates paths that take into account the non-holonomic constraints and generate smoother paths. The results of A* algorithm serve as a guide for FIS planner. The FIS system is initially generated using rules from common sense. Once this model is ready, the fuzzy parameters are optimized by Genetic Algorithm (GA). The GA tries to optimize the distance from the closest obstacle, total path length, and the sharpest turn at any time in the journey of the robot. Many times, the planning algorithm may require a map breakup such that the coarser-level graph also has a high degree of resolution and A* algorithm may not be able to work. Hence, in the second approach, the authors replace the A* algorithm with GA. Both approaches were tested on various complex and simple paths. All paths generated were optimal in terms of path length and smoothness. Their last approach is based on dynamic programming, which, in this implementation, is similar to the use of neurons embedded in the map being planned. In this approach, the authors talk about the use of extra nodes in the planning framework called accelerating nodes. These nodes are less in number and fully interconnected. These nodes transmit information about any map change and blockages to each other for sudden re-planning to be initiated. These nodes further guide the robot until re-planning completes.*

DOI: 10.4018/978-1-4666-2074-2.ch009

INTRODUCTION

The problem of planning for mobile robot has multiple issues which planning algorithms need to solve as far as possible. The same is true with the behavioral planning approaches in which we chiefly studied the fuzzy and the neural approaches. These approaches model the behavior of the robot and make it move from one point to the other in the robotic map. The fuzzy approaches in particular are good for fast and real-time planning. They can work in dynamic maps as well as the generated paths and obey non-holonomic constraints. However, it is possible for the robot to get struck. The optimality of the path cannot be ascertained. These algorithms cannot maneuver maze-like maps by their modeling techniques. The various advantages and disadvantages are summarized in Table 1. The purpose of this chapter is to attempt to improve upon these techniques, making them better for planning for most scenarios.

It the earlier chapters we dealt with the problem of planning by A* algorithm which was shown to generate good results in a variety of maps. However, the algorithm is computationally very expensive. In practical problems, the map may be too large for the standard A* algorithm to perform. Hence, the A* algorithm cannot be applied over the entire problem map. The A*algorithm also generates very sharp paths which the real robot would find it very difficult to follow. This is the case when the number of allowed moves is few and increasing them further increases the computational complexity drastically. The paths generated by the algorithm disobey the non-holonomic constraints of the robot. In this chapter, the aim is to adapt A* algorithm according to these constraints.

In hybridizing the solution, we have usually been making different algorithms run on maps at different resolutions or degree of abstraction. A manner of doing the same is to use probabilistic maps as discussed in chapter 7. In the discussed approach, the map is modeled in a way similar to the quad tree approach (Kambhampati & Davis, 1986). Here at the top, the nodes have a high degree of uncertainty regarding the presence or absence of the obstacles. This uncertainty is measured in terms of grayness of the cell. A completely white cell denotes the complete absence of obstacles from the complete region. On the other hand, full black color denotes the conformation of presence of obstacles in the cell. The grey denotes the intermediate values with the intensity denoting the probability of the cell being free from obstacles.

Table 1. Advantages and disadvantages of fuzzy inference systems

S. No.	Advantages	Disadvantages
1.	Very less computational time. Real time performance	Not-complete. Can get struck in simple maps.
2.	In base forms, works for dynamic maps and sudden obstacles	Cannot solve maze-like maps. There is no global level planning.
3.	Validity of non-holonomic constraints	Not-optimal

Hence, we see that the advantages and disadvantages of the fuzzy and A* algorithms complement each other, which motivates their hybridization. The side-by-side comparisons of fuzzy approach and A* algorithm is provided in Table 2. In the first approach to be discussed in the chapter, the problem of planning is solved by making a hierarchical solution to the problem making use of A* algorithm at coarser level and fuzzy planner at the finer level. The A* algorithm works on a probabilistic map. It tries to find a path of the smallest path length that has a high probability of non-collision. The high probability factor ensures that it is feasible to follow this path and reach the goal without collision. The path length factor ensures that the path generated is short to a good extent. Discussions in introduction, conclusions, and sections covering the entire algorithm

Table 2. Comparative advantages and disadvantages of fuzzy inference systems and graph search algorithms

S. No.	Fuzzy Inference System		Graph Search	
	Advantages	Disadvantages	Advantages	Disadvantages
1.	Very less computational time. Real time performance	Not-complete. Can get struck in simple maps.	Complete	High computational time on medium to high resolution maps
2.	In base forms, works for dynamic maps and sudden obstacles	Cannot solve maze-like maps. There is no global level planning.	Can solve maps of any complexity	Used for static path planning only
3.	Validity of non-holonomic constraints	Not-optimal	Optimal	Non-validity of non-holonomic constraints

are reprinted with few modifications from Kala et al. (2010a) with kind permission from Springer Science+Business Media: Artificial Intelligence Review.

The A* algorithm gives the basic structure of the solution on a low-level detailed map. This solution acts as a guide for the Fuzzy Inference System (FIS)-based planner. The FIS planner generates solution in real time manner. The solutions follow the non-holonomic constraints. In addition, the solution generated by the FIS avoids the robot getting too close to the obstacle. This is analogous to the motion of vehicles in everyday life. The FIS solution works on the detailed map, and the solution generated by them is the final solution.

The initial fuzzy model used by the system was developed by experimentation and best of the understanding. Once this basic prototype was ready, Genetic Algorithm (GA) was used to further optimize the model. There were 3 basic maps of ranging complexity that were used by the GA as benchmark problems. The GA was supposed to optimize the performance with respect to these maps. The performance was measured in terms of total path length, distance from nearest obstacle, and the maximum turn angle. It may be seen that any good path would mean the least total path length, comfortable distance from nearest obstacle and least maximum turn angle at any time in the run on the path. GA fine-tunes the generated model to convert it into the model that could be used for any general case.

This approach seems good to solve most of the problems associated with planning for mobile robot. However, the approach may face a problem in case the map given to A* algorithm is itself of high-resolution. The possibility of a coarser map to be complex always needs to be accounted for. Further, it may be possible that the map cannot be worked over at very coarse level due to its complex nature. Coarser level may not record very fine important detail for which it might be important to work at a map of high resolution or a comparatively finer map at the higher planning or coarser level. Since the coarser map in this representation has a uniform resolution, even if a single segment requires a high degree of details, it would be important to keep resolution high for a complete map. Hence, A* algorithm may be optimal and complete for a range of maps, but not a desired approach for high-resolution complex maps at coarser level. For the same reason, we extend the first approach and use genetic algorithm for planning at the coarser level. This is discussed as the second approach of the chapter. Here, we discuss our work reported in Kala et al. (2010b). Discussions in introduction, conclusions, and sections covering the entire algorithm are reprinted with few modifications from Kala, Shukla, and Tiwari, (2010) with permission from Inderscience Publishers.

The last approach discussed in this chapter deals with mechanism of adding special nodes called accelerating nodes in the dynamic programming formulation of the problem. This formulation as discussed in this chapter is similar to the use of neural networks, where neurons are embedded into the map, which gives information to the robot regarding the distances to goal. The robot traverses as per the same information until it reaches the goal. At every point the decision of the robot is governed by the immediately neighboring nodes. If there is a change in the map in terms of emergence of a new obstacle, the nodes re-compute distances by passing each other updated information. This takes a little time, but towards the end, every node gets its updated value, which the robot may consult for its motion.

The basic motivation behind the work is to discuss the case where the path may have a sudden blockage. In such a case, the optimal path of the robot changes drastically. The numerous nodes may take a long time to convey the information to the robot, within which duration it would need to go by the wrong route, which was previously considered optimal. This problem is solved by using a lesser number of nodes called accelerating nodes in the map. Unlike the normal nodes, these nodes are less in number and can transmit information to each other and complete the computations in a short span of time. These nodes inform the robot about any blockage or other obstacle emergence, which might change the robot's path. These nodes further give a vague idea to the robot regarding the optimal path, which the robot may consider for motion until the normal nodes compute the optimal path.

An interesting application of path planning in real life is Mars rovers (JPL, 2012). These are robots left for exploration on the surface of Mars. The robots are, from a global perspective, guided by human operators who give them broad instructions about the motion to carry out. These robots have inbuilt autonomy, which enables them to avoid obstacles and move in the Mars terrain. The robots are able to perceive the terrain and decide the safest and shortest path. This marks an interesting application of using path-planning algorithms over some general guidance path.

Once again, it would be wise to talk about a few good approaches that may be of interest in regard to this chapter. Readers may leave the rest of this section without any loss of continuity. Potential approaches are extremely powerful modeling techniques where the complete map is interpreted with every obstacle having a negative potential field and the goal having a positive potential field. There is extensive scope to hybrid these methods as well, which we do not discuss in this text. Key works in solving the problem using potential field includes (Pozna, et al., 2009; Tsai, et al., 2001). Hui and Pratihar (2009) gave a comparison between the Potential Field and the Soft Computing solutions.

The basic model of using Dynamic Programming for Real Time Robot Path Planning was displayed in the work of Willms and Yang (2006). Neural networks and related approaches are further extensively used. At every iteration, we try to propagate the distance from the target to all the points of the map. Related works in use of neural networks for solving the problem include the work of Yang and Weng (2000) and Zelingsky (1994).

Fuzzy-based planners are further under constant research. Key works include the work of Shibata et al. (1993) who reported the use of Fuzzy Logic for fitness evaluation of the paths generated. Other approaches using fuzzy logic include Camilo et al. (2008) and Cortes et al. (2008). Chen and Chiang (2003) made an adaptive intelligent system and implemented using a Neuro-Fuzzy Controller and Performance Evaluator. Pradhan et al. (2009) solved a similar problem for unknown environments using fuzzy logic.

Embedded sensor networks (O'Hara, et al., 2008) are further highly related approach where sensors are embedded into the map, which does most of the planning. Each sensor stores information of its distance from goal. These computations

Figure 1. A guided fuzzy planning algorithm*

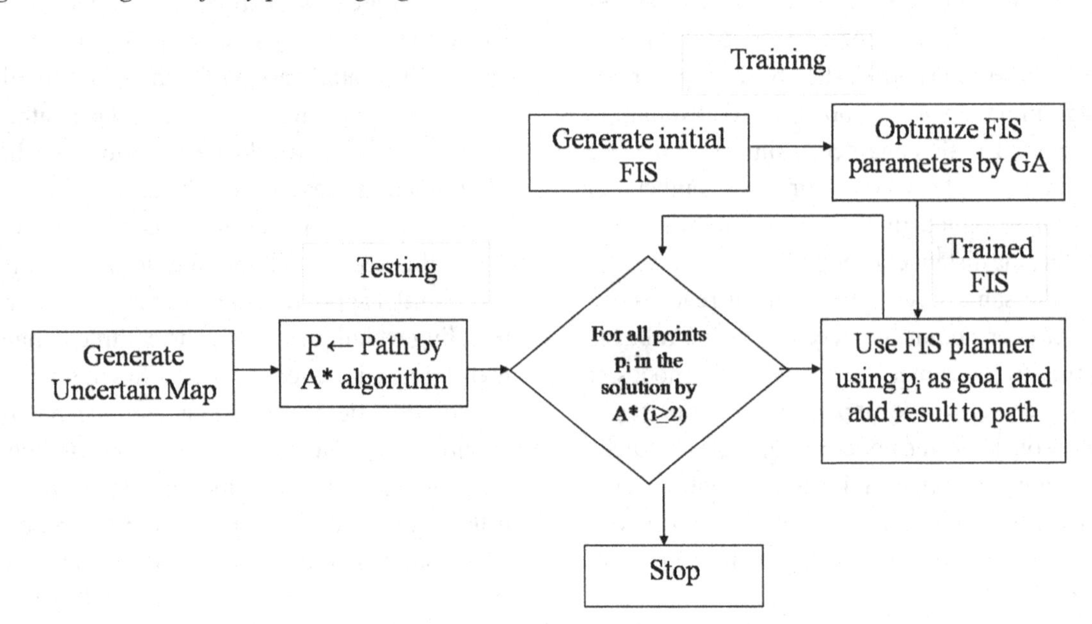

are shared with other sensors for their computations. The robot simply uses the sensor information for its traversal.

The other major issue in the work of this chapter is map representation, where we use hierarchy of graph as a means of solving the problem. The commonly used representations in literature for the problem include Quad Tree (Kambhampati & Davis, 1986), Mesh (Hwang, et al., 2003), Pyramid (Urdiales, et al., 1998). Zhu and Latombe (1991) used the concepts of cell decomposition and hierarchal planning. Hierarchical Planning can also be found in the work of Lai et al. (2007) and Shibata et al. (1993).

FUSION OF A* AND FUZZY PLANNER

The first approach to be discussed fuses A* algorithm and FIS for effective planning of the mobile robot. In this section, we give a general outline to the algorithm. The basic methodology is to use the A* algorithm for a higher end planning and FIS for the lower end planning. The FIS model is optimized by the use of GA. The general structure of the algorithm is given in Figure 1. This section and subsequent sections covering the entire algorithm are reprinted with few modifications from Kala et al. (2010a) with kind permission from Springer Science+Business Media: Artificial Intelligence Review.

The algorithm starts by taking as input the initial graph. This graph may be approximately built by the map-building algorithm. This means that we do not know for sure whether the obstacle lies at some cell or not (Kambhampati & Davis, 1986). The A* algorithm runs on this map to generate a path P. The path P is a collection of points p_i such that p_1 is the cell cluster where the source node is found and p_N is the cell cluster where the goal node is found, N being the number of points in the A* solution.

These points are then used one after the other (other than the source) to act as guide for the fuzzy planner. As soon as the robot enters the region in which the guide cell is found, the next point in P becomes the goal, and the robot has to move a step further. This goes on and on until the robot reaches the final block where the goal node is found. The FIS is used in every block for guiding the robot. After the last iteration, FIS planner is used to make the robot reach the exact cell where the goal is located.

First, the initial FIS is generated and then the FIS parameters are optimized using GA. For this purpose, benchmark maps were used. This training stage needs to be applied only once to generate the FIS. Afterwards the FIS may be used reputedly for all problems. We discuss the various steps of the algorithm one by one.

Map

Map is the representation of the robotic world. The path-planning algorithm uses map to get information about the obstacles and paths that are accessible. In this approach, we have used a probabilistic approach to represent a map that is the same as that employed in chapter 6 with the difference that there are only two hierarchies in the current approach. The map is represented in the form of grids of variable sizes. Each grid is marked with a color from white to black through grey. The color denotes the probability of the obstacles lying in that region. White means no obstacle and black means a confirmed presence of obstacle. The grids are converted into a graph before running the A* algorithm. In the graph, each grid represents vertex. All grids that are accessible by other grids (irrespective of presence or absence of obstacles) are marked as edges. One such representation of the map is given in Figure 2.

The grayness is the measure of the obstacles in the cell. Grayness of any cell lies between 0 and 1 and denotes the probability of collision of the robot in that cell. The probability-based map is used only by the A* algorithm. The FIS planner uses the grid map where each grid is of unit size. The probability of the occurrence of obstacles in this grid is either 0 or 1.

This graph may be generated by decomposition of the probability-based map. The map represents a two-level hierarchy. The A* algorithm runs at the first level and the FIS planner at the second level. The first level is the abstract level, and the second level is the detailed level. A graph at the 1st level (coarser level) consists of a number of grids of the 2nd level (finer level). Let each grid at the 2nd level consist of α x α grids of 1st level, where α is a parameter of the algorithm. If we increase α,

Figure 2. The map at any general point of time

Figure 3. The 2-level hierarchy in map

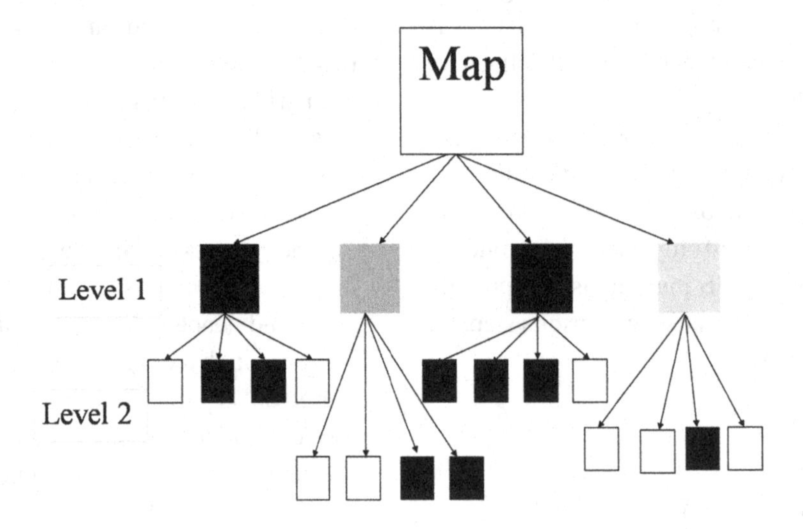

the number of grids at the 1st level would decrease and vice versa. This is shown in Figure 3.

The A* Guidance

The A* algorithm acts as a guide for the FIS. In this section, we discuss this relation between the A* algorithm and FIS algorithm. The A* algorithm returns a set of points p_i at the first level that lead us to the goal. For the sake of convenience, we denote any point by (x_i,y_i) where x and y are the coordinates of the mid-point of the region covered by this point comprising of the finer level points. The source is a finer level point located somewhere in the p_1 region. The algorithm first sets p_2 as the goal for the robot that is still situated at the start position. The robot moves towards p_2 with an aim to reach it. As soon as the robot exits the

Figure 4. The formation of layer 2 grid from layer 1 points

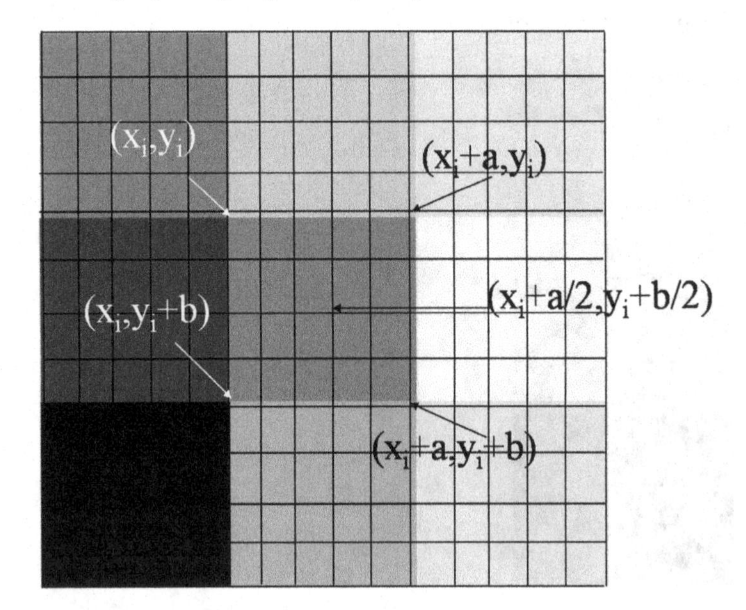

region in which p_1 is located and enters into the region where p_2 is located, the goal is changed to p_3. Now the robot tries to move towards p_3. The robot keeps moving in this manner. In general if it is at any point of time the robot is at position (x_i,y_i) at some region covered by p_i, it would have as its goal the point p_{i+1}. This mechanism of guiding the robot goes on until the robot enters the region of p_N. Now the actual goal guides the robot to complete the left journey. The concept of the level 1 (coarser level) and level 2 (finer level) points is shown in Figure 4.

Figure 5 shows the robotic guidance. The red line shows the solution generated by the A* algorithm. Here the robot starts from the source with the goal as P_2. While it is at area P_3, the goal is P_4. As soon as the robot reaches the point A, the goal is changed to P_5. This way the robot keeps moving.

Fuzzy Inference System Planner

The movement of the robot in the system is done by a FIS Planner. The initial FIS was generated by hit and trial method. The FIS is supposed to guide the robot to reach the goal position as directed by the A* algorithm. The FIS tries to find out the optimal next move of the robot. The FIS is same as discussed in chapter 6, key points of which are being repeated for the sake of understanding.

The FIS takes 4 inputs. There are angle to goal (α), distance from goal (d_g), distance from obstacle (d_o), and turn to avoid obstacle (t_o). The angle to goal is the angle (α, measured along with sign) that the robot must turn in order to face the goal. This is measured by taking the difference in current angle of the obstacle (φ) and the angle of the robot (θ). The result is always between -180 degrees and 180 degrees. The distance from goal (d_g) is the distance between the robot and the goal position. This distance is normalized to lie between 0 and 1 by multiplying by a constant. Similarly, the distance from obstacle (d_o) is the distance between the robot and the nearest obstacle found in the direction in which the robot is currently moving. This is also normalized to lie between 0 and 1.

The turn to avoid obstacle angle (t_o) is a discrete input that is either 'left,' 'no,' or 'right.' These stand for counter-clockwise turn, no turn, and clockwise turn respectively. This parameter represents the turn that the robot must make in order to avoid the closest obstacle. This input is

Figure 5. The robotic guidance

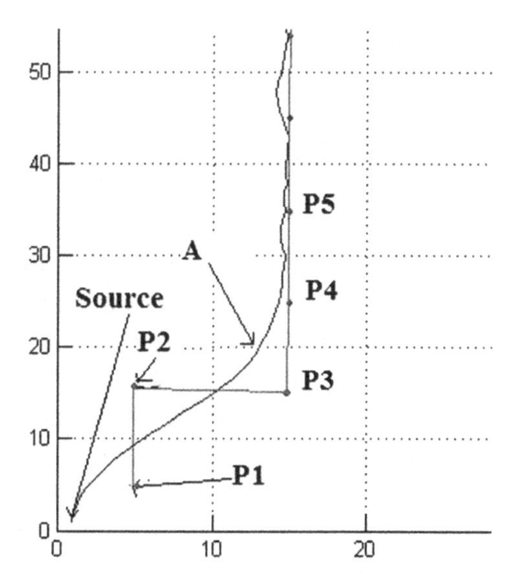

measured by measuring the distance between the robot and the obstacle at three different angles. The first distance is the distance between the obstacle and the robot measured at the angle at which the robot is facing (a). The second distance is the same distance measured at an angle of dθ more (c). The third distance is measured at an angle dθ less (b). Various cases are now possible. The first case is c>b. This means that the obstacle was turned in such a way that turning in the clockwise direction made it even furtherer.

In this case, the preferred turn is clockwise with an output of 'right.' The second case is b>c. This means that the obstacle was turned in such a way that turning in the anti-clockwise direction made it even furtherer. In this case, the preferred turn is anti-clockwise with an output of 'left.' The third case is b=c. This is the case when the robot is vertically ahead of the robot. In such a case we follow a 'left preferred' rule and take a 'left' turn. There is a single output that measures the angle (β) that the robot should turn, along with direction.

Rules are the driving force for the FIS. Based on these inputs, we frame the rules for the FIS to follow. The rules relate the inputs to the outputs. Each rule has a weight attached to it. Further some inputs have been applied with the not operator as well. All this makes it possible to frame the rules based on the system understanding.

The rules can be classified into two major categories. The first category of rules tries to drive the robot towards the goal. The second category of rules tries to save the robot from obstacles. If an obstacle is very near, the second category of rules become dominant. 11 rules are identified.

A* Algorithm

The A* algorithm is responsible for the high-end or coarser-level planning of the robot. A* algorithm generates very efficient solutions in any graph problem. In this algorithm, we have used a probability-based A* algorithm. The reason for this is that the A* algorithm runs at a level, where the exact information about the obstacles is unknown. The A* algorithm tries to find a solution in this probability based map. The algorithm is supposed to minimize the path length and at the same time maximize the probability of reaching the goal from this path without collisions.

The A* uses the cost function for its functioning, the cost decides the goodness of the solution. The better solutions have lower costs. Hence, the approach is generally to find smaller costs and expand them further. The cost function used by the algorithm is same as discussed in chapter 7. The function aimed to maximize the probability of non-collision and minimize the path length. The costs that the algorithm uses are heuristic cost h(n) and the historical cost g(n). For the A*, the total cost f(n) is given by the sum of heuristic cost h(n) and historical cost g(n). In this algorithm, we use the historic cost as distance of the center of the node from the source and the heuristic cost as the cost of the center of node to the goal. The total fitness C(n) in this case is the probability based total cost. This is given by Equation 1.

$$C(n) = f(n)\,Grey(P) + (1 - Grey(P)) \quad (1)$$

Here, P is the path of traversal till node n, f(n) has been normalized to lie between 0 and 1.

The graph of this function is as shown in Figure 6.

The grayness plays a very vital role in deciding the performance of the algorithm. If the A* algorithm tries to maximize the probability too much, it may result in the longer paths being selected, because of the fact they led to higher probability. This would result in the robot opting for longer paths to avoid any possible obstacle. Based on these ideas, we try to modify Equation 1) by introducing a parameter β that controls the effect of the grayness in the cost function. If the value of β is kept as 0, the A* algorithm completely discards the effect of grayness and converts into a normal un-probabilistic A* algorithm. The modified Equation 1) is given by Equation 2:

Figure 6. The plot of fitness function

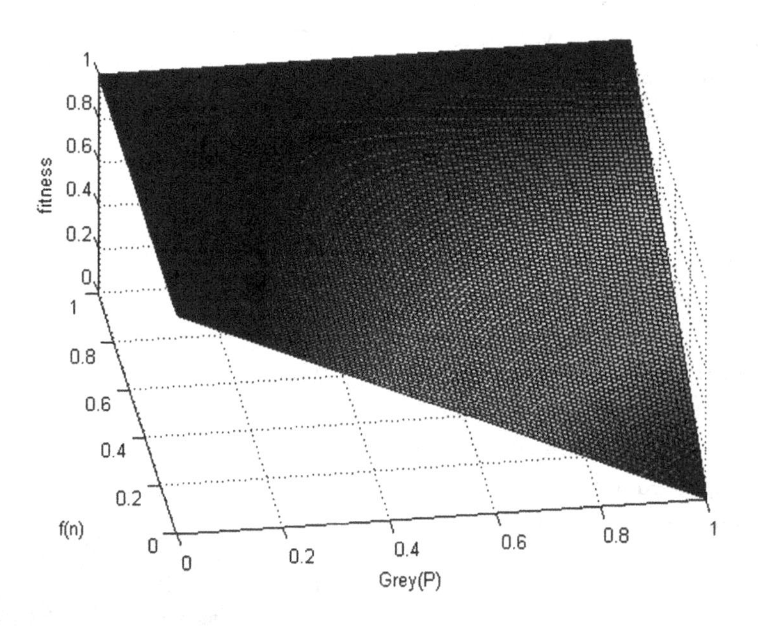

$$C(n) = f(n)\,Grey'(P) + (1 - Grey'(P)) \tag{2}$$

Here, Grey`(P) is given by Equation 3:

$$Grey'(P) = \begin{cases} 1 & Grey(P) > \beta \\ Grey(P) & otherwise \end{cases} \tag{3}$$

The fitness function for β=0.6 is given in Figure 7.

Genetic Optimizations

The fuzzy model discussed in earlier section was an initial fuzzy model that was generated. The

Figure 7. The plot of modified fitness function

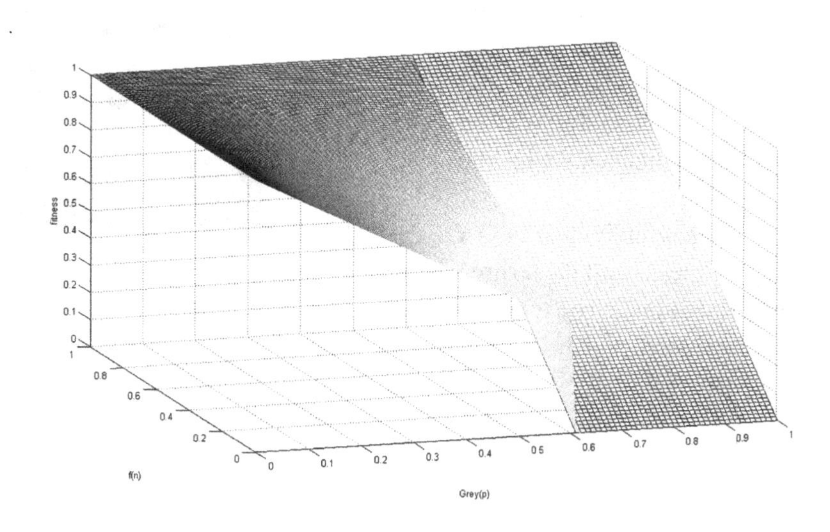

parameters of this model were decided by trial and error. In order to make this a scalable system, there were further optimizations that were necessary. This was done by the means of GA. The GA was supposed to optimize the fuzzy parameters for optimal planning of the robotic path. There were a total of 32 fuzzy parameters that were optimized by the GA.

For this problem, we first created 3 maps that acted as benchmark maps for the GA. The GA was supposed to optimize the performance of the system against these maps. The first map was a simple map with no obstacles. The second map had single obstacle on the way from source to goal. The robot was supposed to avoid this path. The third map had many obstacles and the robot was supposed to find its way out of them. These maps are given in Figure 8.

The fitness function for any of the map i tried to optimize three things (1) The total path length (L_i) (2) The maximum turn taken any time in the path (T_i) (3) Distance from the closest obstacle anytime in the run (O_i). All these were normalized to lie on a scale of 0 to 1. The total fitness for any map i may hence be given by Equation 4:

$$F_i = L_i (1-O_i) T_i \quad (4)$$

The fitness F_i for any graph hence lies between 0 and 1 (lesser preferred). 1 is the maximum fitness a map can attain. If the robot at any time moves out of the map, or collides with an obstacle, we assign it a fitness value of 1. The GA would naturally result in deletion of such solutions in the course of generations. In this manner, we handle infeasible solutions in GA.

The total fitness (F) of any individual in GA is the sum of its score or fitness in all these three maps. This is given by Equation 5. The total fitness can be anywhere between 0 and 3.

$$F = F_1 + F_2 + F_3 \quad (5)$$

Figure 8. Benchmark maps for GA (a) no obstacle, (b) single obstacle, (c) many obstacles

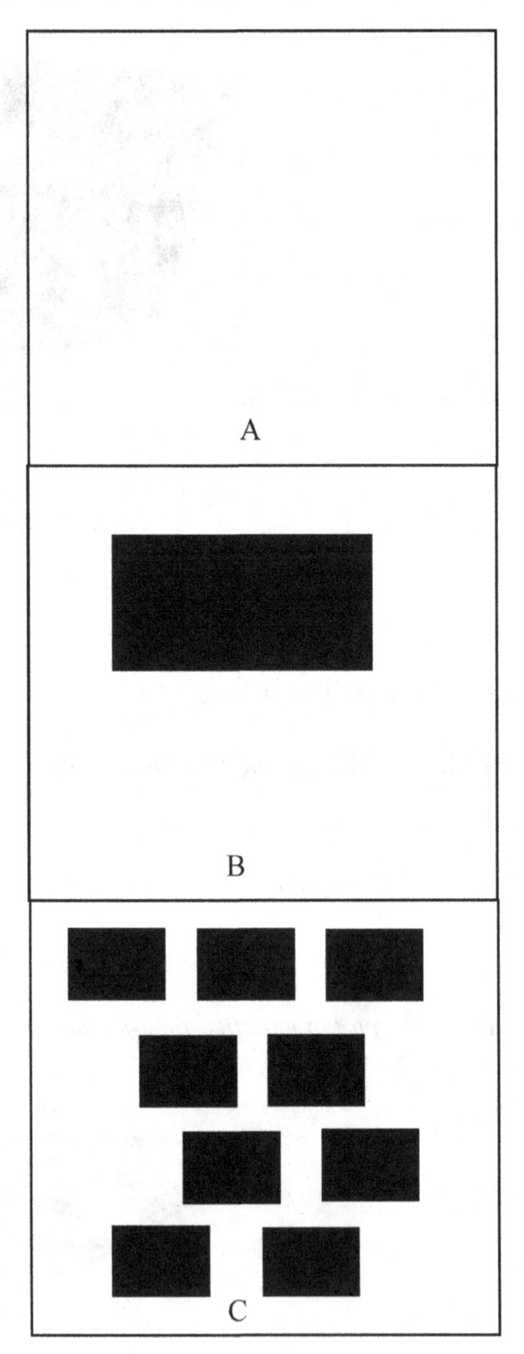

SIMULATION AND RESULTS

In order to test the working of the algorithm, we made a simulation engine of ourselves. Every attempt was made to ensure that the simulation engine behaves in a way similar to actual robot. This would ensure that the algorithm can be easily deployed to the real robot for the purpose of path planning. All simulations were done on a 2.0 dual core system with 1 GB RAM. The initial graph was taken as an input in form of an image. The image depicted the obstacles as black regions and the path as the white region.

The first major task was the Genetic Optimization of the FIS. Then this optimized model was implemented and tried on the benchmark problems. The system was then tried on numerous maps with varying obstacles. These are given in the next sub-sections.

Genetic Optimizations

The purpose of the GA was to find out the exact parameter values for the FIS. 32 such parameters were identified that needed optimizations. The Matlab GA toolkit was applied for the optimization purposes. The GA search space consisted of the region around the value chosen by the hit and trial method of the 32 parameters on both sides by 10%. This means if the value of any parameter was x, the GA was supposed to optimize it in the search region of x-0.1L to x+0.1L where L is the total range of values that x can take. It is natural that on the basis of rules framed, the solution was not likely to be present in any other region. This provided ample of space for GA to locate the global minima and at the same time making the space finite to search for possible solutions.

The population was represented by the double vector mechanism. The population size was fixed to 50. Rank-based scaling was used. Stochastic uniform selection was used. The crossover rate

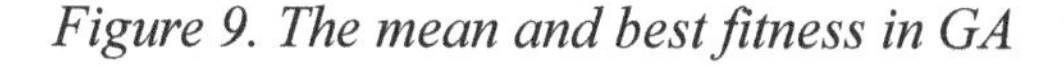

Figure 9. The mean and best fitness in GA

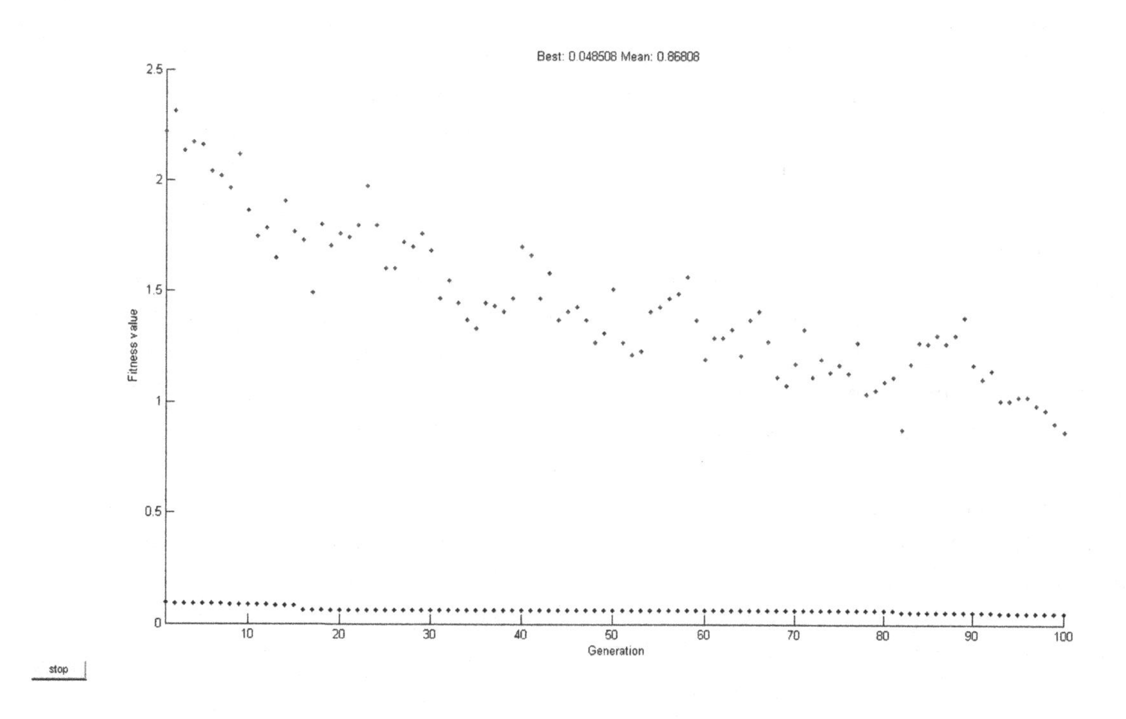

was 0.8 with elite count as 2. Gaussian mutation function was used with a scale and spread of 1 each. The GA was made to run over 100 generations. It took about 1 hr 15 mins for the GA to complete the generations. The fitness curve of GA showing the mean and best fitness is given in Figure 9. The final fitness of 0.04850 was achieved by the GA.

Benchmark Maps

The first tests were applied at the benchmark problems. This was an attempt to see the behavior of the algorithm at the benchmark problems. Since these benchmark problems were applied on low-resolution graph of size 100X100, there was no need to run the A* algorithm. Direct FIS planner was used. The maps are shown in Figure 10. The path traced by the algorithm after optimization is given in Figure 10. Figure 11 shows the same problem being solved by application of A* algorithm alone. In all cases the source is the top left corner of the map and the goal is the bottom right corner of the map.

The solutions generated by the algorithm were all smooth so that the robot can be easily turned as per requirements. They all seemed to be optimal. In each case, the robot was easily able to steer its way by avoiding obstacles and reach the goal node.

In all the cases, the robot was initially facing at the direction of the X-axis and not towards the goal. This is the reason why we see a smooth transition in path and angle at the starting points in the graph. The solutions generated by A* algorithm however do not reveal this. These solutions assume the robot would somehow turn towards the angle of the goal and then it would start moving in a straight line towards it.

Figure 10b shows another very interesting solution where the robot made a smooth transition and started moving in a manner to just avoid the obstacle. When the obstacle ended, it again smoothly changed its direction. The nearness of the obstacle avoided the undue long tracing of path and the turns were smooth enough for the robot to be taken. Similar inferences may be made from Figure 10c.

The very sharp turns present in Figure 11b and 11c clearly show that these solutions are not practical, and further, they cannot be easily converted to smoother paths as well. This exposes the weaknesses of A* algorithm.

Looking at Figure 10c and 11c, we see that our algorithm guided the robot from the top of the obstacle (remember we adopted a 'left preferred' approach). The A* algorithm on the contrary guided the robot from the middle of obstacle. It may be argued that the unequal length may not be optimal. We will see later that the guidance by the A* is responsible for the path length, and it divides the problem in such a way that the optimal path is selected in most of the cases.

More Maps

In order to fully test the behavior of the algorithm, we tested the algorithm against 3 more maps. This time the maps were of larger size of 1000X1000. Many obstacles were places in between the source and the goal. The robot was supposed to reach the goal from the source. The top left corner is the source and bottom left corner is the goal. In all the cases, the source was fixed as the top left corner and the goal was fixed as the bottom right corner.

The first test case was the obstacle avoidance test. Here we placed numerous small and big obstacles that were kept on the path of the robot. We observed that the robot was easily able to guide its way towards the goal. It could smoothly manage to overcome the numerous obstacles of all sizes and shapes that were on its way. The path traversed by the robot seemed optimal. The map and the path traced by the algorithm are given in Figure 12a. The initial path given by the A* algorithm is given in Figure 12b. The path traced only by the application of A* algorithm is given in Figure 12c and only fuzzy planner is given in Figure 12d.

Figure 10. The path traced by algorithm for benchmark maps (a) map 1, (b) map 2, (c) map 3

Figure 11. The path traced by standard A algorithm for benchmark maps (a) map 1, (b) map 2, (c) map 3*

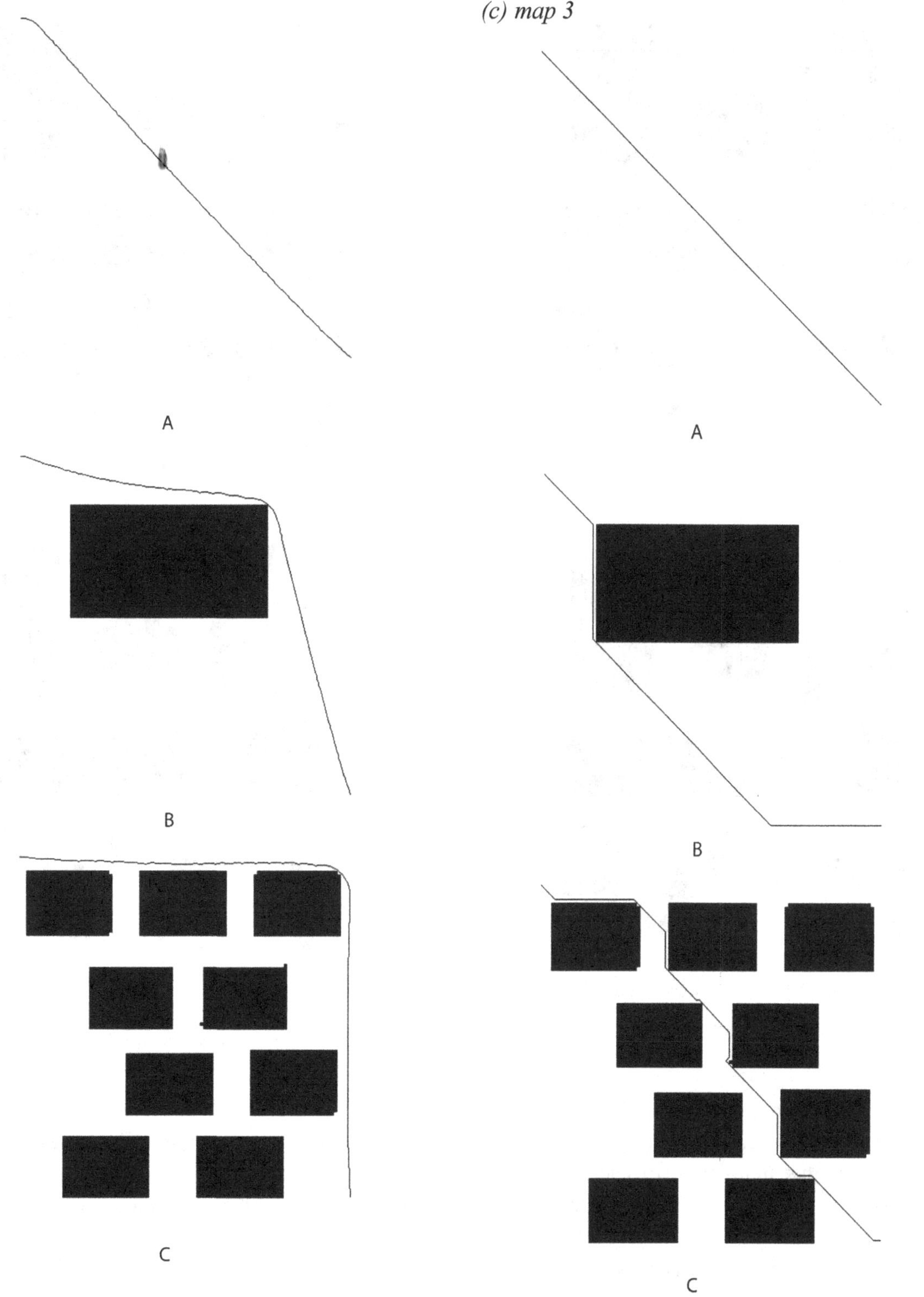

Figure 12. The obstacle avoidance test case (a) this algorithm, (b) result of A planning, (c) only A* algorithm, (d) only fuzzy planner*

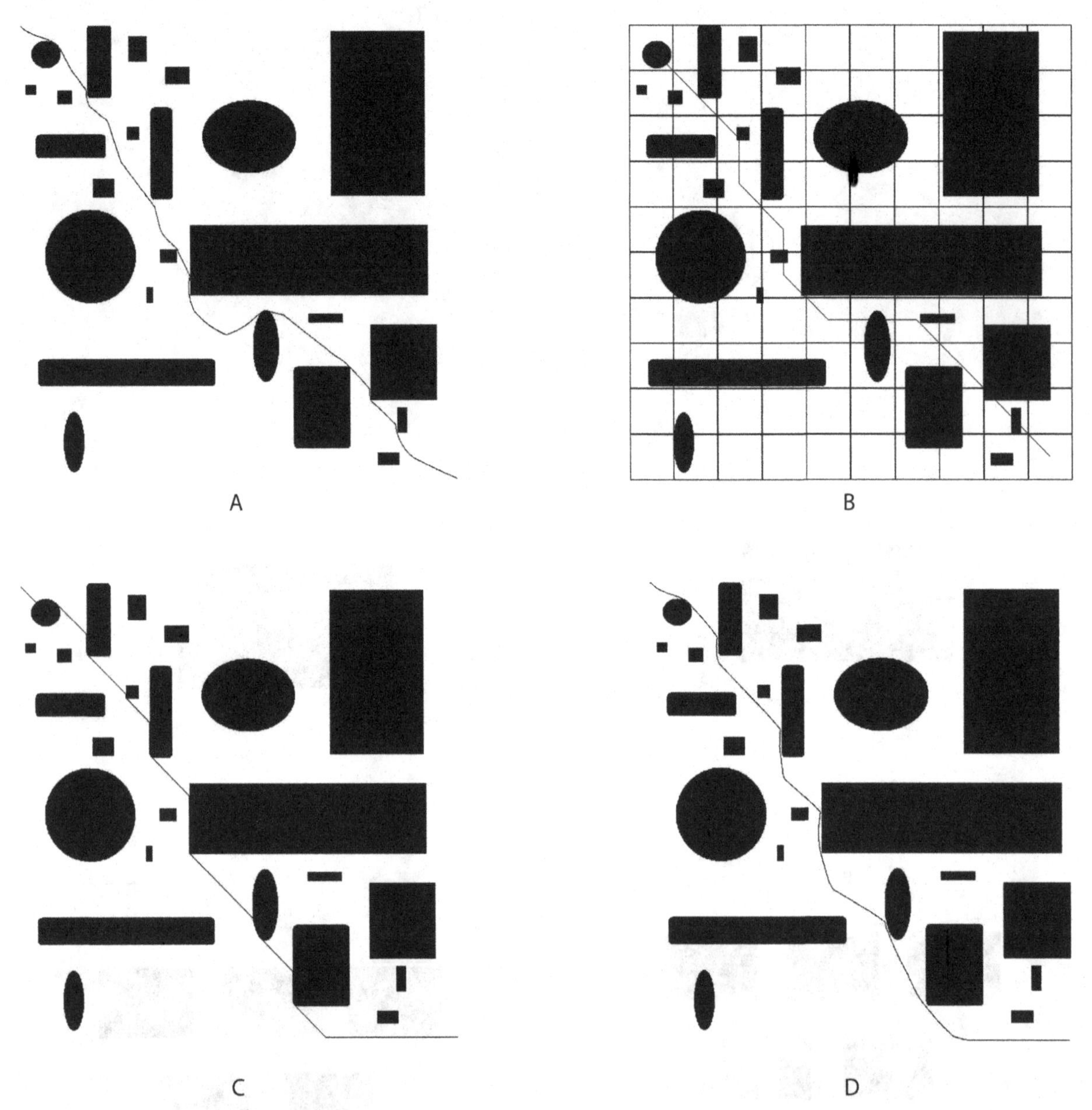

The second test case applied was of scattered obstacles. In this, many small obstacles of abrupt lengths were placed between the source and the goal. In this case as well, the algorithm could smoothly guide the robot. The total path length was not optimal in this case, but was decent enough considering the non-holonomic constraints. The extra path was a result of the probabilistic approach of the A* algorithm that wanted to maximize the certainty. The map and the path traced by the algorithm are given in Figure 13a. The initial path given by the A* algorithm is given in Figure 13b.

The path traced only by the application of A* algorithm is given in Figure 13c and only fuzzy planner is given in Figure 13d.

The third test case applied was of a simple maze-like situation. A complex map was given. The robot was supposed to find its way out. In this case as well, the robot was able to guide itself out. It may be noted that the maze considered was kept simple enough for the fuzzy planner to make out. Had we used a difficult map, the fuzzy planner would have ended up in a collision. The map and the path traced by the algorithm are given in

Figure 13. The scattered obstacle test case (a) this algorithm, (b) result of A planning, (c) only A* algorithm, (d) only fuzzy planner*

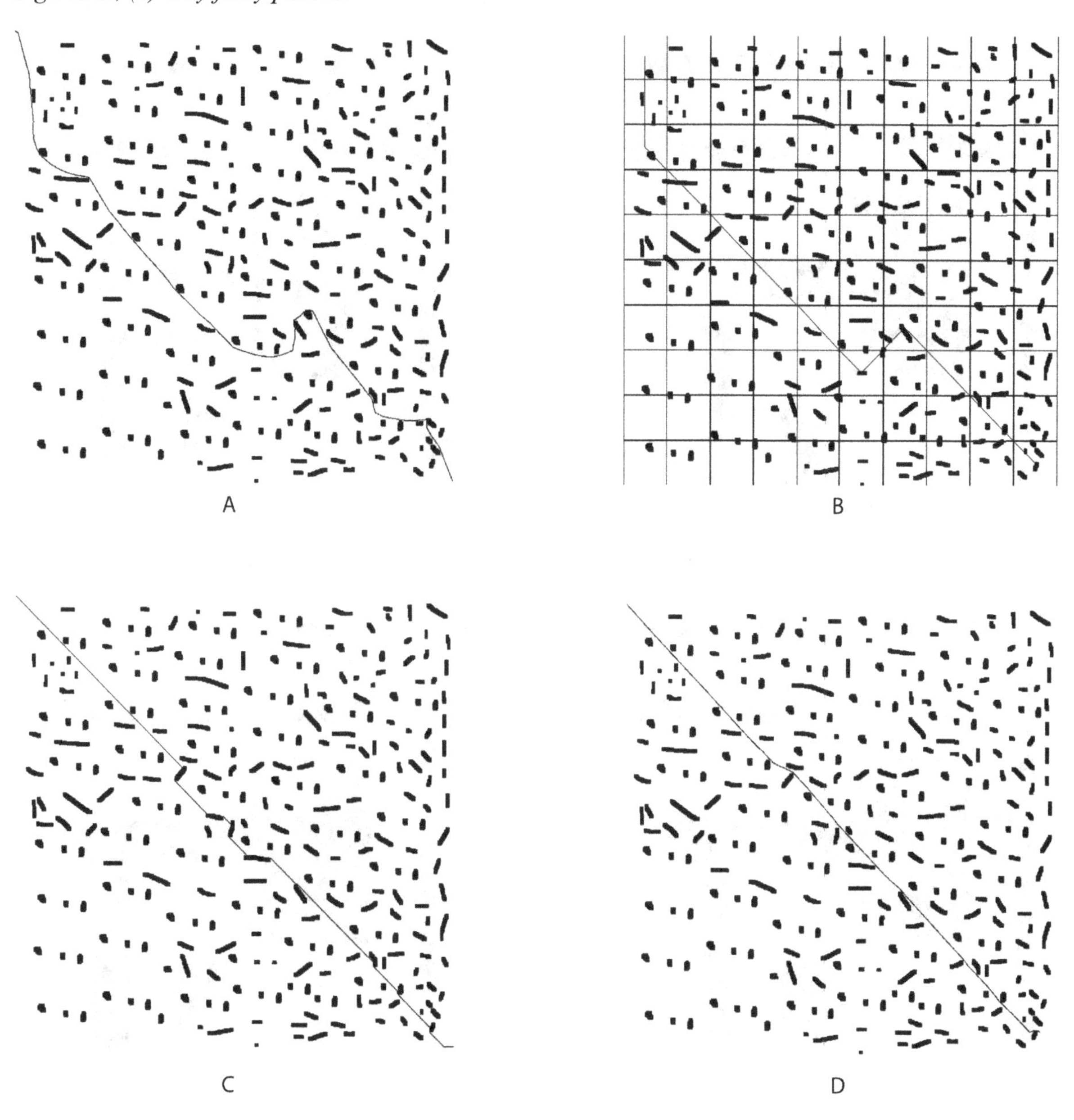

Figure 14a. The initial path given by the A* algorithm is given in Figure 14b. Here also, it tried to maximize the probability. The path traced only by the application of A* algorithm is given in Figure 14c, and only the fuzzy planner is given in Figure 14d.

Effect of Change in Grid Size

In this problem, we had introduced a 2-level hierarchical map. The finer level map was a detailed version of the coarser level map. The coarser level was probabilistic in nature. Here various

Figure 14. The simple maze test case (a) this algorithm, (b) result of A planning, (c) only A* algorithm, (d) only fuzzy planner*

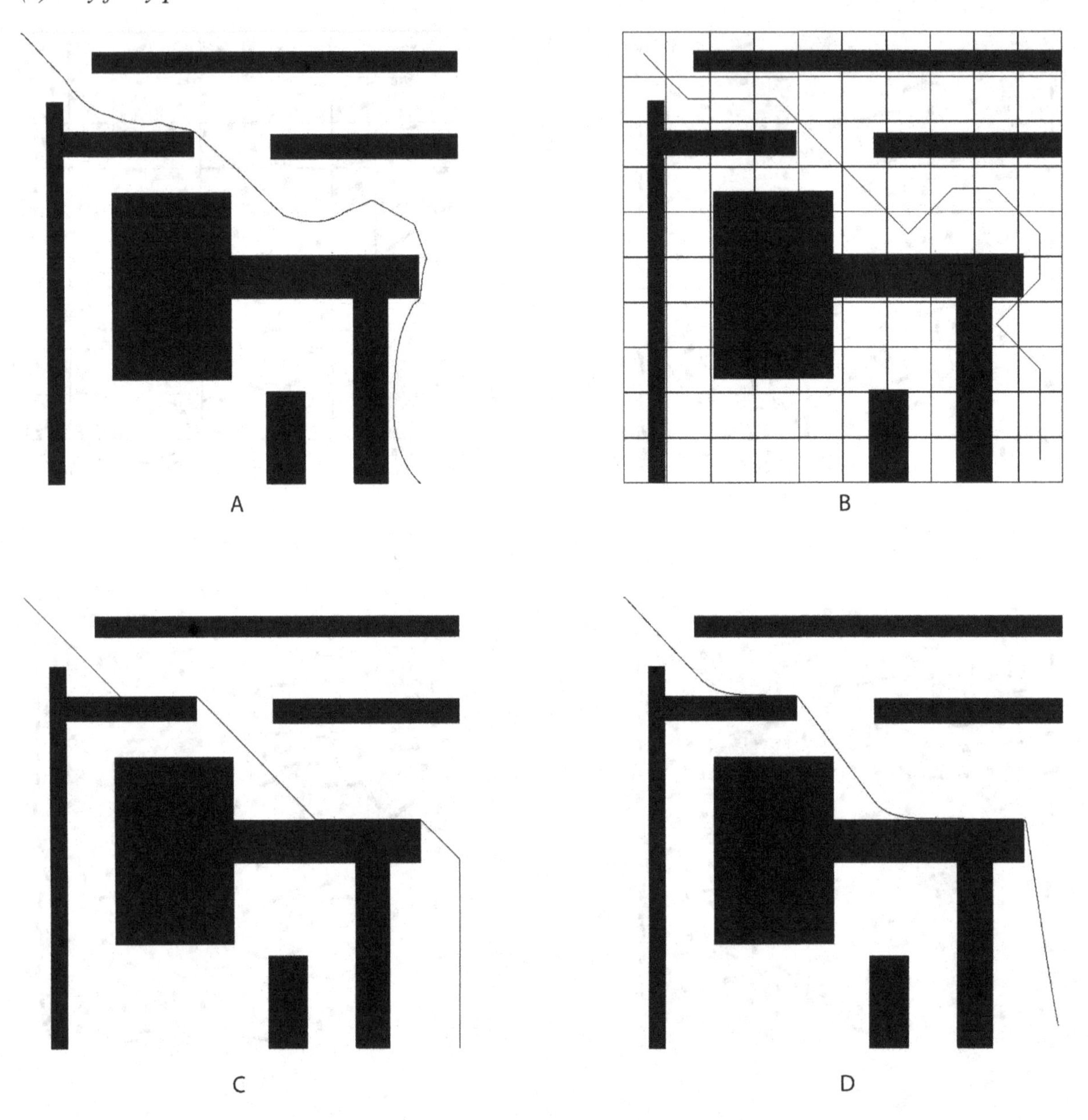

Figure 15. The path traced by robot for various values of α (a) α = 1000, (b) α = 100, (c) α = 20, (d) α = 10, (e) α = 5, (f) α = 1

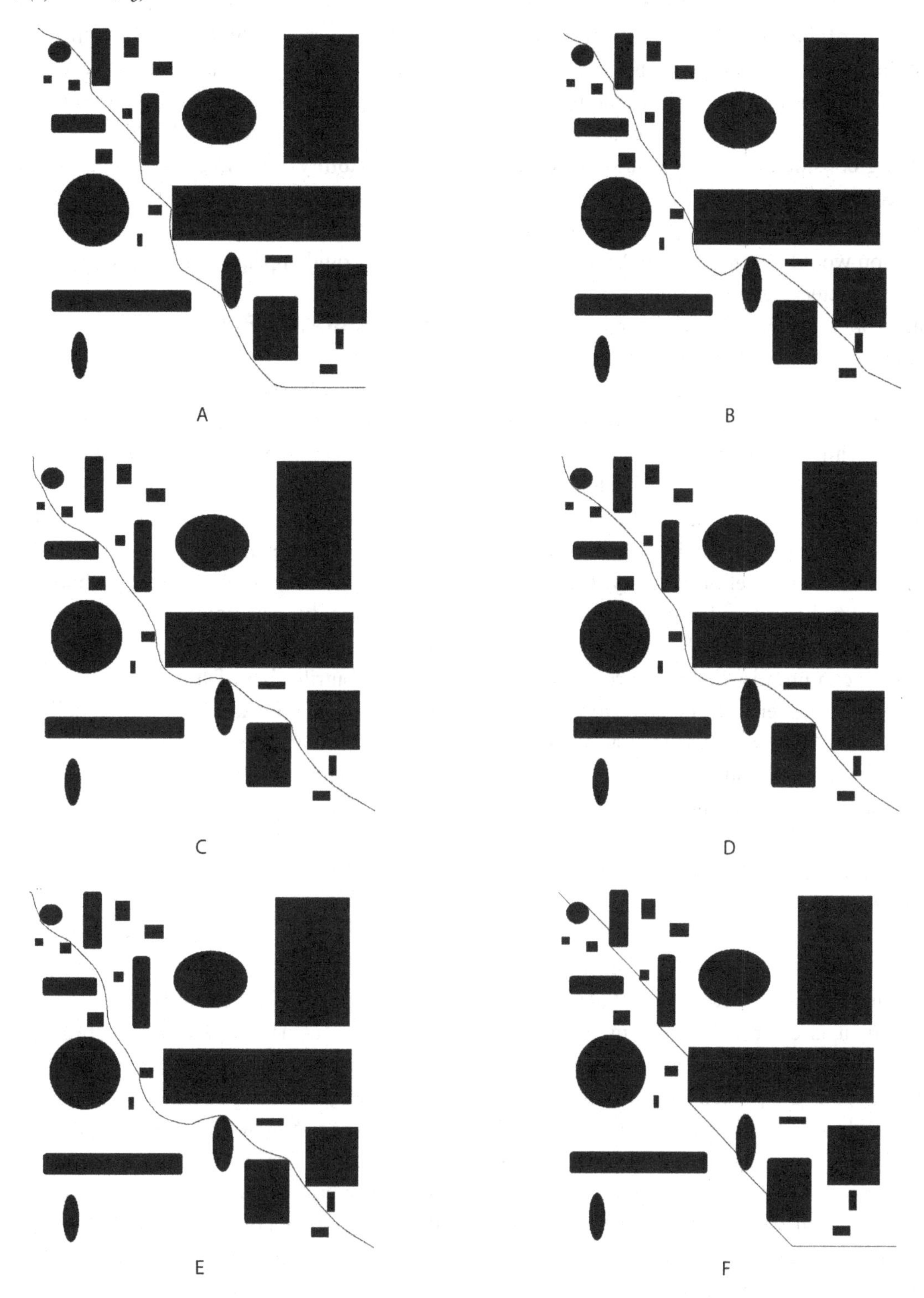

nodes of finer level had been clubbed to form a single grid or cell. The size of this grid (α) decides the working of the algorithm. If α is very small, almost the entire graph would be available at the finer level. The solution would be predominantly A* in nature. If the size is too large, there would be nothing considerable at the coarser level. In that case, the solution would be predominantly of the nature of fuzzy planner. In all other cases, the solution would be a mixture of A* and fuzzy planner characteristics.

The probability-based fitness used in the A* algorithm is the other factor that changes with the change of α. As we increase α, the probability keeps getting vague due to the higher grid size. The A* algorithm tries to find paths that have high probability of reaching to the goal. This in turn ensures that under any circumstances, the fuzzy planner would be able to reach the goal position with the maximum level of certainty. However, the probability has an undue effect on the path length. In order to increase the path probability, the length is compromised. This in turn causes the total path length traversed by the robot to increase.

We studied this effect into the various maps that we had used. The path traced by the robot for one such map is given in Figure 15.

Effect of Change of Grayness Parameter

The other parameter that we try to change is β. This parameter was discussed earlier. As per the discussion, it is clear as this parameter would increase, the conventional non-probabilistic behavior would get predominant in the behavior of the algorithm. We studied the change of this parameter into numerous maps. Figure 16 shows the results on the same graph that was used in our discussion. It may be seen that the algorithm failed for $\beta=0$. This is because the A* algorithm returned a straight path from source to goal which had very less probability. The other inputs could not produce a change in the output of the A* algorithm. It may be observed here that change of β tries to make a change in A* algorithm result. As a result, the change in path would be discrete rather than continuous. Whenever there is a change in path of the A* algorithm result, a discrete change is reflected. This depends upon the probability conversion and the grid size (α). The experiments were conducted at a large grid size. In smaller grids, it may be verified that the change would appear to be more continuous.

Discussion of Parameters

It may easily be seen that the total path is affected by 3 factors: (1) the contribution of the fuzzy planner that makes the path smooth at the same time reducing time of algorithm (it, however, may result in a longer path or the failure in finding path in case of complex maps), (2) the contribution of the A* algorithm in reducing the path length (α), which can solve very complex maps with optimal path length at the cost of computational time, and (3) the contribution of the A* to maximize the probability of the path (β), which usually would increase the path length.

The combined effect of the change in values of α and β would reveal the generation of unique paths. The final path, however, has a strong dependence on numerous factors like map, path simplicity, total map size, etc. Based on the results of Figure 12 and 13, it may be seen that we can generate all kinds of paths between A* algorithm, fuzzy planner and probability-based approach by the adjustment of these parameters. The flexibility, however, puts the constraint that these parameters need to be judiciously set, so as to attain optimal results. The parameters have a strong dependence on numerous factors like map, path simplicity, robot design, total map size, etc. This eliminates the possibility of constant values of these parameters for different scenarios and maps. The task of parameter setting for α and β may be done by considering the same factors. Very simple maps with fewer turns may have more A* contribution.

Figure 16. The path traced by robot for various values of β (a) β = 0, (b) β = 0.2, (c) β = 0.3, (d) β = 0.5, (e) β = 0.6, (f) β = 1

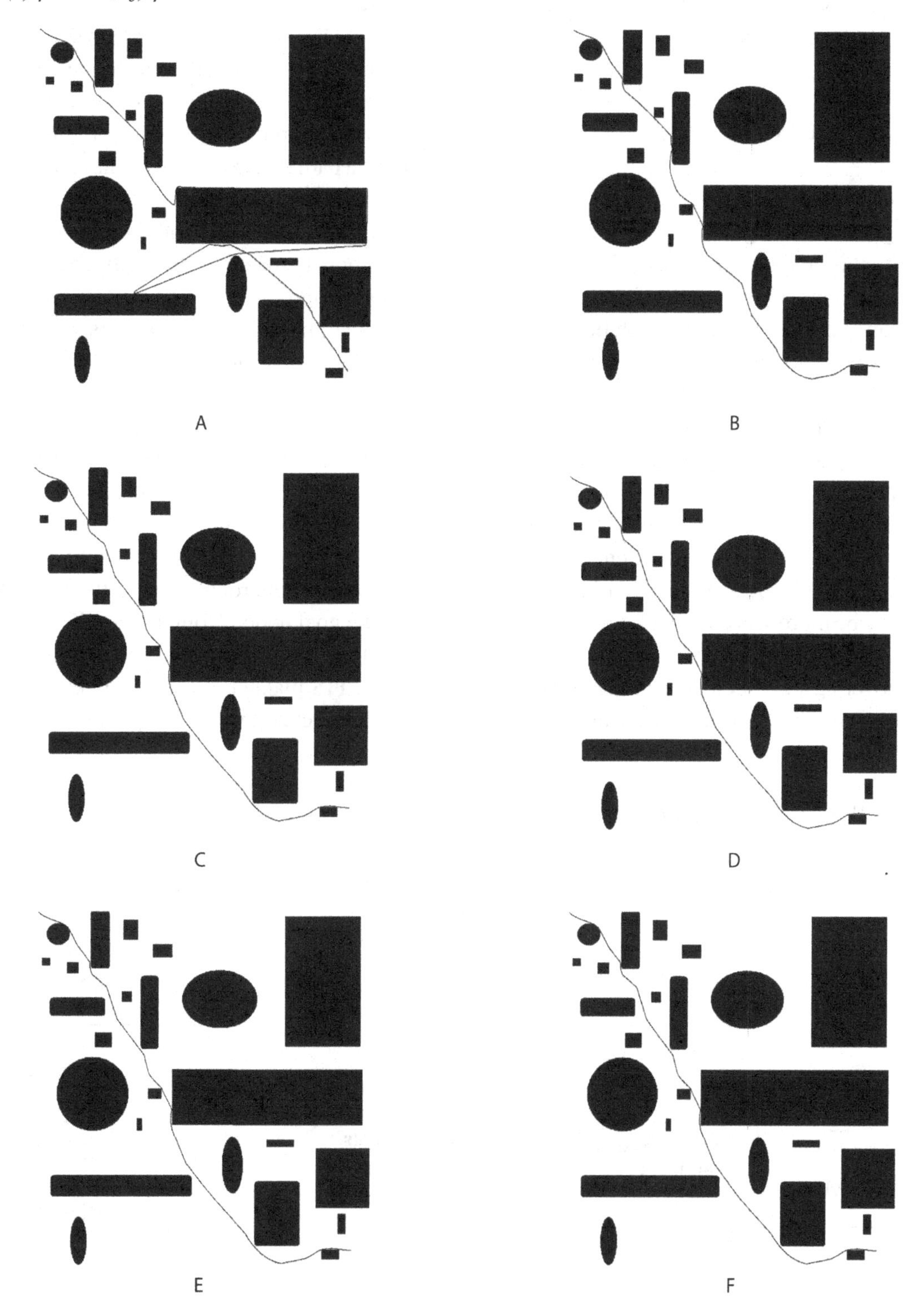

The presence of multiple objects in the optimal path may require a lot of fuzzy planning to escape from the variety of objects.

EVOLVING ROBOTIC PATH

The approach presented in the previous couple of sections gives a fairly good performance over a variety of maps and situations. Consider the A* algorithm serving at the coarser level. It is possible that the map be of extremely high resolution. There might be extremely high details that prohibit from decomposition of the maps into levels such that the coarser level has a low resolution. In such a case, the A* algorithm may struggle to produce results in a finite time span. High resolution may hence prohibit its use. Hence, in this section we replace the A* algorithm with the Genetic Algorithm (GA). The GA by its population-based approach may be more scalable for high-resolution coarser level maps. This section and subsequent sections covering the entire algorithm are reprinted with few modifications from Kala, Shukla, and Tiwari (2010) with permission from Inderscience Publishers.

The general structure of the algorithm is given in Figure 17. The only difference is the use of GA at the coarser level. The algorithm starts by taking as input the graph. The GA runs on this map to generate a path P. Here, we do not have a probabilistic map, but rather a probabilistic map converted into a discrete map. This path is generated using an evolutionary technique that optimizes both the path length and the complexity. The path P is a collection of points p_i such that p_1 is source node and p_N is the goal node. These points are then used one after the other (other than the source) to act as guide for the fuzzy planner. As soon as the robot gets close enough by some predefined threshold amount to the region in which the goal cell is found, the next point in P becomes the goal and the robot has to move a step further. This goes on and on until the robot reaches the final block where the goal node is found. The FIS is used in every block for guiding the robot. After the last iteration, FIS planner is used to make the robot reach the exact cell where the goal is located.

Figure 17. Evolutionary algorithm guided fuzzy planning algorithm

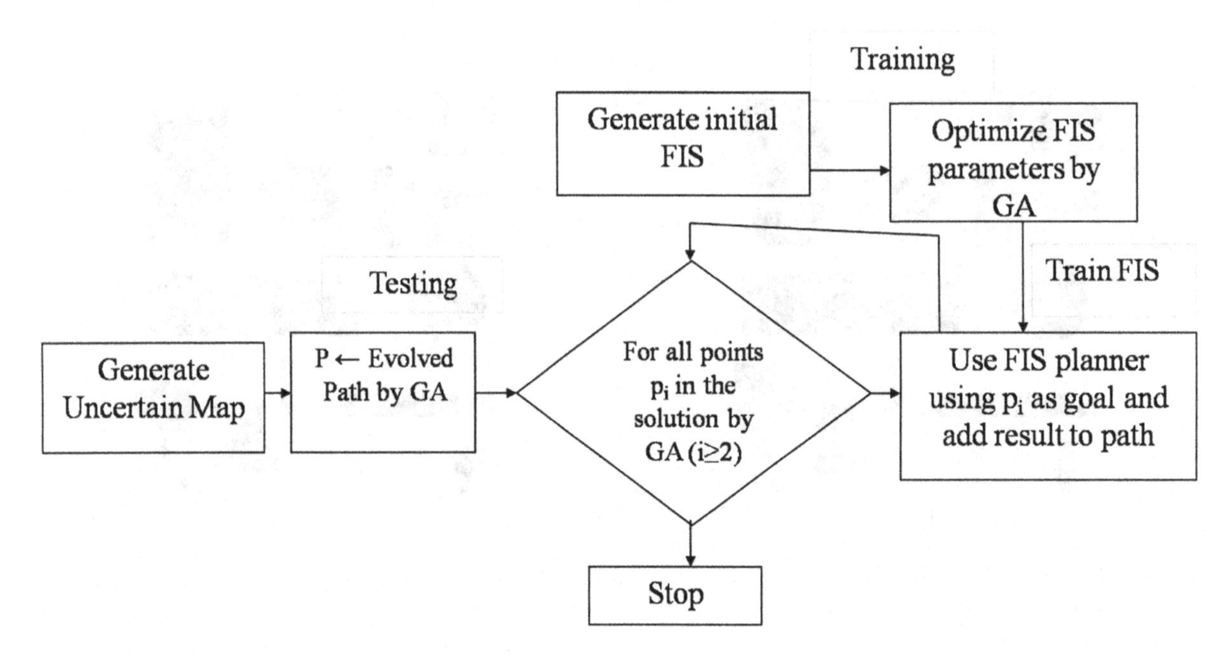

The only point of discussion of the algorithm is the GA framework, which is similar to the framework presented in chapter 8. Key points are highlighted for better understanding of the readers. In this algorithm, we try to evolve the robotic path with the help of GA. It is natural that the better paths that GA can return would be as straight as possible without too many turns. We have already stated that the path returned by the GA is in form of a collection of points P. The total number of points in the path is known as path complexity. It is assumed that the robot can reach the goal from the source by travelling in a straight line between every adjacent pair of points (P_i, P_i+1) in this collection P. This is checked at every fitness call of the GA. If a straight line path is obstructed by some obstacle, the path is regarded as infeasible.

Since we are interested in increasing the straightness, we claim that good solutions have as few points in this collection P as possible. In other words, the complexity needs to be the least. It may be easily visualized that for an empty graph only the source and the destination would be present in P. Similarly, if there is a unit obstacle in the way from source to destination, P may contain one point additional along with the source and the destination. The source and the destination would always be there in the solution. Hence, we do not consider them in the collection set but rather add them before returning the solution.

In order to control the size of the returned solution, we restrict the maximum number of points that the GA can sustain in any individual to α. We start with very less value of α (α =0, this factor was abbreviated differently earlier, change of abbreviation is to ease discussions). As the generations increase, we keep increasing the value of α. Here α increases as per the Gaussian curve. First, the increase in value of α with respect to generations is very rapid. As the generations keep increasing, the rise in the value of α is not very vast. After few more generations α seems to be more or less constant. The relation between α and N, the number of generations is given by Equation 6. Here, g is the number of generation and σ and A are constants.

$$\alpha = A \exp(-g^2/2\sigma^2) \tag{6}$$

The solution pool at any time would be consisting of individuals of different length and fitness. Hence, it is necessary to adapt a strategy for the genetic operators to be used between these chromosomes of unequal length. In addition, many new genetic operators may be required for the purpose of carrying out specialized operations specific to the problem.

The first task is to frame an individual representation technique. Here each individual is a potential solution to the problem of path planning that we are dealing. The individuals are a collection of points that the robot must access in sequential order in order to reach the goal. The source and the goal are not present in the collection of points. Each point has x coordinate and y coordinate. We place a constraint on the individuals. Suppose a line is drawn from the source to the destination. All points represented by the solution need to be sorted in the order of this line. This means that every movement of the robot would take it closer to the destination. If any individual generated during any operation disobeys this rule, it needs to be repaired via a repair operation where this sorting takes place. The chromosomes in this case would be of varying sizes.

Crossover is the major evolutionary operator of use. Here, we derive a strategy that carries out crossover operation between two chromosomes that are of unequal length. The first task that we do is to make the length of the chromosomes equal. This may be done by duplicating random points in the smaller chromosome, until the lengths of both chromosomes are same. Duplication creates the new point at the same index to keep the sorted order same. Now one point crossover is carried out in which half the points are taken from first parent and half from the other. Duplicate points from the generated child are deleted. This many times would result in killing of smaller solutions, which do not generate results.

Mutation selects points in the individual at random and moves them over the graph by some arbitrary percentage. The percentage depends upon the mutation rate used in the algorithmic implementation. A custom genetic operator used is Add Points that, when applied over an individual, cause it to request for an addition of a point in its collection. If it is less than α, then the request may be granted with a probability p. The probability p is inversely proportional to the chromosome length. Similarly, another custom operator is used called Add Individual, which inserts new large-sized solutions in the population pool. As the generations grow the need of larger population size would be necessary. This operator inserts individual at a rate R that increases similar to that of α. The last custom operator used is Flush Population. This operator that is active mainly at start of the algorithm when the total number of points (α) is much less that the least number of points to solve the problem. This operator deletes all small infeasible solutions and replaces them with relatively big feasible solutions. Elite passes the best individuals of lower generation to the higher generation. This operator ensures that the best individuals do not get deformed with generations due to genetic operations.

The fitness function is simply the length of the total path that may be computed by the point wise traversal. If traveling at some path results in a collision, then the maximum fitness function is assigned to the individual and the individual is regarded as infeasible. An extra penalty of Lβ is added where L is the path complexity or the number of points in the chromosome representation and β is a constant.

SIMULATIONS AND RESULTS

The population was represented by the discussed technique. In all simulations, the population size was fixed to 700. Rank-based scaling was used. Stochastic Uniform selection was used. The crossover rate was 0.7 with elite count as 2. Mutation rate was fixed to be 0.06. The GA was made to run over 500-1000 generations depending upon the map used. The maximum value of the complexity (A) was fixed to be 6 and associated σ was fixed to 0.3 times the total generations. β was fixed to 0.15 times the total map perimeter.

We tested the algorithm against 3 maps. This time the maps were of larger size of 1000X1000. Many obstacles were places in between the source and the goal. The robot was supposed to reach the goal from the source. The top left corner is the source and bottom left corner is the goal. In all the cases, the source was fixed as the top left corner and the goal was fixed as the bottom right corner. The results along with the maps for these cases are shown in Figure 18a, b, and c.

In all the cases, we can easily see that the algorithm tried to optimize not only the total path length but also the complexity. Both these parameters have different roles to play in the entire algorithm. The total path length tries to optimize the net traversal of the robot. At the same time, the complexity factor tries to return a path that would be simple to work with using the FIS planner. Further lesser turns would mean more straightness and a higher speed of travelling. This would in turn give a lot of advantage to the robot to run at a high speed and reach the goal. The FIS planner did a lot of smoothening of the path generated by the GA. This is clearly seen in the results that show a smooth change in angles as the robot travels. The path thus generated has the discussed advantages.

Figure 18. The path traced by robot for various maps (a) map 1, (b) map 2, (c) map 3

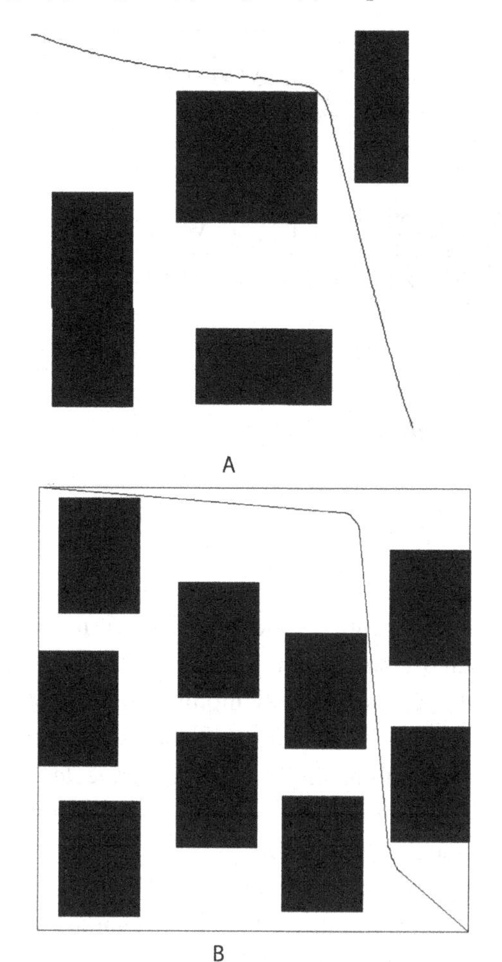

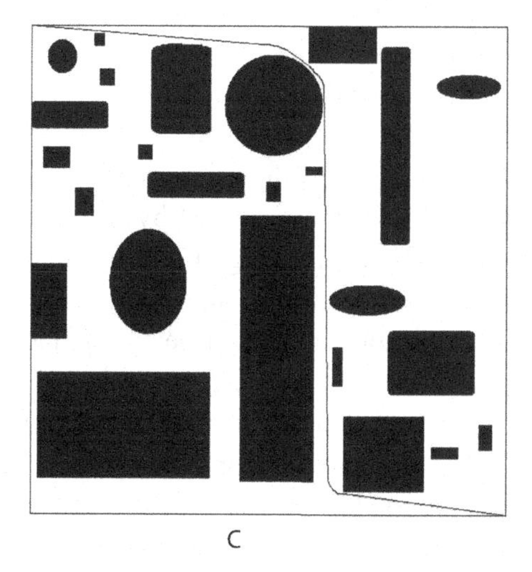

PLANNING BY ACCELERATED NODES

This section presents the work of planning for a mobile robot by using Dynamic Programming as a base planning algorithm. The implementation of the algorithm is similar to the implementation of neural networks where neurons are embedded into the map and exchange information with each other regarding distances to the goal. The dynamic programming also uses nodes with continuous exchange of information between them. However, the dynamics of the node is governed by the recurrence relation of the dynamic programming. Basics of the technique and the method to solve the problem were discussed in chapters 2 and 3. For more details regarding the approach, please refer to Kala et al. (2012).

It is assumed that the entire map is embedded with nodes, which we take to be uniformly distributed in a grid pattern for simplicity. Hence, the nodes of dynamic programming are similar to the grids used in graph-based approaches. Similarly, we model the allowable moves of the robot signifying from which node the robot can move. Recurrence relation of the dynamic programming is used in which every node computes its shortest distance to goal without encountering any obstacle. In this manner, there is an initial phase where all nodes compute their distances to the goal. The parent information is also kept which records the path, which makes this shortest distance. Once this stage is over, the robot simply moves towards the goal by iteratively following the parent from the current node.

Any change in the environment produces a change in the distances to goal, which must be made in the algorithm using the dynamic programming recurrence relation. These changes require time for the calculations to complete and the time depends upon the shortest distance of the goal from robot's current position in the modified map. During this time, the robot does not have

Figure 19. Sudden blockage situation

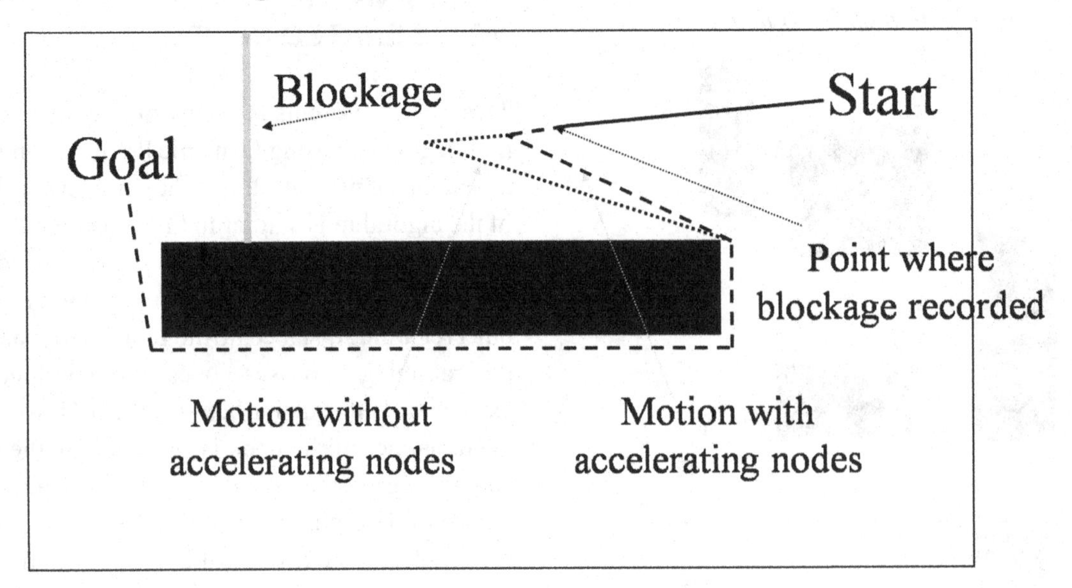

an idea of the optimal path. If there is no or little change in the optimal path in the changed map, the robot may continue moving as per the old non-updated values. However, condition of sudden blockage is different. In such a situation, the path may completely change and moving on the non-updated path may take the robot away from the goal. This situation is represented in Figure 19. This forms one of the major limitations of the approach. Since updating may require a long time, the robot continues to march in the wrong direction for quite long, until the values are updated and then it needs to go back and follow the correct path.

The manner in which we solve this problem is by using additional nodes in the map called the accelerating nodes. These nodes are much less in number and form another dynamic programming framework at a coarser level. These nodes along with the normal nodes for a map are shown in Figure 20. An accelerating node can talk to few accelerating nodes in its neighborhood and exchange information with them. Since these nodes are lesser in number, their updates are very fast in the entire map at the coarser level. Every accelerating node knows the distance to all neighboring accelerating nodes which is an information extracted from planning at the finer level. This distance is continuously monitored. If there is a small change in the map, such that two neighboring accelerating nodes are still connected, this value being monitored increases by a small amount and then becomes stable.

The increase in the value is equal to the increase in the path length by the small change. If the change produces a better path, this value decreases. Since the two nodes are still connected to each other, no action is taken, and while the robot travels in the map, all nodes at finer level attain their optimal values. Since the change is small the change in path may normally be small. However a blockage is sensed by a continuous increase in this value being monitored. In case a blockage occurs between any two neighboring nodes, the separation between them keeps increasing up to very large value. This can be obtained from the recurrence relation of dynamic programming. In such a case, the main algorithm, operational at the final level, is withheld, and a substitute algorithm is initiated. The change only takes place if the blockage affects the optimal path of the robot. For this, it is essential that the robot

Figure 20. Normal and accelerating nodes

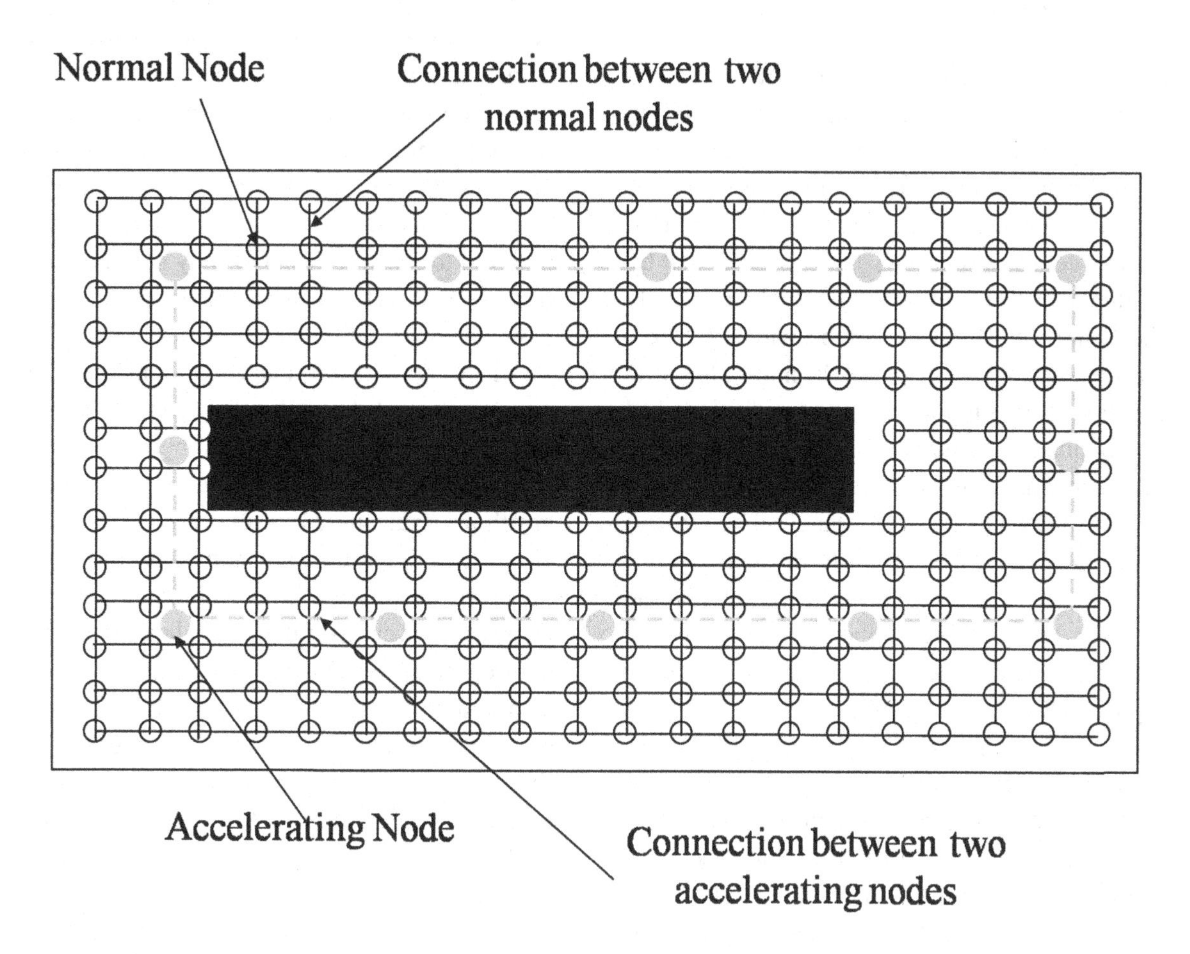

should be traveling in the region around the two nodes between which the blockage is recorded.

Upon sensing any blockage that affects the robot's plan, the first task is to decide the immediate action of the robot. Every accelerating node knows the neighboring accelerating nodes to which it is connected. This forms a graph at the coarser level between the accelerating nodes. A* algorithm is used to find a path from the accelerating node nearest to the current position of the robot to the goal at the coarser level consisting only of accelerating nodes. This path at the coarser level forms a guiding strategy for the robot's motion in a manner similar to the use of A* algorithm for robot traveling with FIS as discussed earlier in the chapter. In this manner, the robot may immediately change its course of action if required and move towards the goal. Note that this may not be the optimal path to goal, as at every iteration, the next accelerating node is used as immediate target. Therefore, the robot basically follows one accelerating node from the other to reach the target. This may not ensure overall optimality.

After a few robotic steps or some iterative cycles, however, the dynamic programming at the finer level has all the updated values recording the blockage. Hence, the robot now leaves execution by the substitute algorithm and continues moving by the basic dynamic programming algorithm.

CONCLUSION

In this chapter, we first proposed a method to solve the problem of path planning using a combination of A* algorithm and fuzzy planner. We tested the algorithm for various test cases. In all the test cases, we observed that the algorithm was able to find the correct solution. The solutions generated by the algorithm were optimal. They took a relatively short time to solve for each of the maps presented. Further, the solutions obeyed the non-holonomic constraints. In this approach, we had further used a probabilistic approach in the map representation. The map was implemented at two levels for the A* algorithm and the fuzzy planner. The FIS was genetically optimized for performance.

The results convey a very unique relation between the working of the A* algorithm and the FIS planner. The A* algorithm works over a probabilistic map trying to optimize the path length and at the same time making the probability of reaching the goal high. The FIS was used primarily for giving smooth paths according to the non-holonomic constraints and ensuring a timely response to the system. The A* algorithm reduced the total path length by selecting the solutions that have less total length. At the same time, the increase of probability resulted in selection of longer paths at some positions.

We further solved each of the paths with all three algorithms: (1) presented algorithm, (2) only A* algorithm, and (3) only fuzzy planner. By results, it may be seen that this algorithm is a unique mix of the other two algorithms. It tries to combine the best features of path length and smoothness in the solution it generates.

The contribution of each of the A* algorithm and fuzzy planner may be varied by changing the grid size (α) and the probability factor (β). This would make the algorithm behave more like one or the other. If the grid size is very small, the algorithm would be primarily A* in nature. This may induce very sharp paths. Further, if the grid size is very large, the resulting algorithm would be dominated by the fuzzy planner properties. It may not be able to solve complex maze like paths. Similarly, if β is very large, the total path length might be more due to the effort of A* to optimize the path length. Reducing β may result in shorter paths, but at the same time the robot may fail to reach the goal or may get obstructed from the path as the path it is trying to follow may have low probability.

In the second approach, we considered the complexity and the high resolution of the map. The A* algorithm was replaced by GA for planning at the coarser level. We again tested the algorithm for various test cases. In all the test cases, we observed that the algorithm was able to find the correct solution. The evolutionary technique adopted was very useful in path planning in static environments where the planning led to very small paths being generated, which had a broad area that was highly straight. The straightness of the path was mainly due to the preference given to the shorter solutions with lesser complexity.

In the last approach, we floated the concept of accelerating nodes. These were special nodes added into the conventional dynamic programming architecture that resulted in a good performance in case of sudden blockage. These nodes could quickly sense the blockage and communicate the same so that the robot can re-plan accordingly. These nodes further provided some idea of optimal path to the robot while the normal nodes update themselves.

REFERENCES

Camilo, O., Collins, E. G. Jr, Selekwa, M. F., & Dunlap, D. D. (2008). The virtual wall approach to limit cycle avoidance for unmanned ground vehicles. *IEEE Transaction on Robotics and Autonomous Systems*, *56*(8), 645–657.

Chen, L. H., & Chiang, C. H. (2003). New approach to intelligent control systems with self-exploring process. *IEEE Transactions on Systems, Man, and Cybernetics. Part B, Cybernetics*, *33*(1), 56–66. doi:10.1109/TSMCB.2003.808192

Cortes, J., Jaillet, L., & Simeon, T. (2008). Disassembly path planning for complex articulated objects. *IEEE Transactions on Robotics*, *24*(2), 475–481. doi:10.1109/TRO.2008.915464

Hui, N. B., & Pratihar, D. K. (2009). A comparative study on some navigation schemes of a real robot tackling moving obstacles. *Robotics and Computer-integrated Manufacturing*, *25*(4-5), 810–828. doi:10.1016/j.rcim.2008.12.003

Hwang, J. Y., Kim, J. S., Lim, S. S., & Park, K. H. (2003). A fast path planning by path graph optimization. *IEEE Transactions on Systems, Man, and Cybernetics. Part A, Systems and Humans*, *33*(1), 121–128. doi:10.1109/TSMCA.2003.812599

JPL. (2012). In-situ exploration and sample return: Autonomous planetary mobility. *Jet Propulsion Laboratory*. Retrieved on January, 2012, from http://marsrover.nasa.gov/technology/is_autonomous_mobility.html

Kala, R., Shukla, A., & Tiwari, R. (2010a). Fusion of probabilistic A* algorithm and fuzzy inference system for robotic path planning. *Artificial Intelligence Review*, *33*(4), 275–306. doi:10.1007/s10462-010-9157-y

Kala, R., Shukla, A., & Tiwari, R. (2010b). Evolving robotic path with genetically optimized fuzzy planner. *International Journal of Computational Vision and Robotics*, *1*(4), 415–429. doi:10.1504/IJCVR.2010.038196

Kala, R., Shukla, A., & Tiwari, R. (2012). *Robot path planning using dynamic programming with accelerating nodes*. Paladyn Journal of Behavioural Robotics. *3*(1), 23-34. doi:10.2478/s13230-012-0013-4

Kambhampati, S., & Davis, L. (1986). Multiresolution path planning for mobile robots. *IEEE Journal on Robotics and Automation*, *2*, 135–145. doi:10.1109/JRA.1986.1087051

O'Hara, K. J., Walker, D. B., & Balch, T. R. (2008). Physical path planning using a pervasive embedded network. *IEEE Transactions on Robotics*, *24*(3), 741–746. doi:10.1109/TRO.2008.919303

Pozna, C., Troester, F., Precup, R. E., Tar, J. K., & Preitl, S. (2009). On the design of an obstacle avoiding trajectory: Method and simulation. *Mathematics and Computers in Simulation*, *79*(7), 2211–2226. doi:10.1016/j.matcom.2008.12.015

Pradhan, S. K., Parhi, D., & Panda, A. K. (2009). Fuzzy logic techniques for navigation of several mobile robots. *Applied Soft Computing*, *9*(1), 290–304. doi:10.1016/j.asoc.2008.04.008

Shibata, T., Fukuda, T., & Tanie, K. (1993). Fuzzy critic for robotic motion planning by genetic algorithm in hierarchical intelligent control. In *Proceedings of 1993 International Joint Conference on Neural Networks*, (pp. 77-773). IEEE.

Tsai, C., Lee, J., & Chuang, J. (2001). Path planning of 3-D objects using a new workspace model. *IEEE Transactions on Systems, Man and Cybernetics. Part C, Applications and Reviews*, *31*(3), 405–410. doi:10.1109/5326.971669

Urdiales, C., Bantlera, A., Arrebola, F., & Sandoval, F. (1998). Multi-level path planning algorithm for autonomous robots. *IEEE Electronic Letters, 34*(2), 223-224.

Willms, A. R., & Yang, S. X. (2006). An efficient dynamic system for real-time robot-path planning. *IEEE Transactions on Systems, Man, and Cybernetics. Part B, Cybernetics*, *36*(4), 755–766. doi:10.1109/TSMCB.2005.862724

Yang, S. X., & Meng, M. (2000). An efficient neural network approach to dynamic robot motion planning. *Neural Networks*, *13*(2), 143–148. doi:10.1016/S0893-6080(99)00103-3

Zelinsky, A. (1994). Using path transforms to guide the search for findpath in 2d. *The International Journal of Robotics Research*, *13*(4), 315–325. doi:10.1177/027836499401300403

Zhu, D., & Latombe, J. (1991). New heuristic algorithms for efficient hierarchical path planning. *IEEE Transactions on Robotics and Automation*, 7(1), 9–20. doi:10.1109/70.68066

KEY TERMS AND DEFINITIONS

Accelerating Nodes: Additional nodes added in the map that represent the map at a coarser level. These nodes sense blockage. In case a blockage is detected, an alternative path is quickly built for immediate motion of the robot. This path is followed until the optimal path is re-computed as per modified map.

Evolutionary Fuzzy Inference System: Tuning the fuzzy rules, membership function parameter values, and sometimes even the entire design of fuzzy inference system by Evolutionary Computing algorithms. In the presented approach, a small search space across the hit and trial solution was cut out in which genetic algorithm carried optimization.

Fusion with Fuzzy Inference System: Attempt to use the real time and non-holonomic properties of fuzzy inference system, by adding a planning algorithm at coarser level that ensures completeness and optimality.

Path Blockages: Scenario where current path of the robot to goal may be obstructed by some obstacle. The optimal path in the modified map after blockage may be very different from the original optimal path.

Path Guidance: Computing a rough path consisting of some points or milestones from source to goal, which may be used for directing a lower-level planning algorithm. The immediately next milestone may become goal for the lower-level planner, which may be changed to next milestone as per strategy.

Chapter 10
Multi-Robot Systems

ABSTRACT

While the concepts of robotics and planning may be easily understood by the taking a single robot, it is not necessary that the problems we solve have a single robot in the planning scenario. In this chapter, the authors present systems with multiple robots, each robot attempts to coordinate and cooperate with the other robots for problem solving. The authors first look at the specific problems where multiple robots would be a boon for the system. This includes problems of maze solving, complete coverage, map building, and pursuit evasion. The inclusion of multiple robots in the scenario takes all the concepts of single robotic systems. It also introduces some new concepts and issues as well. They look into all these issues in the chapter which include optimality in terms of computational time and solution generated, completeness of planning, reaching a consensus, cooperation amongst multiple robots, and means of communication between robots for effective cooperation. These issues are highlighted by specific problems. The problems include multi-robot task allocation, robotic swarms, formation control with multiple robots, RoboCup, multi-robot path planning, and multi-robot area coverage and mapping. The authors specifically take the problem of multi-robot path planning, which is broadly classified under centralized and decentralized approaches. They discuss means by which algorithms for single robot path planning may be extended to the use of multiple robots. This is specifically done for the graph search, evolutionary, and behavioral approaches discussed in the earlier chapters of the book.

DOI: 10.4018/978-1-4666-2074-2.ch010

INTRODUCTION

The basic notion of the book was to explain the basic concepts in planning for mobile robotics. With the same aim, we discussed a number of algorithms from simple to complex that solve the problem. While this completes the book to a large extent in terms of path planning, it is important to understand other key problems in robotics, which may be of interest to the readers. Hence, we devote this chapter entirely to the complementary topics, which could not find a place in the rest of the book. The broad domain of discussion is multi-robot systems. While a lot has been said and discussed about the use of robots in industry and household, it may be easily realized that many times the work to be performed can be done quickly by using multiple robots in place of a single robot. However, to do the work effectively all the participating robots must cooperate and coordinate with each other, ensuring that the total work is performed efficiently. This opens an entire pool of issues, which we briefly touch upon.

The agenda of the chapter is two-fold. First, we intend to present the broad domain of planning, which goes beyond the task of path planning as discussed throughout this text. Much of the problems presented in this chapter open new dimensions to planning, where intelligent techniques can be used. The entire domain of path planning introduced plenty of heuristics and intelligence specific to the problem modeling, which helped to better understand the base methods and how to use them for solving problems in a better way. The new problems further introduced this approach, adding new thoughts and problem-solving techniques. Second agenda of the chapter is to discuss the notion of multi-robot motion planning in depth. Having multiple robots in the system requires intelligent techniques at multiple stages of planning. Every level presents some challenges, which may be effectively tackled by sophisticated algorithms. All this adds to the understanding of the base methods and their intelligent use for problem solving.

We first discuss some of the other planning problems and later extend the discussion to the issues of multi-robot systems. Specific problems to multi-robotics are also taken and explained. The problem of path planning, which is the major problem tackled in the book, is later extended to the presence of multiple robots.

PLANNING PROBLEMS IN ROBOTICS

The domain of robotics is extremely wide. There is increasing use of robotics from industry to personal use, from military to space, and from healthcare to entertainment. A large number of domains are benefited by robotics. While the automation experts are working towards making systems as autonomous as possible, research continues to raise new issues, concerns, and solutions. While each of these domains has problems from the complete hierarchy of robotics, our focus is specifically onto the problem of planning. Path planning is not the only problem that these domains demand. There may be other levels of planning needed, depending upon the domain and the requirements. Further research sometimes creates benchmark problems that have physical relevance. The intention is to research what would be needed in future, as well as to encourage innovative ways of problem solving which may have relevance to existing problems.

While the focus of the book was the task of motion planning, we cannot neglect the other problems. In this section, we briefly describe some of the research problems in robotic literature, at the planning level, which are under constant research. All these problems in some manner are similar to the problem of robot path planning as per the ways we modeled it in the earlier chapters; however, these introduce some characteristic behaviors that make them challenging to solve

Figure 1. Maze solving problem

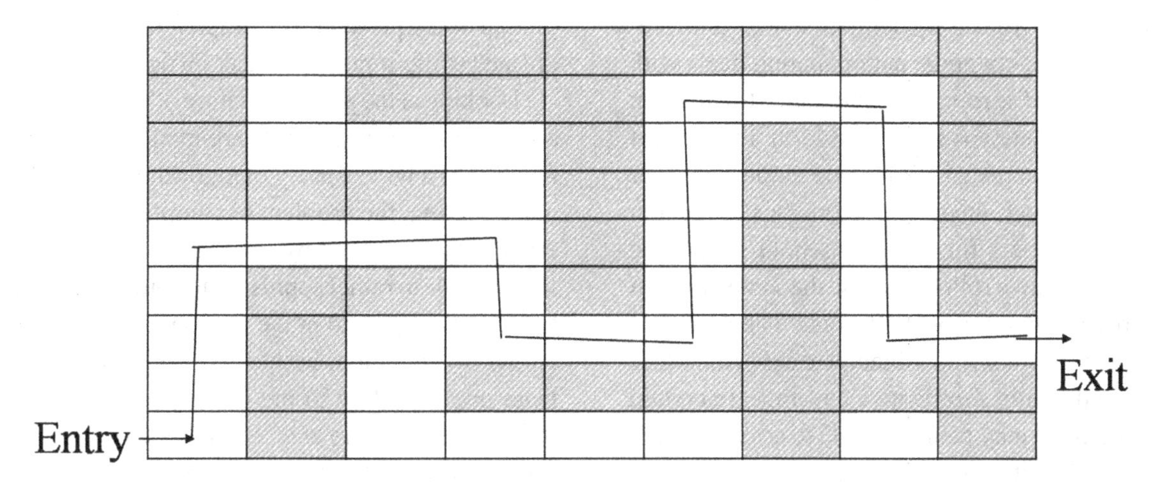

as well as expose new set of issues. We restrict our discussion to three such problems, which are maze solving, complete coverage problem, and pursuit evasion. These are discussed in the next sub-sections.

Maze Solving

The basic motivation for this problem comes from the use of robots in dangerous remote operations (Dracopoulos, 1998; Sadik, et al., 2010). Robots may be needed to be sent in remote places and explore the same. Lack of signals or large delay may make it impossible for humans to control robots. Further, if environment is complex, computational power of robots may make the operation easier. In the problem of maze solving robots need to navigate an unknown or semi-known environment. The robot, once given the source and the goal, must be able to figure out the complete path and reach the goal safely. The environment may not be known. In this manner, the problem differs from conventional path planning problem. MicroMouse robotics competition is a well-known competition where robots made by numerous research groups compete with each other for maze solving. A sample problem is given by Figure 1.

A large number of algorithms are designed for the problem which may be broadly classified into the categories whether the environment is known or unknown. For known environments, this problem reduces to problem of robot path planning. For unknown environments, a simple technique is to make random movements in hope to reach the goal. Another simple technique that works a number of times is to simply follow a wall. Here, the assumption is that the different walls connect to the outer boundary of the maze, or else the robot may find itself trapped in a loop. A more complex pledge algorithm keeps track of its angle and makes use of it to decide the motion, attempting to move towards a direction. A different modeling is done by the Tremaux's algorithm, which marks the way it visits. Markings are used for decision making regarding the way to follow at any juncture. The aim is to visit places with lesser and lesser marks, which represent possibility of a solution.

Complete Coverage Problem

Complete coverage represents a very similar problem as compared to maze solving or general path planning (Choset, 2001; Enders, et al., 1998; Hedberg, 1995; Luo, 2002). Unlike conventional path planning, where the attempt is to make the robot move from a predefined source to a predefined goal, coverage path planning deals with the ability of a robot to completely traverse an area. Complete coverage is used in a variety of application areas which include automatic floor cleaning, which attempts to automatically vacuum clean the floors for both household and office purposes. Lawn rollers also use the same problem for regular maintenance of grass. The complete coverage problem must attempt to cover the entire area with minimal repetitive coverage. In certain situations, it may be impossible to cover the area efficiently with no overlap. In many other cases, algorithmic limitations might lead to the same. The knowledge of the environment also plays a major role. Completeness in this problem deals with the ability of the algorithm to insure complete area is covered. Many times the characteristic nature of map or the presence of complex framework of obstacles might make the task difficult.

Looking at the practical nature of the problem, a large number of attempts are made, which has resulted in a wide variety of algorithms. Just like maze solving, the problem of complete area coverage may be solved by a variety of techniques from simple to rather sophisticated ones. A simple technique may be to make random moves at the corner of the region. In this manner by bouncing from one boundary to the other, eventually one may cover the entire area. Simple behaviors may further be modeled to enable robot to overtake obstacles, move to a point, etc. In case the map is known, there is a lot more that can be done.

Similar to the problem of planning, a technique may be to have the map represented in multiple

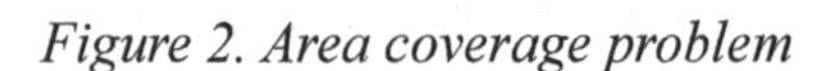

Figure 2. Area coverage problem

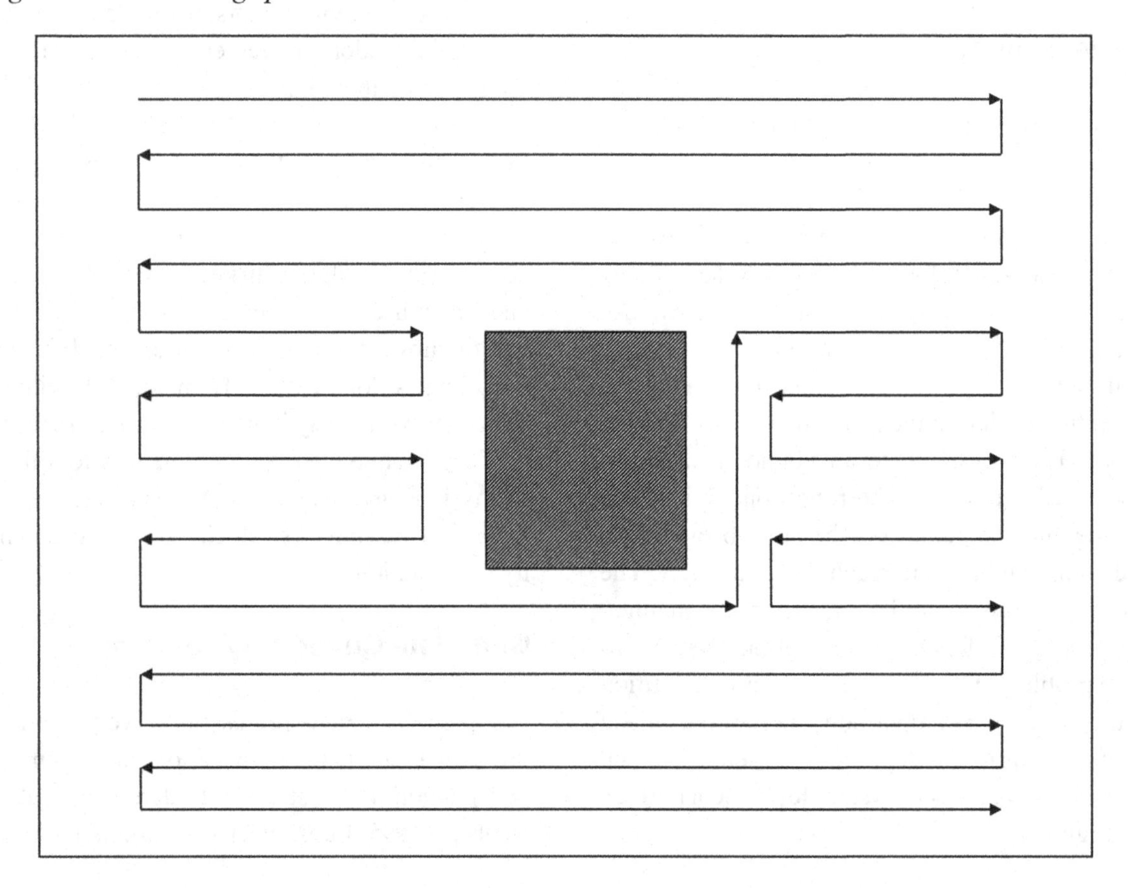

granularities with the initial planner at coarser levels. Here the complete area to be covered needs to be represented as a collection of cells that together make up the map. The task is to formulize an order in which the different cells are visited. Common technique is to follow the entire map in a manner similar to manual clearing of a room, where the entire length of room is traversed for fixed segments of breadth. Regions behind obstacles may be left for latter traversal, or intercepted as we find a path to reach the area covered by obstacle. Similarly, traversal may be circular in nature in place of being aligned to room length or breadth. A simple strategy for solving the problem is shown in Figure 2.

Pursuit Evasion

Pursuit Evasion is a game that is used as a benchmark problem these days due to the various research challenges that it presents along with the potential practical applications of it in real life equivalents (Brooks, et al., 2008; Cheng, 2003; Guibas, et al., 1985; Isaacs, 1965; Vidal, et al., 2002). The basic game structure consists of two players called as the pursuer and evader. The task of pursuer is to capture the evader as early as possible, while the evader attempts to escape from being captured. Both the pursuer and evader might represent groups, in which case any pursuer may catch any evader. The challenging part of the game is that the pursuer needs to think over the potential evader moves and act accordingly so that capture may be performed. It may be assumed that the pursuer possesses a higher speed, failing which it would be very simple for the evader to escape.

Potential applications include defense and police applications, search and rescue operations, etc. The motion of evader makes this problem very different from the path-planning problem, where the source and goal are completely static. In this problem, there is little or rather no use to reach the point presently occupied by the evader, since the duration in which the pursuer moves the evader would already have left. The task here for the algorithm is to develop a strategy that may be used for accomplishing the task. The game may be illustrated by Figure 3.

For this problem, also a number of efforts have been made which has resulted in a variety of

Figure 3. Pursuit evasion problem

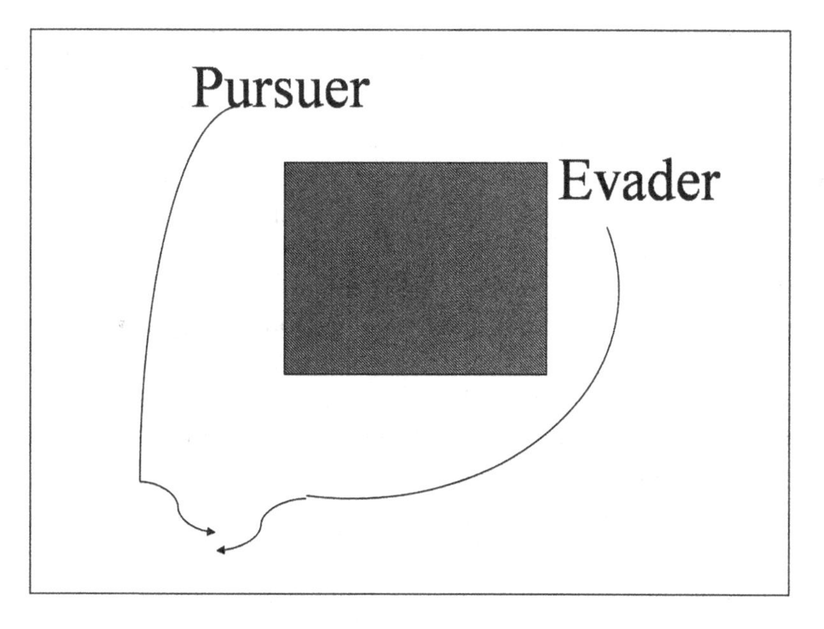

solutions of varying complexities and slightly varying in the manner of problem modeling and assumptions. Use of differential equations for modeling the motion of both the pursuit and the evader are one class of solutions. These equations then make it a mathematical problem formulation where equilibrium denotes the capture. Another common technique uses a probabilistic formulation. The technique uses probabilistic techniques to build map and move the pursuit and evader. Bayesian rules are use to guess the potential moves and the same are used for planning ones movements.

MULTI-ROBOT SYSTEMS

Robots do many of the tasks ideally done by humans and are constantly being used for a large number of domains. However, using a robot for a task may not always solve the problem. It may be impossible to design a robotic framework that solves the problem in totality. This especially happens when a significant amount of tasks need to be performed in numerous stages that are not necessarily related. In other scenarios, a single robot may be inefficient to carry forward the entire task on its own. In these scenarios, the intent is to use multiple robots in place of a single robot to solve the entire task. This is similar to use a diverse workforce for carrying forward a task in place of a single human doing the complete task. However, the notion of using multiple robots creates a new set of issues that must be addressed to effectively solve the problem. Much of the issues are related to creating a proper planning framework for the various robots, and hence the same is of our interest.

The notion presented here is the use of multiple robots for problem solving that creates a multi-robot system. Each of the systems is a single robot. This robot has its own robotic architecture, which we broadly defined to be consisting of sensors for sensing the robotic world, information processing modules, localization and mapping modules, planning modules, and control modules. Until now, we were interested in one such system of a single robot. However, the multi-robot systems constitute a system of such systems, which denotes a system of higher complexity. In addition, the multiple systems communicate with each other, which opens possibilities for better and efficient ways in which the task in hand may be performed.

Coordination is an important aspect, and the multiple robots must cooperate and coordinate with each other to make the task in hand possible. This is again similar to the manner in which humans work in teams. Any collective task involves a lot of discussion, mutual understanding, leadership, cooperation, etc. for it to be performed successfully. Teams do not constitute individual humans, each doing any part of the task it pleases in any manner it likes. While in humans, many of the aspects may come naturally or by common sense, in case of multiple robots things need to be clearly defined for the entire system to perform flawlessly. We describe few issues that arise in the design and operation of multiple robotic systems in this section.

Time and Solution Optimality

From a planning perspective, we know that the requirement is to enable the individual robots making up the multi-robot system to decide what is to be done and how. The planning may again be at various levels with different assumptions and complexities. The basic purpose here is to comment upon the planning perspective of multi-robot systems from the planning perspective of single robot systems. The time required for the generation of the plan as well the optimality and completeness of the plan are important indicators for selecting any method or algorithm for the generation of this plan.

The multi-robotic systems may have different robots behaving in different manners, and all combination of these basic possible movements at every instance of time and situation make up the solution space. It is evident that the solution space of multiple robots is highly complex as compared to the solution space of a single robot. Consider the discussion over planning with a single robot with graph search algorithms, where we studied algorithm behavior for a different number of allowable moves, which if increased, led to increase of solution space complexity. A multi-robot system equivalent would mean multiplication of the number of allowable moves, which means a significant increase in complexity with very little increase in number of robots. This makes planning for multiple robots a lot more difficult than planning of single robots. It may be noted that the effectiveness of the work to be performed is more in multi-robot systems than a single robot; however, generation of an effective plan that does the effective work is more difficult in multi-robot systems than the case of a single robot. As the solution space increases, the time required to generate the optimal plan also increases.

The multi-robot systems are indeed used in real time scenarios as well, for which we use different algorithms, which do not go into the complexity of the solution space. However, these algorithms may not guarantee completeness or optimality. We further delve into these concepts with specific notion of path planning for multiple robots later in the chapter.

Reaching a Consensus

The multi-robot systems have a large number of robots working over their own part of the problems. The problem may many times require the robots to make some kind of decision. The decision may affect the working of the other robots as well. Most times the decision may involve deciding the mechanism in which the problem needs to be tackled. It is evident that since all robots are building same solution to the problem, they must all agree to a common plan of solving the problem (Ishiguro, et al., 1997; Olfati-Saber, et al., 1997; Ren, et al., 2005; Sheng & Pan, 2009). It would be wrong for each of the different robots to solve its part of problem taking its own decisions, without considering the manner in which the decisions may affect the other robots work. It is possible that all robots do something entirely contradictory to each other's work, and hence, the solution to the problem being solved is unacceptable.

The problem of reaching a consensus in a multi-robot system deals with designing of mechanisms in which all the different robots reach to a common plan of action or decision for problem solving. Many times these decision need to be made in real time, as the solution is build up. The mechanisms are all dependent upon the problem of study. Briefly speaking, the multi-robotic system may be centralized or decentralized. The centralized system has a leader that makes all decisions and all the slave robots have to follow the decision. In such a case, reaching the consensus is simple; all computations for decision-making are done by a single master. The slave robots may however provide useful information for decision making, which may not be available with the master. The decentralized systems, however, do not have a central leader. In such a case, many problems may have predefined policies, which may be used by all the individual robots for decision making. In other systems, the decisions may be made to vote by the different robots or may alternatively be decided based on the robot priority that made the decision along with its confidence towards the decision made.

While we state the consensus in centralized system as trivial, it is important to understand about the manner in which a leader may be selected out of the available robots in the system. Initially, any random leader may be specified by the system. This robot may continue to take decisions and act accordingly; however, it is possible that the leader robot collapses before the task is completed. This may be a mechanical failure, software failure, leader stuck in some deadlock, etc. In such a case, the other robots need to complete the task

Figure 4. Types of systems on basis of cooperation

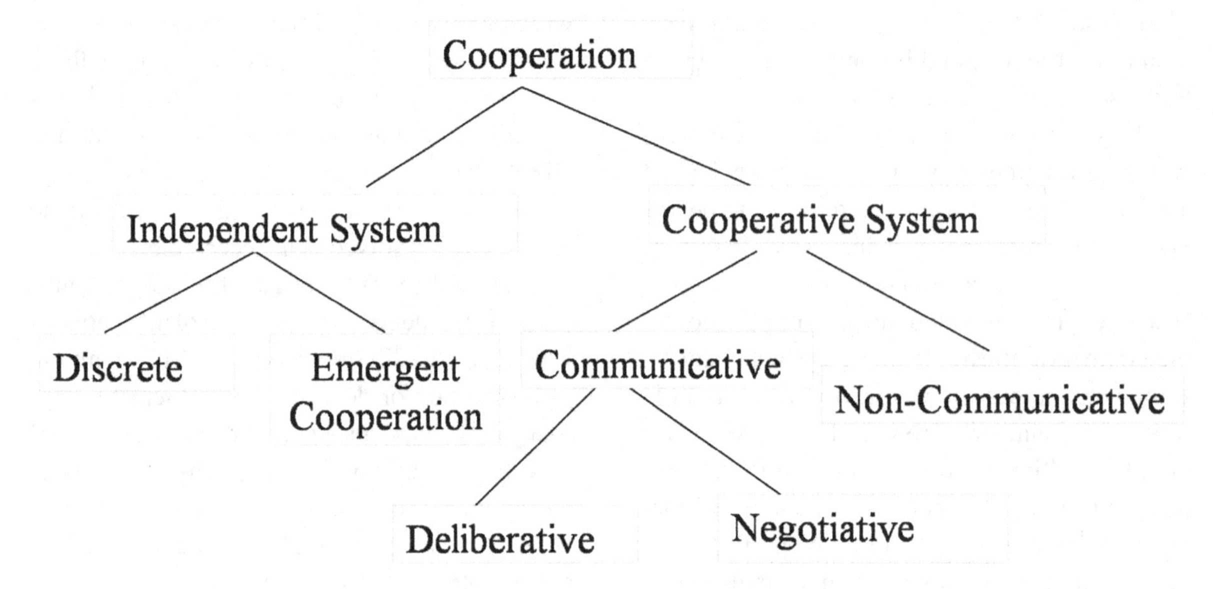

for which they require a new leader. Since the job is to be done autonomously, human intervention is not allowed. In such a case, again, robots may vote for a new leader, or one may be randomly set that attains consensus. This robot would continue the task of decision-making.

Cooperation

Another major issue that has importance in the multi-robot system is cooperation (Beckers, et al., 1994; Fierro, et al., 2002; Noreils, 1990; Tuci, et al., 2006). It is unlikely that the problem in hand would be simple enough that it could be cleanly divided amongst the robots by a master robot, and the different robots could simply do their part of the problem, catering to all the instruction given by the master. In a decentralized system, as well, it is unlikely that the different robots would easily reach a consensus and act accordingly. Systems usually require a continuous communication between robots in order for them to perform the task in hand optimally.

Robots working together should not get each other's way, but should ensure that, if a robot requires something from the other robot, it would be catered to. The problem of cooperation especially catches importance from the fact that cooperation with one robot might mean sacrificing one's optimality or comfort. In multi-robot systems, it is not important that individual robots act in the best manner for the work allocated for them; it is also important that they cooperate with each other. This is especially important as optimal plans of different robots may collide, making it possible for only one of them to be implementable. In such a case, different robots need to cooperate and decide which robot sacrifices and till what limit. In the case that such decisions are needed in real time, cooperation may become even more difficult to carry out. The general hierarchy of cooperation in case of multiple-agent systems is given in Figure 4.

Independent systems do not have any element of cooperation amongst robots, which all act in the manner they like, without considering each other. This strategy may work in the case that the task to perform is completely independent and non-interrelated between robots. The system is discrete when the individual tasks of different robots do not relate at all to each other. The system shows emergent cooperation behavior when the task is somewhat dependent; however, the robots are not explicitly planned to cooperate with each other for accomplishing the same. From the robot's point of view, no cooperation is desirable, even though a glimpse of the entire task may look as if the robots are actually cooperating.

The cooperating systems are further classified into communicative systems and non-communicative systems. The communicative systems have a formal communication mechanism in which the different robots can talk to each other via the designed communication protocol. This is what happens when robots transmit their plans, and attempt to agree upon them. However, in non-communicative cooperation, no formal communication is available. The robots need to observe each other and make decisions. The communicative systems may further be classified into deliberative systems and non-deliberative systems. This classification is more to do with the mechanism in which the systems communicate to reach a consensus. In deliberative technique, interchange of information takes place to decide the action, while in negotiating systems, decisions are made by competition.

Communication

Communication in multi-robot systems plays a great role in order to ensure robots can communicate with each other and coordinate to effectively solve the problem (Farinelli, et al., 2004; Parker, 2003; Zhang, et al., 2006). It is natural for such a system in which one robot makes an observation, which forms a useful piece of information for another robot. This robot may further make some plan or decision, which involves another set of robots, to which the plans are communicated. In this way, constant interchange of information may be required in the entire cycle of the multi-robotic systems. The communication between the robots takes place by the communication protocol, which is specified in the design phase. It is important for the communication to be extremely quick with as little overheads as possible. The communication may demand an assurance that it is received safely at the other end. These are naturally contradicting demands as a more secure communication may have higher overheads and making connections, communicating, and safely terminating may take more time.

A major decision in communication between the multiple robots is to decide whether the communication needs to be in broadcast manner or in one-to-one manner. In broadcast technique, all robots broadcast the messages they need to communicate. In case the message to be communicated was common for all the robots, all these robots accept the message and act accordingly. However, if the message was intended for a particular robot, only the intended recipient accepts the message, whereas all the other robots simply neglect it. This technique is good because only an ad-hoc network needs to be formed between robots after the time in connection formation and breaking is minimized.

Further, the method is very quick in case the same message needs to be transmitted to all robots, especially in a typical master-slave centralized architecture. However, in case a lot of interaction between pairs of robots is needed, this technique results in too many broadcasts and a lot of processing time for every robot is wasted in listening to messages. The other technique is the one-to-one technique in which a robot connects to the robot to which the message is intended to be delivered and transfers the message. In case the message is intended to multiple recipients, it needs to be communicated to all individually, which may take more time.

PROBLEMS IN MULTI-ROBOTIC SYSTEMS

We have stated numerous times that multi-robot systems pose the potential to solve problems more efficiently and effectively than single robot systems. While the application areas are wide and span across domains, it would be considerable to study few general problems as an example. This would enable us to understand how the multiple robots add to efficiency as compared to a single robot, as well as open new issues and challenges, which are dealt as benchmark problems in multi-robot systems. In this section, we do not give the specific application areas but rather address the problems in multi-robotics, which are under constant research.

Task Allocation

One of the most fundamental problems in multi-robotics is task allocation (Gerkey & Mataric, 2004; Viguriaand Howard, 2009; Zheng, et al., 2008). The robots collectively may be needed to perform a wide stream of tasks that are given to the robotic system as a whole. As time passes, these tasks keep coming and need to be performed. Every task may be performed by either of the available robots. The problem is to decide which robot performs which task. Hence, the problem modeling is like a centralized system where there is a master that does the role of task allocation, as they are demanded from the system. The problem is described by Figure 5. We need to formulate an efficient algorithm that can enable the master to decide which task goes to which robot. The tasks here may be discrete in nature wherein there is a predefined start and end to it, and the robot is expected to complete it in totality. The tasks may also be continuous in nature, which deal with continuous monitoring, surveillance, etc. The problem of task allocation differentiates between robots. Different robots may have different capabilities. Further the tasks are differentiated, which means some tasks may be easy to perform and may require less time and skills than others. The problem is widely studied by the research community.

The problem may broadly be classified depending upon the modeling scenario. The first classification is based on the robots, which classifies on the basis of whether the robots can perform a single task at a time, or they can perform multiple tasks at the same time. Similar classification may be done on the basis of tasks. This classification specifies whether a task needs to be performed by a single robot, or there are tasks which multiple robots would need to perform collectively. In case tasks require some skill set, which may not be found in a single robot, the system may

Figure 5. Multi robot task allocation problem

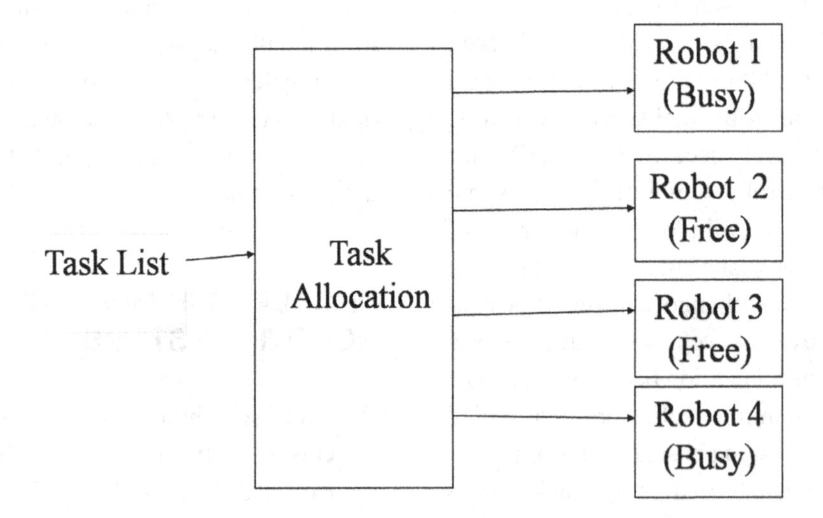

chose to allocate multiple robots for the same. The third classifications technique is more to do with method and specifies whether the allocation is done instantaneously as the task arrives into the system, or a list of task is available and algorithm needs to carry forward the assignment of tasks of the lists only. In the former case, it is unknown what tasks may come in the future, while in the latter case, all information is known in advance.

The decision must be given to which robot is the basic motive of the algorithm. If all the tasks are known before hand, the problem converts into an optimization problem, where every combination of robot and task is a potential solution to the problem. Any optimization algorithm may be formulated to map the tasks to robots available. However, the algorithm would not be considered real time.

Utility is defined which denotes the ease or effectiveness with which a robot may solve a particular task. Utility depends upon both robot and task, as every robot may not solve every task with the same ease and vice versa. The utility broadly depends upon two factors, which are quality of task execution and expected resource cost. The utility is defined as the difference between these quantities. For a robot to be able to perform a certain task, its utility value must be positive. In order to make decisions while allocating a task, every robot attempts to predict its utility which denotes if selected to perform the particular task, with what effectiveness it would be able to solve the task. Since utility depends upon both task and robot, every robot comes up with some different utility value when asked for every task. The decision to which the robot must be selected for performing the task may be done by optimization of the utility if complete set of tasks is available. Alternatively, for systems where the task list is not in hand, heuristic measures may be chosen for selecting the robot. A simple technique is to select the idle robot having the highest utility value for the task. However, consideration needs to be given that this robot would become busy until it completes the job. In the meantime, if a task arrives which can only be solved by this robot, the allocation plan would be regarded sub-optimal, as the newly entered task would need to wait till the robot becomes free.

Robotic Swarms

The concept of robotic swarms has emerged from the basic inspiration taken from the natural swarms (Iocchi, et al., 2001; Yogeswaran & Ponnambalam, 2010). The natural swarms like schools of fishes, flocks of birds, etc. are to be always found together. Consider the motion of a school of fishes. In the entire motion, the group of the fish stays together. Even if a variety of obstacles come in front of the school, they are able to coordinate with each other such that no fish is hurt from the obstacle as well as all fishes stay together with the group. All these are interesting behaviors, which have always attracted the interests of multiple researchers from multiple domains. The intent is to study the dynamics of this behavior and to possibly replicate it in the artificial systems. In the robotic domain, we find that many times a large number of robots may be easily deputed for a task. This considers that robots are small mobile devices which are inexpensive, and hence, even using a significant number of them, does not add up to the system cost. With little increase in financial costs, these robots may give a good performance boost to the entire system. A sample swarm is given in Figure 6.

The challenge lies in the manner in which the different robots may be moved about and the manner in which they work together in cooperation and coordination. Since the number of robots is very high, centralized techniques of planning may not be useful for their planning. Here again we turn to the natural counterparts. The ease with which the school of fish moves is highly motivating. Similarly, the ants easily do their tasks. This motivates the use of highly simplistic techniques.

Figure 6. Moving swarm of robots

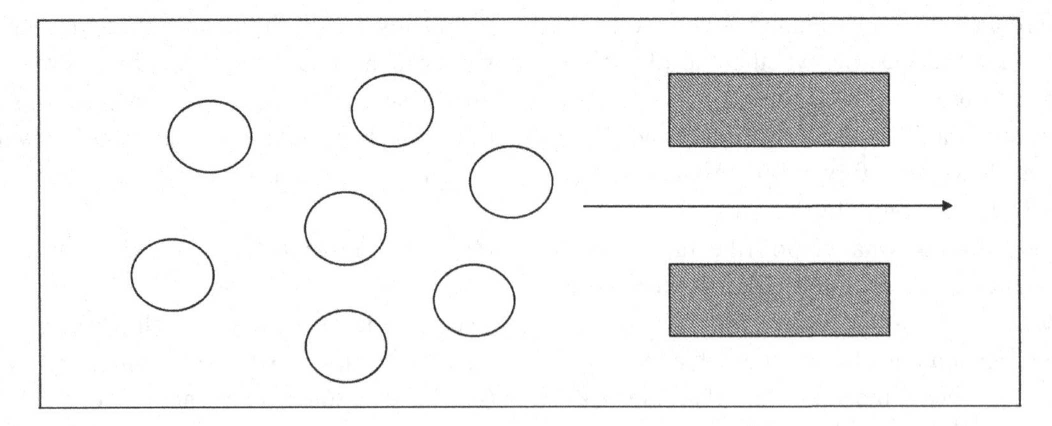

For motion related problems, a simple technique is to make one robot the leader to which all the other robots follow. They may follow by keeping some relative distances to each other. Similarly, heuristics may be defined for tasks that a large group of robots collectively does.

Formation Control

Another interesting problem that is commonly studied under the head of multi-robot systems is robotic formation (Balch & Arkin, 1998; Das, et al., 2002; Fredslund & Mataric, 2002; Yun, et al., 1997). In this problem, we have a number of robots, which we attempt to arrange in some kind of geometric shape or pattern. The arrangement of robots deals with planning the motion of each one of the robots, such that every robot reaches a position without collision with each other. The position of robots is not known and needs to be found by the algorithm itself. What is known to the algorithm is the pattern or the shapes that the robots need to make. The robots may be variable in number and the algorithm must consider the optimal positions such that the shape is represented. The problem for making a circular shape is shown in Figure 7.

Here, there are three major steps that need to be performed for the robots to have the formation. The first step is to specify the shape that we wish the robots to make. Mostly, the shape is predefined. The second step involves measures for robots to compute the positions that they need to take in the final shape that they collectively make. The robots are all considered similar and need to be placed at key positions as per the desired shape. The approaches may be centralized or decentralized. In centralized approaches, a master is nominated, which knows the current position of all the robots. It computes and tells all other robots which position they need to take. In the decentralized approaches, no leader may be available. Every robot may make its own computations and decide accordingly. The next task is physical motion of the robot. The initial positions and the positions that the different robots need to take are all known. The physical motion is performed by a control algorithm, which also ensures no mutual collisions.

RoboCup

RoboCup is many times seen as a benchmark problem for planning and control of mobile robots (Kitano, et al., 1997, 1999; Reis & Lau, 2001). The attempt is to make a robotic team play a match of soccer with another robotic team. RoboCup stands for Robot Soccer World Cup and the complete design of the competition is like a soccer championship. Search and Rescue, and

Figure 7. Formation control problem for making circular shape

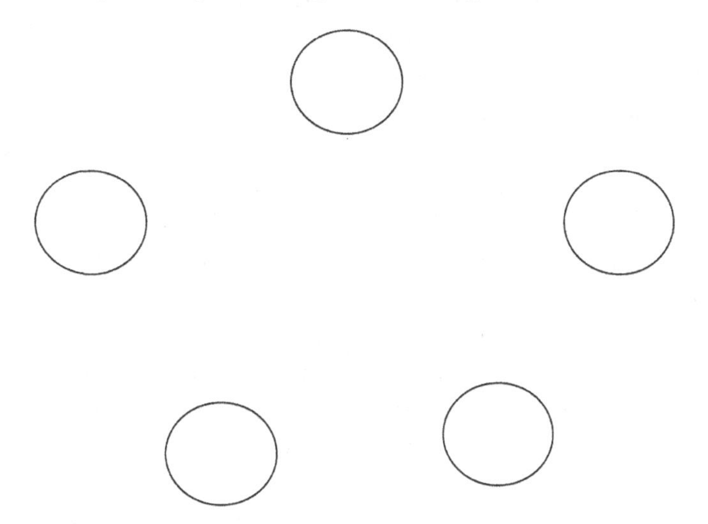

Robot Dancing competitions are also organized. The aim of a robotic team is to win the match by scoring higher number of goals as compared to the opponent.

The contest has always attracted the interests of numerous research groups who all make their robot players and enable them to participate in the mega event, where the prepared robots need to operate autonomously without any external human aid. Making effective teams worthy of winning the cup requires intelligent robots. At the level of individual robots, the architecture is similar to that studied in chapter 1. All the levels are found comprising of modules of sensing, recognition, sensor fusion, planning, control, etc. The different robot players, however, should not play their own game. They must rather cooperate with each other and put up a team performance. This means that at the highest level of planning there needs to be a game strategy based on which the complete team operates. Making and following an effective strategy are many times the keys to winning. Further, the different robots must ensure that they help each other to operate as per the set strategy. All robots going to fetch the ball is highly unlikely as is not passing the ball to another player and only attempting to hit the goal. Different players may occupy different positions and may play differently as per their role in the game strategy. Further, each robot must plan itself. It is essential to avoid any collisions with other robots, as well as to successfully fetch the ball, and pass or strike. All this makes the task of planning and control very difficult.

The problems become further difficult considering the highly dynamic environment that the game results in. Decisions need to be made very early so that the robotic players can react swiftly to situations. Many times this should lead to being the first player to get the ball, or successfully blocking the ball. In this problem, global view of the game is not available with the robotic players. Hence, the players can only look at a part of the field and based on the same information they need to make all of their decisions.

Path Planning

Multiple robots for use in a system for problem solving may need to move about in the entire workspace. Many times the robots come from different sub-systems that revolve around the workspace for whatever are their requirements. This creates a potential problem that while navigating one robot may collide with another robot. This is dangerous for the system as collision may make the robots unusable, or in other cases may leave them lost and un-localized in the robotic map. When the optimal paths of two robots happen to collide with each other, it becomes mandatory for a robot to leave way for other robot, or both robots must change their paths. In such a case, the central issue comes up in path planning of the robotic system. This type of planning is the multi-robot path planning where we have multiple robots that have their own source and goal and need to be moved accordingly. It may not be possible for all the robots to move by their optimal paths. However, the goal of the algorithm is that the travel plan as a whole must be optimal and no robot must collide with any other robot. The problem is summarized by Figure 8.

The issue of multi-robot path planning is similar to the issue of manipulation by multiple manipulators, wherein each of the manipulators must be planned so that the resultant task is done. The additional manipulators make the planning task more difficult by the increase in the complexity of the problem.

Area Coverage and Mapping

The problem of area coverage was discussed earlier in the chapter. The coverage of the complete area that is given in the problem shall become extremely difficult and time consuming task with the use of a single robot. It may be difficult for a single robot to roll over a big lawn or a robot to clean the entire workshop floor. Many times the work may be needed to be completed early. In such cases, it is natural to use multiple robots for carrying forward the task, rather than a single robot (Aulinas, et al., 2008; Choset, 2001; Enders, et al., 1998; Hedberg, 1995; Luo, 2002). The use of multiple robots stresses upon the fact that there needs to be a proper division of work. Further proper coordination and cooperation amongst the individual robots is necessary. The efficiency of the problem lies in ensuring that every robot carries forward the task as per its efficiency and no robot sits idle for long. The efficiency of the problem may be measured by the time in which the entire area is covered. Real time obstacle avoid-

Figure 8. Multi robot path planning problem

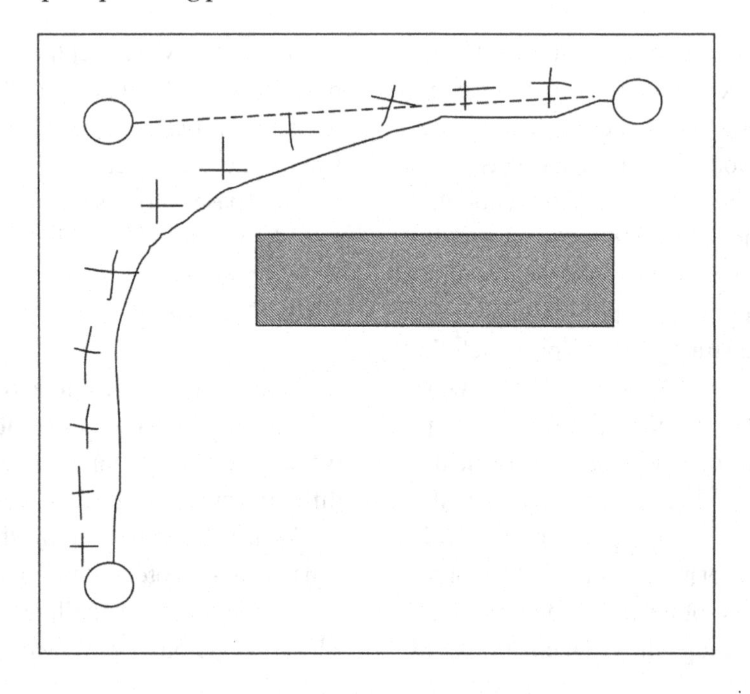

ance and fallback strategies in case of failures of some robots is always useful. The problem is given in Figure 9.

Another similar problem is of mapping or map building. Here the area is unknown, which needs to be explored, and a map needs to be built accordingly. This eliminates the use of humans to do the job. Many times, it may not be possible for humans to enter the area, as it may be dangerous. Further use of humans is also costly task for which robots may be cheap and effective alternatives. Again single robot may take too long a time for the complete area to be covered. The use of multiple robots may help in this regard. A common problem here is that the area is unknown and, hence, a prior division of task is difficult. This is unlike the area coverage problem where the complete area is usually known, at least to some extent. Individual robots may only be required to cover the allocated area with obstacle avoidance. In this problem, all the robots need to explore the surroundings. As the robots explore, they make the map, which is shared between robots and all the robots build a unanimous map, every part of which is made by some or the other robot. As the map is explored, robots further get to know areas where more exploration is required. With this knowledge of map, heuristic techniques may be made to send robots to specific places of map, once their part of exploration is over.

A common problem that may arise in this problem or any other problem is loss of communication. In single robot systems when the mobile robot reaches a place where it does not have communication with the ground system, the robot may be stated lost. In such a case, only predefined programming within the robot can enable the robot to get out or come to a region where communication is available. Many times robot running on GPS may enter into region where no GPS connectivity is available. In such cases the locally fitted redundant sensors may help to rescue.

This problem is more important in a multi-robot system. Even if a robot is in connection with another robot which has connection to the robot group, routing principles may be used for the robot to communicate with any other robot. However if a single robot or a robot group loses

Figure 9. Multi robot area coverage problem

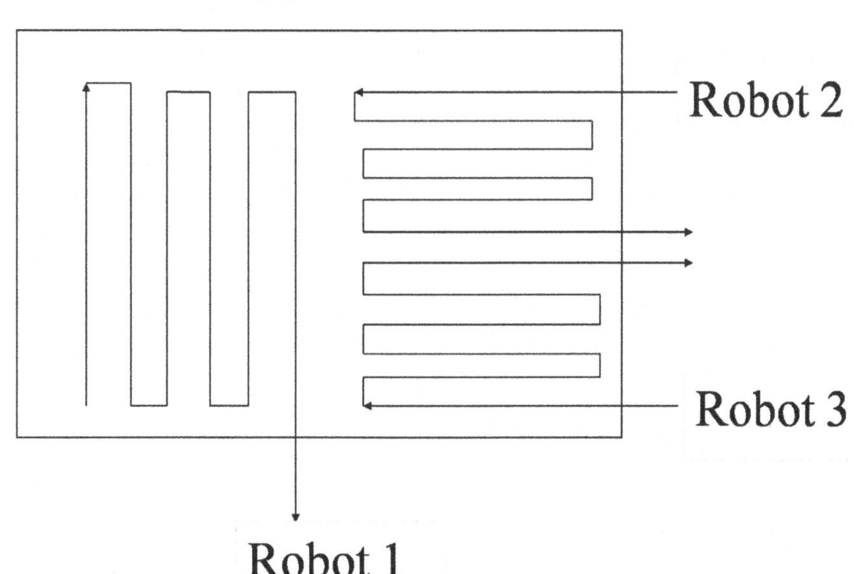

its contact with the other robots, it is stated to be lost. In such a case, the robotic group does not have any information of the lost robot or robot groups. Both systems behave independently of each other. Randomness may be expected until communication is restored. This again depends upon the robotic capabilities.

MULTI ROBOT PATH PLANNING

In the previous chapters of this book, the focus was to plan a single robot from the source to the goal. The scenarios in hand comprised of use of a robot for a task, which required the robot to move from one location to the other. In such a case, it was important to model the environment as a map, and to use the map for the motion of the robot. Here we classified the environment as static or dynamic. Static maps do not change with time, while the dynamic maps have some obstacles that are moving around the map. Both the types of maps have their own relevance in the problem. Most of the home and office environments are dynamic, where people keep moving around and placing objects here and there, which may suddenly appear in front of the robot. However, most tasks where robots are sent for exploration do not have dynamic obstacles around and the robot may easily navigate itself in the complex map. Based on the planning algorithms proposed, we saw that it was fairly easy for the robot to use any simple control algorithm to navigate itself in the robotic map. For complex maps, the control mechanisms may be difficult, which was not necessarily discussed in the algorithms presented, though the smoothness was attempted to be added, or provision of smoothness was made.

Now, since we have upgraded the systems from single robot to multiple robots, it is essential to understand how it impacts the problem of path planning specifically (Warren, 1990; LaValle & Hutchinson, 1998; Erdmann & Lozano-Perez, 1996; Carpin & Pagello, 2001; Azarm & Schmidt, 1997; Yegenoglu & Stephanou, 1989). Most of the concepts presented in the preceding sub-sections are true for the problem of path planning as well. The path planning algorithms need to show a high degree of cooperation, ensure that every robot has a near optimal travel path, while the overall travel plan is optimal. This means a high degree of problem complexity at the planning level, while the results may still be needed in real time. The additional robots open a complete pool of ideas that may be implemented in the problem of multi-robot motion planning. Designing effective algorithms that give optimal results, are complete in nature, and take less computational time. While we do attempt to ensure optimality and completeness in real time systems, it may be considered based on discussions over planning of a single robot that these may be practically impossible. Multiple robots representing a still higher degree of problem complexity as compared to single robots make it a difficult task, even in case of low resolution maps.

The problem of planning for multiple robots may sometimes be seen as the task of simple planning with the design of effective coordination strategies between the different robots. By simple planning, we point to all the planning algorithms that we discussed in the text. These algorithms may be used taking into account the multiple robots, or more simply on a single robot itself. However, it is evident that if all the robots are planned independently, these would come up with a navigation plan in which a collision may occur. This happens when two or more robots occupy a position at the same time as per their plan. Since the entire travel plan consisting of travel plans of individual robots is prone to collision, it cannot be used. This means that the travel plan needs to be mended, which would mean mending the individual travel plan of robots.

Here the factor of coordination strategy comes into the picture. This strategy tells how to mend the individual travel plan to ensure no collision. By changing travel plan of one or more robots, we certainly make its travel sub-optimal; however, the aim is that the overall travel plan must be optimal. Every algorithm used for the problem of planning of multiple robots differs in the manner in which the cooperation is carried out amongst multiple robots. We discuss some of the concepts associated with multi-robot path planning in this section.

Problem

The problem statement of multi-robot path planning states that we are given a number of robots. The initial position of all these robots is given. Each robot intends to travel to a specific place or the goal. The goals of all the robots are also known in advance. We further have a map, which is common for all the robots. The map consists of static obstacles, and may in addition also have some dynamic obstacles. The task is to construct the trajectories for every robot such that all the robots reach their goal, as well as no robot collides with any other robot or obstacles as they move. Completeness of the problem in this context is defined whether all the robots reach their respective goals, in case it was possible to come up with a travel plan that makes every robot reach the goal safely. Optimality of the problem further does not stress that each and every robot must reach the goal with the optimal route. It rather stresses upon the fact that the overall route of all robots combined must be optimal.

Since all the robots move simultaneously in the map, the speeds of the robots are of concern. In solving for a single robot, we did not discuss much about the speed, since it was assumed that the controller would take the robot by the maximum permissible speed possible with the generated trajectory. Alternatively, the robot might be moved at some convenient constant speed. Hence, speed was not much of a concern in the planning stage. However, speed of robots is an important aspect in multi-robotic motion planning. Increasing or decreasing speed of a robot may result in a collision between two robots or may have the collision avoided. Hence, speed decides which robots may collide, and at what space in the map they may collide. Many algorithms may take the speed of the vehicles as constant and may then aim to work the robot paths. Collisions are checked assuming each robot travels with the constant speed in the planned path. Alternatively, many algorithms try to change the speeds as well so that collisions may be avoided. The maximum permissible speed is usually fixed. The solutions to the problem may be broadly divided into centralized or decentralized solutions. Each of these is discussed.

Centralized Planning

The centralized planning approaches look at the planning problem as a whole and attempt to come up with a plan that includes the travel plan of all the vehicles. These approaches attempt to ensure completeness and they need to hence try out all possibilities, which mean computing every possible combination of move by all robots at every instance of time. Hence, the problem is to search a solution, if it exists, in the entire problem space, which includes all possible moves of all robots. The solution must further be optimal. Now consider the discussions in the earlier chapters where we attempted to use graph search or evolutionary approaches. The solution space in application of any of these methods was very complex and had high dimensionality. This made it impossible to generate solutions in a real time manner. As a result, an assumption had to be the static nature of the map, which made the entire search space static. Hence, even though the search space was large for these algorithms, it was still possible to invest high amount of time. Investing such a high amount of time ensured completeness as opposed to other algorithms, which could be executed in real time.

The search space for a problem of multi-robot motion planning is simply the multiplication of search spaces of all the individual robots. If we increase the number of robots from one to two or maybe three, we can expect the algorithms to give results after a large span of time as compared to the single robot counterpart. However, as we further increase the number of robots, it is evident that significantly large span of time may not also give a result. Hence, even though these solutions add to the completeness and optimality, the time required is very large. This is especially true in cases where the number of robots is quite high.

Decentralized Planning

This planning technique believes it is not possible to deal with a search space, which is extremely complicated. This is especially the case when there are large numbers of robots. Hence, the problem of having a complex search space is to be broken down into simpler problems. In decentralized solutions, each robot has its own search space (Azarm & Schmidt, 1996). Hence, each and every robot is planned individually. This would result in plans that have collisions. These collisions are handled by framing a coordination strategy. In these approaches the complexity of the search space is restricted to the complexity of search space of single robot, which can be solved in fair amount of time, depending upon the algorithm used for planning. Completeness and optimality in this class of algorithms cannot be guaranteed; however, it may be possible to generate solutions in real time. This makes it possible to plan multiple robots in real time, ensuring no collision occurs.

One of the commonly used coordination strategies amongst robots is prioritization. Here different robots are assigned different priorities at the start of the planning process. Then planning takes place strictly in the order of priorities. The highest priority robot is the one that first plans. While planning, it only considers the obstacles in the scenario and not the other robots. Every subsequent robot that plans takes the earlier planned robots as obstacles and tries to avoid them as well, along with the other obstacles in scenario. Since the travel plans of all higher priority robots are well-known, they may be easily treated as mobile obstacles whose dynamics are well-known. The higher priority robots simply need to avoid these robots as obstacles, while they plan their individual algorithms. In this way, we can ensure that collision would not happen.

A major issue is the mechanism by which the priorities are assigned. Randomly assigned priorities work well; however, they may even lead to loss of completeness and highly sub-optimal results in many cases. A lower priority robot may have to travel a significantly long way because its optimal way is blocked by some high priority robot and no other way is possible besides traveling very long. In such a case, solutions could change if the lower priority robot had higher priority, which highlights the importance of prioritization scheme. Genetic Algorithms are commonly used for assigning priority.

APPROACHES

The approaches discussed for the single robots may be easily extended to the multi-robot counterparts. The method of application may be centralized or decentralized, which solely depends upon the complexity and resolution of map and the number of robots in the planning scenario. The centralized approach would demand that we extend the notion of planning to account for the additional robots in the solution formulation. The decentralized approaches do not affect the planning algorithm of the individual robots but do demand that they would account for the presence of other robots in the planning scenario. In this section, we briefly comment upon the extension of the studied basic techniques of path planning to the presence of multiple robots.

Graph-Based

The graph-based approaches for single robot basically relied upon existence of some state or position of the robot, which was extended to produce more positions or states that the robot could connect to. The total number of moves were defined which gave the total number of states to which a state expands to. This made the graph consisting of vertices and edges. Every expanded state is used only if it is feasible. In extending the problem to multiple robots using a centralized architecture we simply use a state or a vertex specification as the states or positions of all robots combined. Hence, a vertex of the graph in this case is the positions of all robots at a particular time. This further stresses on the fact that the total number of vertices become excessively large in the presence of multiple robots.

Let every robot R_i be denoted by its position $<x_i, y_i>$. In case of single robot the total number of vertices was m x n, the size of map. However, in case of multiple robots, the vertex is denoted by $<x_1, y_1, x_2, y_2, \ldots x_r, y_r>$, where r is the number of robots. The total number of vertices becomes $(m \times n)^r$. For high r, this is clearly too complex to handle. Further, we need to specify the edges. In case of single robot, we simply took the permissible unit move of a single robot. Here, however, we need to take every combination of permissible move of all robots. This clearly states that the number edges follows the same trend in increase from single robot as the number of vertices. This gives the total graph that needs to be searched. The source of all robots constitutes the source state. The goal state is also when all the robots reach their goals. Alternatively, if the number of robots is high, we may simply follow a priority-based planning, where every robot of lower priority considers the robots of higher priority as obstacles.

Evolutionary Planning

The evolutionary techniques encoded the path of a single robot in a genotype, which was optimized by the evolutionary process. The genotype that we used for the optimization consisted of specification of multiple points, which were to be traced one after the other to get to the goal. The source was always the first of these points while the goal was the last. Genotype of a single robot R_i may hence be written by $<P^1_i, P^2_i, P^3_i, \ldots P^s_i>$. Here s is the maximum number of turns. The complexity of the fitness landscape is clearly $(m \times n)^s$, since each point x coordinate can vary from 1 to m and y coordinate from 1 to n. The centralized path planning techniques would emphasize on taking the optimization problem as a whole. Hence, the genotype must contain specifications of all the robot paths.

The complete genotype in case of centralized planning becomes $<P^1_1, P^2_1, P^3_1, \ldots P^s_1, P^1_2, P^2_2, P^3_2, \ldots P^s_2, \ldots, P^1_r, P^2_r, P^3_r, \ldots P^s_r>$. We assume that maximum number of turns of all robots is the same. r is the number of robots. The task in fitness computation is to compute the average travel time, given the specification of all the robotic paths. Here, as per specification of the path, all robots must be simulated to watch for collision. In case of collision, extra penalty is added. The complexity of the resultant fitness landscape increases to $(m \times n)^{sr}$. This clearly represents significant increase if the number of robots is very high.

The other technique may be to use priority-based planning, where each robot may be given a priority. Alternatively, a co-evolutionary algorithm may be used for the task of optimization. In this algorithm, each module represents the path of a single robot, the optimization of which is carried by a separate evolutionary algorithm. The various evolutionary algorithms attempt to cooperate with each other for generating paths. Cooperation in this technique is built in the evolutionary algorithm itself. Individuals are assigned a high fitness not only if the individual robot's path is optimal, but also when the individual encourages some other robot to have an optimal path.

In this manner, we encourage individuals that help other robots attain the optimal path. In our case, it would mean that the optimal path of a robot, which leads to collision with some best paths of other robots, is given a poor fitness value and its generation is discouraged. In this manner, we may evolve better and better paths for individual robots, which do not have collisions with each other. A master evolutionary algorithm is taken that generates best combination of the various individuals of the individual robot's evolutionary algorithm. Each individual represents a path. Hence, this algorithm mainly attempts to evolve best combination of paths, which is the purpose of the problem.

Behavioral Planning

The last technique we discuss is the behavioral approach. We saw in the earlier chapters that the behavioral algorithms compute the optimal path by simply looking at the world around. Complete map may not be required, and no intensive computations are made. Hence, centralized approaches are not used in this class of algorithms. Further, no specific strategy may be needed while working with multiple robots. The robots are planned as they move around in the planning scenario. For planning any robot, we consider all the other robots as obstacles. In this manner, the robot may naturally overcome collisions in a manner similar to obstacle avoidance. In this way, all the robots keep moving from source to goal until they reach the goal position. Planning is done in real time. Alternatively, techniques may be developed to avoid moving robots considering their motion dynamics. This would enable robots to compute the best possible manner of collision avoidance.

CONCLUSION

The focus of the chapter was on multi-robot systems. The basic notion is that the multiple domains where robots are used may benefit if multiple robots are employed in place of a single robot. As a result, a far greater efficiency in performance is expected, which means a higher degree of automation. The inclusion of the additional robots opens an entirely new set of issues and problems, which was the focus of the chapter.

We first presented a number of problems where planning played a major role. This included the problems of maze solving, complete coverage, map building, and pursuit evasion. Each of these problems presents a unique challenge to planning as a whole. The discussion was later extended to the real issues associated with the use of multiple robots. The problem of planning for multiple robots adds up a lot of possibilities in which the multiple robots may behave. This means a significantly high complexity in terms of solutions possible. This makes the problem highly complex to solve in real time, especially if completeness and optimality are key factors. Many times the individual robots may themselves solve for their part of the problem, which constitutes the decentralized approaches. Here it is important for the various robots to agree with each other and reach a consensus. Coordination is the key with which multiple robots may work together and as a team solve the problem effectively.

Communication mechanism was discussed, which is the basis for the robots to show cooperation and coordination. A number of problems were disused to understand the working of multiple robots together. The problems include multi-robot task allocation, robotic swarms, formation control with multiple robots, RoboCup, multi-robot path planning, and multi-robot area coverage and mapping. Path planning is one of the specific topics of interest in the entire book. We specifically take this problem and discuss the issues of multi-robotics. Centralized solutions to the problem look at the entire problem as a whole and build up a complex search space in which the solution needs to be searched. The decentralized solutions on the other hand plan each and every robot independently. Collision between the robots may be avoided by building effective coordination strategies. The mechanism of extending the single robot planning algorithms to multiple robots are discussed for the graph search, evolutionary-, and behavioral-based planning algorithms.

REFERENCES

Aulinas, J., Petillot, Y., Salvi, J., & Llado, X. (2008). The SLAM problem: A survey. In *Proceedings of the 11th International Conference of the Catalan Association for Artificial Intelligence,* (Vol. 184), (pp. 363-371). IEEE.

Azarm, K., & Schmidt, G. (1996). A decentralized approach for the conflict-free motion of multiple mobile robots. In *Proceedings of the IEEE Conference on Intelligent Robots and Systems 1996,* (Vol. 3), (pp. 1667-1675). IEEE Press.

Azarm, K., & Schmidt, G. (1997). Conflict-free motion of multiple mobile robots based on decentralized motion planning and negotiation. In *Proceedings of the IEEE International Conference on Robotics and Automation,* (pp. 3526-3533). IEEE Press.

Balch, T., & Arkin, R. C. (1998). Behavior-based formation control for multirobot teams. *IEEE Transactions on Robotics and Automation*, *14*(6), 926–939. doi:10.1109/70.736776

Beckers, R., Holland, O. E., & Deneubourg, J. L. (1994). From local actions to global tasks: Stigmergy in collective robotics. In Brooks, R., & Maes, P. (Eds.), *Artificial Life IV*. Cambridge, MA: MIT Press.

Brooks, R. R., Jing-En, P., & Griffin, C. (2008). Game and information theory analysis of electronic countermeasures in pursuit-evasion games. *IEEE Transactions on Systems, Man, and Cybernetics. Part A, Systems and Humans*, *38*, 1281–1294. doi:10.1109/TSMCA.2008.2003970

Carpin, S., & Pagello, E. (2001). A distributed algorithm for multi-robot motion planning. In *Proceedings of the Fourth European Conference on Advanced Mobile Robotics*, (pp. 207-214). IEEE Press.

Cheng, P. (2003). *A short survey on pursuit-evasion games*. Champaign, IL: University of Illinois at Urbana-Champaign.

Choset, H. (2001). Coverage for robotics - A survey of recent results. *Annals of Mathematics and Artificial Intelligence*, *31*(1), 113–126. doi:10.1023/A:1016639210559

Das, A. K., Fierro, P., Kumar, V., Ostrowski, J. P., Spletzer, J., & Taylor, C. J. (2002). A vision-based formation control framework. *IEEE Transactions on Robotics and Automation*, *18*(5), 813–825. doi:10.1109/TRA.2002.803463

Doran, J. E., Franklin, S., Jennings, N. R., & Norman, T. J. (2002). On cooperation in multi-agent systems. *The Knowledge Engineering Review*, *12*(3), 309–314. doi:10.1017/S0269888997003111

Dracopoulos, D. C. (1998). Robot path planning for maze navigation, In *Proceedings of the IEEE International Joint Conference on Neural Networks,* (pp. 2081-2085). Anchorage, AK: IEEE Press.

Enders, H., Feiten, W., & Lawitzky, G. (1998). Field test of navigation system: Autonomous cleaning in supermarkets. In *Proceedings of IEEE International Conference on Robotics and Automation,* (pp. 1779–1781). IEEE Press.

Erdmann, M., & Lozano-Perez, T. (1986). On multiple moving objects. In *Proceedings of IEEE International Conference on Robotics and Automation,* (pp. 1419 – 1424). IEEE Press.

Farinelli, A., Iocchi, L., & Nardi, D. (2004). Multi-robot systems: A classification focused on coordination. *IEEE Transactions on Systems, Man, and Cybernetics. Part B, Cybernetics*, *34*(5), 2015–2028. doi:10.1109/TSMCB.2004.832155

Fierro, R. D., Spletzer, A., Esposito, J., Kumar, J., Ostrowski, V., & Pappas, J. P. (2002). A framework and architecture for multi-robot coordination. *The International Journal of Robotics Research*, *21*(10), 977–995. doi:10.1177/027836490202101098 1

Fredslund, J., & Mataric, M. J. (2002). A general algorithm for robot formations using local sensing and minimal communication. *IEEE Transactions on Robotics and Automation*, *18*(5), 837–846. doi:10.1109/TRA.2002.803458

Gerkey, B. P., & Mataric, M. J. (2004). A formal analysis and taxonomy of task allocation in multi-robot systems. *The International Journal of Robotics Research*, *23*(9), 939–954. doi:10.1177/0278364904045564

Guibas, L., Latombe, J.-C., LaValle, S., Lin, D., & Motwani, R. (1985). A visibility-based pursuit-evasion problem. *International Journal of Computational Geometry & Applications*, *4*(2), 74–123.

Hedberg, S. (1995). Robots cleaning up hazardous waste. *AI Expert*, 20–24.

Iocchi, L., Nardi, D., & Salerno, M. (2001). Reactivity and deliberation: A survey on multirobot systems. In *Proceedings of Balancing Reactivity and Social Deliberation in Multi-Agent Systems* (pp. 9–32). IEEE. doi:10.1007/3-540-44568-4_2

Isaacs, R. (1965). *Differential games*. New York, NY: Wiley.

Ishiguro, A., Watanabe, Y., Kondo, T., Shirai, Y., & Uchikawa, Y. (1997). A robot with a decentralized consensus-making mechanism based on the immune system. In *Proceedings of the Third International Symposium on Autonomous Decentralized Systems*, (pp. 231-237). IEEE.

Kitano, H., Asada, M., Kuniyoshi, Y., Noda, I., & Osawa, E. (1997). Hitoshi matsubara robocup: A challenge problem for AI. *AI Magazine*, *18*(1), 73–85.

Kitano, H., Tadokoro, S., Noda, I., Matsubara, H., Takahashi, T., Shinjou, A., & Shimada, S. (1999). RoboCup rescue: Search and rescue in large-scale disasters as a domain for autonomous agents research. In *Proceedings of the 1999 IEEE International Conference on Systems, Man, and Cybernetics*, (vol. 6), (pp. 739-743). IEEE Press.

LaValle, S. M., & Hutchinson, S. A. (1998). Optimal motion planning for multiple mobile robots having independent goals. *IEEE Transactions on Robotics and Automation*, *14*, 912–925. doi:10.1109/70.736775

Luo, C., Yang, S. X., Stacey, D. A., & Jofriet, J. C. (2002). A solution to vicinity problem of obstacles in complete coverage path planning. In *Proceedings of the IEEE Conferences on Robotics and Automation*, (Vol. 1), (pp. 612-617). IEEE Press.

Noreils, F. R. (1990). Integrating multirobot coordination in a mobile-robot control system. In *Proceedings of the IEEE Conferences on Intelligent Robots and Systems 1990: Towards a New Frontier of Applications*, (Vol. 1), (pp. 43-49). IEEE Press.

Olfati-Saber, R., Fax, J. A., & Murray, R. M. (2007). Consensus and cooperation in networked multiagent systems. *Proceedings of the IEEE*, *95*(1), 215–233. doi:10.1109/JPROC.2006.887293

Parker, L. E. (2003). Current research in multirobot systems. *Artificial Life and Robotics*, *7*(1-2), 1–5. doi:10.1007/BF02480877

Reis, L. P., & Lau, N. (2001). FC Portugal team description: RoboCup 2000 simulation league champion. *Lecture Notes in Computer Science*, *2019*, 29–40. doi:10.1007/3-540-45324-5_2

Ren, W., Beard, R. W., & Atkins, E. M. (2005). A survey of consensus problems in multi-agent coordination. In *Proceedings of the 2005 American Control Conference*, (pp. 1859-1864). IEEE.

Sadik, A. M. J., Dhali, M. A., Farid, H. M. A. B., Rashid, T. U., & Syeed, A. (2010). A comprehensive and comparative study of maze-solving techniques by implementing graph theory. In *Proceedings of the IEEE Conference on Artificial Intelligence and Computational Intelligence,* (Vol. 1), (pp. 52-56). IEEE Press.

Sheng, L., & Pan, Y. J. (2009). Distributed control for consensus of networked multi-agent robotic system with time delays. *International Journal of Information and Systems Science*, *5*(2), 161–178.

Tuci, E., Gross, R., Trianni, V., Mondana, F., Bonani, M., & Dorigo, M. (2006). Cooperation through self-assembly in multi-robot systems. *ACM Transactions on Autonomous and Adaptive Systems*, *1*(2), 115–150. doi:10.1145/1186778.1186779

Vidal, R., & Shakernia, O., Kim, Shim, D., & Sastry, S. (2002). Probabilistic pursuit-evasion games: Theory, implementation, and experimental evaluation. *IEEE Transactions on Robotics and Automation*, *18*, 662–669. doi:10.1109/TRA.2002.804040

Viguria, A., & Howard, A. M. (2009). An integrated approach for achieving multirobot task formations. *IEEE Trasactions on Mechatronics*, *14*, 176–186. doi:10.1109/TMECH.2009.2014056

Warren, C. W. (1990). Multiple path coordination using artificial potential fields. In *Proceedings of the IEEE International Conference on Robotics and Automation*, (pp. 500-505). IEEE Press.

Yegenoglu, F., & Stephanou, H. E. (1989). Collision-free path planning for multirobot systems. In *Proceedings of the IEEE Symposium on Intelligent Control,* (pp. 537-542). IEEE Press.

Yogeswaran, M., & Ponnambalam, S. G. (2010). Swarm robotics: An extensive research review. In Fuerstner, I. (Ed.), *Advanced Knowledge Application in Practice*. New York, NY: InTech Publishers.

Yun, X., Alptekin, G., & Albayrak, O. (1997). Line and circle formation of distributed physical mobile robots. *Journal of Robotic Systems*, *14*(2), 63–76. doi:10.1002/(SICI)1097-4563(199702)14:2<63::AID-ROB2>3.0.CO;2-R

Zhang, Z., Ma, S., Lu, Z., & Cao, B. (2006). Communication mechanism study of a multi-robot planetary exploration system. In *Proceedings of the 2006 IEEE International Conference on Robotics and Biomimetics*, (pp. 49-54). Kunming, China: IEEE Press.

Zheng, T., Li, R., Guo, W., & Yang, L. (2008). Multi-robot cooperative task processing in great environment. In *Proceedings of the IEEE Conference on Robotics, Automation and Mechatronics,* (pp. 1113-1117). IEEE Press.

Chapter 11
Conclusion

ABSTRACT

Having covered a large number of concepts, algorithms, issues, and challenges in robot motion planning, there is little left to state as the conclusion. In this chapter, the authors simply jot down some last words, final thoughts, and closing remarks. While a large number of algorithms exist for the same problem, it is hard to pick out one algorithm and use it upright for a given modeling scenario. It becomes important to understand what features the algorithm has, what is the level of each of these features, and whether these features suit the problem in hand. Optimality and completeness are important factors provided by the algorithm in real time, which may, however, not always be possible. Hybrid algorithms are more tailored in the sense that these algorithms can be customized as per needs. Again, the choice of algorithm plays an important role, as the limitations get added up along with the advantages which we need by design. Hence, it is always important to know what is desired from the robot and the planning algorithm and what scenarios it may face. Most practical scenarios may be simple, such that these can be simply solved by any general algorithm. Other aspects of robotics may be more challenging than a planning algorithm. However, it is certain the ever-rising use of robotics shall bring forth challenging scenarios for robots to work in, in which case the planning algorithms would be tested to best of their capability.

CONCLUDING REMARKS

Choices play a major role in our everyday life. Be it grocery, cuisine, or mode of transport, we always have a lot to choose from. Picking any one of the available choices, maybe even randomly, may not sound like a bad strategy, since all choices are more or less the same, they mean the same thing to us, or they are short lived—we do not spend our lifetime with the same choice. However, having a large variety of choices in a robotic scenario may not mean the same. Here, the algorithms differ in

DOI: 10.4018/978-1-4666-2074-2.ch011

capabilities; a wrong choice might mean a total system failure. Further, changing an algorithm is always a costly venture, especially having understood that the planning algorithm is ultimately knit to all the other modules of mobile robotics. Now it becomes a critical issue of knowing all about the problem or the modeling scenario in hand which may vary to great extent.

Robot operating for cleaning a house may not be the same as the robot doing postal delivery in homes, which may again not be same as the ones working in industry and public places. Robots used for exploration type tasks may be different from those used in oceans, which may again be very different from those used in rescue operations. For many scenarios, it is possible that default or simple choices of algorithms work well. This leaves space to concentrate on the other issues in robotics, rather than only on the planning aspect.

In most scenarios, if the algorithm fails to find a feasible path, it may be possible to manually or remotely move the robot, which becomes operational again. In many other scenarios, it does not matter whether the robot takes the optimal route as long as it does the desired work. Hence, these simple scenarios may be easily solved by any algorithm, knowing there is always a backup plan for rescue. This methodology is not something entirely new. Software in its initial cases might be created for the normal inputs, later versions being more sophisticated to deal with the critical or tricky inputs, security aspects, and all. Robot operating for a rescue operation in a cave might have no possibility of user interference due to lack of signals inside. In such a case, it becomes important for the robot to be on its own, and any breakdown means a mission failure.

Even in cases where choice of algorithm is critical, simple algorithms may effectively do the job without any error as per the user requirements. Other issues in robotics like the choice of sensors, physical make of the robot, control algorithm, etc. might be more important issues. However, two aspects need to be seen. First, whether the design is made on the basis of the normal inputs that the system has, or based on the exceptional inputs that might someday come into. Robots operating in pre-known environments may not be capable of handling blockage of a path if exceptional cases are not encountered. Second is the importance of optimality of the path. For most life-critical robotic operations, even fractions of seconds can make a difference. For simple scenarios, as well, optimal routes taken over prolonged period of robotic deployment can mean better response of the system as a whole.

This puts forth an interesting issue, whether to look at the modeling scenario in hand at its extremity and devise a proper planning algorithm or to just use a simple algorithm that solves the problem in most cases. The first case is the academic school of thought, which demands the algorithm to give results in all kinds of maps. The results need to be optimal and algorithms need to be complete. Real time nature is an important aspect for operations. The basic behavior is simple, account for everything possible, even with the slightest chance. The academic school of thought basically attempts to create a planning framework that best solves the problems with every factor an important indicator to measuring the worth of the algorithm.

The other case is the industrial school of thought where the attempt is basically to solve the problem in hand for most cases that one may. It is important that the robot physically performs as per the expectations of the user, users are satisfied and are able to use the system created, robot is able to perform with as little assistance as possible, and overall the system does work. Here, the solution provided may not be proven to be optimal or complete. Clearly, the approach of the book was based on the academic school of thought, unlike the industrial school, which is more relevant when discussing cases of physical robots deployed in problem solving.

The basic problem with the academic school of thought is simply that it may be impossible to engineer a solution that is perfect in the sense that solutions are optimal, algorithm is complete, most dynamic scenarios are taken, cooperation of other people in workspace is not assumed, robot delivered is able to perform in all types of environment, there is no user assistance required, there is minimal effort required to install the robot at a new place, and that the robot works considering all uncertainties into the system. It is then that we need to re-think the complete system and come up with algorithms that tradeoff between the requirements. This is exactly when choice of algorithm comes into play. Having said that, compromises might have to be made, depending upon the scenario, and it is important to understand what is needed by the algorithm and what is not so desirable. The emphasis is on understanding the problem scenario.

A simple robot in household may not see too rapid obstacles on the way, which is not the case with robots operating inside some industry with huge machinery movements all around. Robots operating in roads and pedestrian environments need to be cautious of mobile obstacles. However, it is simply difficult to mine out the desirable and not so desirable aspects needed by the algorithm over the modeling in hand. In most scenarios, the designer is tempted to go with some simple algorithm knowing the kind of performance it would give, which might be reasonably acceptable. While the academic school of thought keeps proposing newer and newer algorithms tested for a variety of diverse scenarios, mapping these to real life and making a wise decision in the choice of an algorithm is implementation specific and not an easy choice. Understanding of tradeoffs between algorithms is still an important issue to decide the applicable algorithm over the scenario. In every chapter, we presented the advantages and the disadvantages of the various algorithms, which is something to be used in decision-making.

The problem of path planning has been solved by a large variety of algorithms. The various models may be fundamentally studied in three broad categories. All these categories have some different modeling scenarios, assumptions, and executions of the algorithm. The first category consists of the planning algorithms that model the problem as a graph. The problem is solved as a graph search problem where source is the initial node and goal is the destination node. This includes algorithms like Breadth First Search, A* algorithm, Dynamic Programming, D* algorithm, etc. These algorithms are known to be computationally very expensive, especially in case of high-resolution maps and do not cater to the non-holonomic constraints. However, the solutions are complete. For the allowed number of moves, the solutions are also optimal in nature.

The second class includes the behavioral approaches where we studied Fuzzy Logic and Artificial Neural Networks. These approaches model the robotic behavior to react to obstacles. This class of algorithms has its analogy to the general manner in which the humans move. We are aware of the manner in which we escape from static and dynamic obstacles. We are further able to make turns and make our way out of any situation, without knowing the complete map as a whole. These approaches are used for real time robot path planning. The path generated obeys the non-holonomic constraints. These techniques are very quick, and hence, real time performance is guaranteed. However, these approaches have two problems. First, they do not regard optimization of path at all. They make the robot travel as per the path visibility. Second, they completely fail in complex or maze-like maps where there may be multiple ways but only one of them is correct. The solutions are neither optimal nor complete.

The other class of algorithms is the evolutionary approaches. Here, some path representation technique is used and evolutionary algorithms are used to optimize the robotic path. One of the advantages of these methods is that the algorithm is iterative in nature. The solution improves with time, and at any time the optimization may be terminated and best individual returned as solution. Non-holonomicity can be added to the algorithm.

However, they work only for simple maps. As the map complexity increases and the maps become more maze-like, these algorithms fail to perform as none of their solutions are feasible. This is due to the complex and massive size of the evolutionary search space. Further, the solution is not real time and intensive computation needs to be performed. The computation for fairly high-resolution graph is less than the graph-based approaches. The results are probabilistically complete and optimal. However, the algorithm does not give results in cases where optimal results have robot passing through a narrow corridor like situation.

A solution to most of the problems with algorithms is using Multi-Resolution Path Planning. This includes the hierarchical planning algorithms. In these approaches, maps may be represented in multiple resolutions at various levels. In most of experiments in this book, we either used a multi-layered approach or a strict 2-layer approach. These approaches make use of same algorithm at multiple layers, or different algorithms at different layers.

Another notion presented was the fusion of algorithms or hybridization of algorithms. Here different algorithms were hybridized with each other or with themselves. The vested interest was that doing so adds the advantages of the individual algorithms and reduces their limitations. In other words, the expectation is that we would be able to remove the limitation of one algorithm by the limitation of the other and vice versa. In these approaches, the tradeoff is governed by the individual algorithm taken and the model used and hence cannot be discussed in depth.

Whatever be the case, the fusion of different algorithms does not mean that the limitations of the algorithms would be completely removed and that that advantages would be completely add up. It rather means that to some extent the limitations of the individual algorithms would be removed. At the same time, this would also mean that the basic limitations of the two algorithms are added up, along with the addition of advantages of the base algorithms. Hence, a fused solution of A* algorithm and Fuzzy Inference System is more complete than a fuzzy planner, better in optimality in terms of route selection than a fuzzy planner; however, it also means that the resultant solution is lesser real time that the fuzzy planner and path would not be optimal in the most simple maps. This emphasizes on caution in selection of algorithms for fusion. A bad manner of fusion would only highlight limitations and not add up advantages. In many scenarios, the base algorithms may be better as compared to the fused counterparts.

Even after much research and stress laid over simple and hybrid algorithms for robot planning, it may yet not be possible to come up with a perfect algorithm as per the academic school of thought. While this may be seen as a limitation to robot design, it is not necessarily one. Most simple mobile robots operating in simple pre-known environments might not need sophisticated algorithms to work with, and the exceptional cases might be too un-probabilistic to encounter, with some backup mechanism that caters to these. Hence, simple algorithms, which give decent looking results, are a simple requirement. It is not surprising that a large variety of tasks can be done by simple algorithms, which are mostly to be seen as planning algorithms of mobile robots.

Our personal interest towards these planning algorithms is always wide, which is also the motive behind the book. Into the domain of artificial intelligence and soft computing, we find many concepts and algorithms at one end and applications at the other end. Planning happens to be one of the applications where most of these algorithms can be used in a manner so as to maximize the understanding of the base algorithms. Looking at the problem of robot planning at the very first instance gives a massive view where unimaginable algorithms are all well used and extensively studied. The incentive is that using this problem as a base, all concepts and algorithms can be easily understood and appreciated. It is always believed that artificial intelligence and soft computing is best understood by examples from applications; robot path planning is the one application that can be easily used as a basis for learning all the various tools and techniques of artificial intelligence and soft computing. Further, this problem domain is wide open to logic, ideas, concepts, and innovations.

The problem sometimes emerges out as a tricky puzzle solving exercise, which has its own joy to deal with. Understanding the manner in which humans solve problems, move around, navigate, and make their way out in most simple to complex environments, and using this as a logic or heuristic to problem solving is challenging. More challenging is to note the limitations of the various approaches, to come up with a set of maps or scenarios, and to attempt to embed some logic to solve the limitations over a variety of scenarios. While all this keeps one busy, it further results in understanding of the complexity of the problem and the various issues associated. There is no limit to the creativity of designing challenging maps, or creativity to actually solve them. Here we do not mean the creativity to design toughest mazes, which are simple from an algorithmic point of view, but the creativity exploiting the limitations of planning algorithms. This is where individual algorithmic limitations are more clearly understood, but in addition, this does lead to a better understanding of the entire algorithm.

It is certainly not true that realistic maps are simple. The increasing use of robotics is bringing more challenges especially at the level of Artificial Intelligence, in which the hard task is to make the robot as intelligent as possible. The ultimate aim is to make robots as intelligent as one's imagination can think of, or the ones widely shown in movies. While the day is far off that these robots get so intelligent, the process is always on. It would certainly be interesting to think of the Artificial Intelligence architecture that these robots have, especially in regard to how these robots are able to do all the tasks so well, while we struggle in a small motion planning algorithm.

The complete architecture is certainly modular, operating at multiple layers of abstractions, and computationally intensive, at the same time operating in real time mode. What is more interesting to observe is that in place of humans aiding the robot for exceptional inputs, uncertainties, etc., it may sometimes be the robot aiding humans. This makes it more important to think over the software architecture, with emphasis to the uncertainties involved. Consider only the motion planning part of this intelligent robot. We know that the challenge may be tough and may require the robot to navigate through any scenario and any situation, which demands a complete and real time operating planning algorithm. While at other modules intelligence may need time to grow, it is certainly motivating to keep the motion-planning front as strong as possible.

It is further of interest that requirements would be that most intelligent robots be able to navigate easily through any kind of environment using optimal routes. At this juncture, it would be wise to think the kind of planning algorithm out of all the various possibilities that would be optimal. Further, it is important to realize that optimal results lie in the use of a particular algorithm or a collation of algorithms that work together. Whichever be the case, algorithms might many times perform well based on certain parameters that may be map dependent. Now, imagining an intelligent robot with a human operator attempting to optimize its operational parameters is certainly not a good thought. We instead imagine intelligent robots to be installed with a planning algorithm that operates in all circumstances. Adaptability, hence, becomes an important issue which has so far been a little discussed issue in robotics. The robot must be able to analyze the situation and tune in its own parameters. The notion may be even further increased to the choice of algorithms for the problem or subset of problem. It is more like having a robot with all skills, and enabling it to decide the skill needed as per assessed situation.

Computation and uncertainty are presently the greatest limitations to planning. Computation limits the calculations that the planning algorithm may do, and hence, the big task is to effectively utilize any calculation in light of how it contributes towards the total result. Uncertainty at all levels results in invalidation of the computed paths, for which the planning algorithm has to leave scope. The build map may not be perfect due to values reported by sensors being error prone, while the control algorithm may not be able to make a robot move by the desired path with the desired speed. All this puts a limitation, which the planning algorithm must overcome. It is natural that a mobile robot may not always be in a position to carry big computational units or a large array of sensors for its motion.

Further assumption of computation being off-board and not on-board inside the robot is not a good design. With constant development in technology, computation is becoming cheap and abundant, while the sensors are becoming more accurate. This makes it possible to carry a lot of computation in a small unit of time, which might eventually even result in smaller resolution versions of offline algorithms working well in online mode. Extremity is certainly having a robot capable of computing all possibilities in fractions of seconds for every situation in which all algorithms are needed to be real time.

For readers interested in this problem domain or who are already working in this problem domain, there is a lot which may be done in the future. In all experiments of the book, our concentration was primarily on creating a robotic trajectory, for which we used simulations. Simulations enabled to easily draw challenging maps. Insertion, deletion, and motion of obstacles were easy to produce. However, experiments may be performed with physical robots. A number of commercially available physical robots may be used which have a differential drive mechanism, a steering mechanism, or likewise. For localization of the robot, a lab environment may be built, where a number of localization points can be fixed. An overhead video camera is widely used, which also supplies the detailed robotic map. Physical experimentation is likely to open new issues.

The basic assumption of robot, which we took in the entire book, was a car-like robot with some non-holonomic constraints. However, the robots vary largely, and along with the type of robots, the constraints that they impose also vary. Robots are widely used both on water and underwater. In both cases, the water currents are important factors. Aerial robots further face air currents. The planning of these robots would differ in case it is a robot driven by wings and wing motion, or a helicopter-like robot. Both types are able to make different moves as per their kinematic constraints. The gait motion or the walking motion of humanoid robots may seem simple but is hard to produce due to problem of balancing. The surface may further not be smooth but inclined. Researchers are further interested in snake-like robots, tree climbing robots, lizard-like robots, wall climbing robots, etc. All these have different applications. Each of these robots implies different constraints and, hence, a different planning. Autonomous vehicles are another type of robots. These are cars that drive themselves, and are being increasingly experimented these days. These may be seen as bigger robots, which operate in the constraints of the road traffic.

From the algorithm point of view, a large number of algorithms have been tried. Current interests are more towards the real time algorithms. Attempts are being made to devise mechanisms to quickly detect changes in map, which invalidate the robot path being followed, quickly detecting blockages, which adversely affects the robotic path, and devising a fast re-planning technique for such scenarios, or to alternatively mend the robotic path. Planning in absence of a complete map is further an important aspect of research, as a complete map may not be initially known to the robot. Algorithms may try to generate a path based on partial information, which can be rectified as the robot moves and becomes aware of new obstacles. The robot is equipped with capabilities to enable it to decide how to surpass unknown obstacles. In the recent past, the RRT class of algorithms have gained a large popularity due to their quick operation and probabilistically complete nature. Due to the same reasons, sensor-based planning is gaining importance where a map may be embedded by sensors. Sensors can quickly detect map changes and communicate the same and further result in fast re-planning.

In this book, we briefly touched upon the problem of multi-robot path planning, which is an interesting problem of research. The methods and algorithms to solve the problem with multiple robots are very wide with new concepts and issues. It may take a book of its own to discuss all the different algorithms. The different robots may or may not be connected to each other via some communication channel. Part of the map may be known to various robots, which may share their information to get a broader idea of the robotic map. Many times different robots may supply supplementary information to the other robot for planning. All these are interesting problems to deal with.

While much is happening into the domain of robotics on both the academic and industrial fronts, it is certainly interesting to see the newer challenges that are put forward at the planning perspective, until we attain the most intelligent robots that operate in any kind of environments. An immense amount of work is required first to look at the adoption of robots in all aspects of life and to note the planning challenges, and second to study all these challenges into formal heads along with the entire research community. As we enter into the era where adoption of robots is becoming common, it is important to have the community to speed up the process. Above all, this requires a strong motivation of researchers, which we hope this book was able to do for the readers to some extent.

Compilation of References

ABB. (2011). *IRB 6400RF*. Retrieved on November, 2011, from http://www.abb.com/product/seitp327/6331470730f0261fc12570e70046bb3a.aspx

Aguirre, E., & Gonzalez, A. (2000). Fuzzy behaviors for mobile robot navigation: Design, coordination and fusion. *International Journal of Approximate Reasoning*, *25*(3), 255–289. doi:10.1016/S0888-613X(00)00056-6

Altenberg, L. (1994). The evolution of evolvability in genetic programming. In *Advances in Genetic Programming* (pp. 47–74). Cambridge, MA: MIT Press.

Alvarez, A., Caiti, A., & Onken, R. (2004). Evolutionary path planning for autonomous underwater vehicles in a variable ocean. *IEEE Journal of Oceanic Engineering*, *29*, 418–429. doi:10.1109/JOE.2004.827837

Antonelli, G., & Chiaverini, S. (2004). Experiments of fuzzy lane following for mobile robots. In *Proceedings of the IEEE Conference on American Control,* (vol 2), (pp. 1079 2004). IEEE Press.

Arnold, D. V., & Beyer, H.-G. (2001). Local performance of the (μ/μI,λ)-ES in a noisy environment. In Martin, W., & Spears, W. (Eds.), *Foundations of Genetic Algorithms* (*Vol. 6*, pp. 127–141). San Francisco, CA: Morgan Kaufmann. doi:10.1016/B978-155860734-7/50090-1

Aulinas, J., Petillot, Y., Salvi, J., & Llado, X. (2008). The SLAM problem: A survey. In *Proceedings of the 11th International Conference of the Catalan Association for Artificial Intelligence,* (Vol. 184), (pp. 363-371). IEEE.

Azarm, K., & Schmidt, G. (1996). A decentralized approach for the conflict-free motion of multiple mobile robots. In *Proceedings of the IEEE Conference on Intelligent Robots and Systems 1996,* (Vol. 3), (pp. 1667-1675). IEEE Press.

Azarm, K., & Schmidt, G. (1997). Conflict-free motion of multiple mobile robots based on decentralized motion planning and negotiation. In *Proceedings of the IEEE International Conference on Robotics and Automation,* (pp. 3526-3533). IEEE Press.

Back, T., Hoffmeister, F., & Schwefel, H. P. (1991). A survey of evolution strategies. In *Proceedings of the Fourth International Conference on Genetic Algorithms*, (pp. 2–9). IEEE.

Back, T. (1996). *Evolutionary algorithms in theory and practice*. Oxford, UK: Oxford University Press.

Back, T., Hammel, U., & Schwefel, H. P. (1997). Evolutionary computation: Comments on the history and current state. *IEEE Transactions on Evolutionary Computation*, *1*(1), 3–17. doi:10.1109/4235.585888

Badran, K. M. S., & Rockett, P. I. (2007). The roles of diversity preservation and mutation in 700 preventing population collapse in multiobjective genetic programming. In *Proceedings of the 9th Annual Conference on Genetic and Evolutionary Computation*, (pp. 1551–1558). IEEE.

Baker, J. E. (1985). Adaptive selection methods for genetic algorithms. In *Proceedings of the First International Conference on Genetic Algorithms and their Applications*. Mahwah, NJ: Lawrence Erlbaum.

Balch, T., & Arkin, R. C. (1998). Behavior-based formation control for multirobot teams. *IEEE Transactions on Robotics and Automation*, *14*(6), 926–939. doi:10.1109/70.736776

Baluja, S. (1994). *Population-based incremental learning: A method for integrating genetic search based function optimization and competitive learning. Technical Report: CS-94-163*. Pittsburgh, PA: Carnegie Mellon University.

Banzhaf, W., Nordin, P., Keller, R. E., & Francone, F. D. (1998a). *Genetic programming: An introduction on the automatic evolution of computer programs and its applications*. San Francisco, CA: Morgan Kaufmann.

Banzhaf, W., Nordin, P., Keller, R. E., & Francone, F. D. (1998b). *Genetic programming: An introduction*. San Mateo, CA: Morgan Kaufmann.

Banzhaf, W., Poli, R., Schoenauer, M., & Fogarty, T. C. (1998c). Genetic programming. *Lecture Notes in Computer Science*, *1391*, 97. doi:10.1007/BFb0055923

Barbehenn, M., & Hutchinson, S. (1993). Efficient search and hierarchical motion planning using dynamic single-source shortest paths trees. In *Proceedings of the IEEE Conference on Robotics and Automation,* (Vol. 1), (pp. 566-571). IEEE Press.

Barbehenn, M., & Hutchinson, S. (1995). Efficient search and hierarchical motion planning by dynamically maintaining single-source shortest paths trees. *IEEE Transactions on Robotics and Automation*, *11*, 198–214. doi:10.1109/70.370502

Bartels, R. H., Beatty, J. C., & Barsky, B. A. (1998). Bézier curves. In *An Introduction to Splines for Use in Computer Graphics and Geometric Modelling* (pp. 211–245). San Francisco, CA: Morgan Kaufmann.

Baturone, I., Moreno-Velo, F. J., Sanchez-Solano, S., & Ollero, A. (2004). Automatic design of fuzzy controllers for car-like autonomous robots. *IEEE Transactions on Fuzzy Systems*, *12*(4), 447–465. doi:10.1109/TFUZZ.2004.832532

Beckers, R., Holland, O. E., & Deneubourg, J. L. (1994). From local actions to global tasks: Stigmergy in collective robotics. In Brooks, R., & Maes, P. (Eds.), *Artificial Life IV*. Cambridge, MA: MIT Press.

Bellis, M. (2011). *Timeline of robots*. Retrieved November, 2011, from http://inventors.about.com/od/roboticsrobots/a/RoboTimeline.htm

Bonev, I. (2011). *Delta parallel robot — The story of success*. Retrieved on November, 2011, from http://www.parallemic.org/Reviews/Review002.html

Brandt, D. (2006). Comparison of A and RRT-connect motion planning techniques for self-reconfiguration planning. In *Proceedings of the IEEE Conference on Intelligent Robots and Systems,* (pp. 892-897). IEEE Press.

Brooks, R. R., Jing-En, P., & Griffin, C. (2008). Game and information theory analysis of electronic countermeasures in pursuit-evasion games. *IEEE Transactions on Systems, Man, and Cybernetics. Part A, Systems and Humans*, *38*, 1281–1294. doi:10.1109/TSMCA.2008.2003970

Bruce, J., & Veloso, M. M. (2003). Real-time randomized path planning for robot navigation. *Lecture Notes in Computer Science*, 288–295. doi:10.1007/978-3-540-45135-8_23

Brunette, E. S., Flemmer, R. C., & Flemmer, C. L. (2009). A review of artificial intelligence. In *Proceedings of the IEEE Conference on Autonomous Robots and Agents*, (pp. 385-390). IEEE Press.

Buehler, M., Iagnemma, K., & Singh, S. (2007). *The 2005 DARPA grand challenge: The great robot race*. Berlin, Germany: Springer-Verlag. doi:10.1007/978-3-540-73429-1

Camilo, O., Collins, E. G. Jr, Selekwa, M. F., & Dunlap, D. D. (2008). The virtual wall approach to limit cycle avoidance for unmanned ground vehicles. *IEEE Transaction on Robotics and Autonomous Systems*, *56*(8), 645–657.

Carbone, A., Finzi, A., Orlandini, A., & Pirri, F. (2008). Model-based control architecture for attentive robots in rescue scenarios. *Autonomous Robots*, *24*(1), 87–120. doi:10.1007/s10514-007-9055-6

Carpin, S., & Pagello, E. (2001). A distributed algorithm for multi-robot motion planning. In *Proceedings of the Fourth European Conference on Advanced Mobile Robotics*, (pp. 207-214). IEEE Press.

Carpin, S., & Pagello, E. (2009). An experimental study of distributed robot coordination. *ACM Robotics and Autonomous Systems*, *57*(2), 129–133. doi:10.1016/j.robot.2008.07.001

Cavalcanti, A., & Freitas, R. A. Jr. (2005). Nanorobotics control design: A collective behavior approach for medicine. *IEEE Transactions on Nanobioscience*, *4*(2), 133–140. doi:10.1109/TNB.2005.850469

Chakraborty, U. K., & Dastidar, D. G. (1993). Using reliability analysis to estimate the number of generations to convergence in genetic algorithm. *Information Processing Letters*, *46*, 199–209. doi:10.1016/0020-0190(93)90027-7

Cheng, P. (2003). *A short survey on pursuit-evasion games*. Champaign, IL: University of Illinois at Urbana-Champaign.

Chen, L. H., & Chiang, C. H. (2003). New approach to intelligent control systems with self-exploring process. *IEEE Transactions on Systems, Man, and Cybernetics. Part B, Cybernetics*, *33*(1), 56–66. doi:10.1109/TSMCB.2003.808192

Choset, H., & Burdick, J. (1995). Sensor based planning: The generalized Voronoi graph. In *Proceedings of the IEEE International Conference on Robotics and Automation,* (vol. 2), (pp. 1649-1655). IEEE Press.

Choset, H. (2001). Coverage for robotics - A survey of recent results. *Annals of Mathematics and Artificial Intelligence*, *31*(1), 113–126. doi:10.1023/A:1016639210559

Clerc, M., & Kennedy, J. (2002). The particle swarm-explosion, stability, and convergence in a multidimensional complex space. *IEEE Transactions on Evolutionary Computation*, *6*, 58–73. doi:10.1109/4235.985692

Coppin, B. (2004). *Artificial intelligence illuminated.* Boston, MA: Jones and Barlett Publishers.

Cormen, T. H., Leiserson, C. E., Rivest, R. L., & Stein, C. (2001). *Introduction to algorithms* (2nd ed.). Cambridge, MA: MIT Press.

Cortes, J., Jaillet, L., & Simeon, T. (2008). Disassembly path planning for complex articulated objects. *IEEE Transactions on Robotics*, *24*(2), 475–481. doi:10.1109/TRO.2008.915464

Cox, E. (1994). *The fuzzy systems handbook.* London, UK: Academic Press.

Das, A. K., Fierro, P., Kumar, V., Ostrowski, J. P., Spletzer, J., & Taylor, C. J. (2002). A vision-based formation control framework. *IEEE Transactions on Robotics and Automation*, *18*(5), 813–825. doi:10.1109/TRA.2002.803463

Davis, L. (1987). *Handbook of genetic algorithms*. Reinhold, NY: Van Nostrand.

de Boor, C. (1978). *A practical guide to splines*. Heidelberg, Germany: Springer Verlag.

Dittrich, P., Burgel, A., & Banzhaf, W. (2006). Learning to move a robot with random morphology. *Lecture Notes in Computer Science*, *1468*, 165–178. doi:10.1007/3-540-64957-3_71

Doitsidis, L., Tsourveloudis, N. C., & Piperidis, S. (2009). Evolution of fuzzy controllers for robotic vehicles: The role of fitness function selection. *Journal of Intelligent & Robotic Systems*, *56*, 469–484. doi:10.1007/s10846-009-9332-z

Doran, J. E., Franklin, S., Jennings, N. R., & Norman, T. J. (2002). On cooperation in multi-agent systems. *The Knowledge Engineering Review*, *12*(3), 309–314. doi:10.1017/S0269888997003111

Dorigo, M., & Caro, G. D. (1999). The ant colony optimization metaheuristic. In Corne, D. (Eds.), *New Ideas in Optimization* (pp. 11–32). London, UK: McGraw Hill.

Dorigo, M., Maniezzo, V., & Colorni, A. (1996). Ant system: Optimization by a colony of cooperating agents. *IEEE Transitions on System, Man, and Cybernetics – Part B*, *26*(1), 29–41. doi:10.1109/3477.484436

Dracopoulos, D. C. (1998). Robot path planning for maze navigation, In *Proceedings of the IEEE International Joint Conference on Neural Networks,* (pp. 2081-2085). Anchorage, AK: IEEE Press.

Dubois, D., & Prade, H. (1985). A review of fuzzy sets and aggregation connectives. *Information Sciences*, *36*, 85–121. doi:10.1016/0020-0255(85)90027-1

Eberhart, R. C., & Shi, Y. (2001). Tracking and optimizing dynamic systems with particle swarms. In *Proceedings of the IEEE Congress Evolutionary Computation,* (pp. 94-97). Seoul, South Korea: IEEE Press.

Ellery, A. (2000). *An introduction to space robotics*. New York, NY: Springer-Verlag.

Enders, H., Feiten, W., & Lawitzky, G. (1998). Field test of navigation system: Autonomous cleaning in supermarkets. In *Proceedings of IEEE International Conference on Robotics and Automation,* (pp. 1779–1781). IEEE Press.

Erdmann, M., & Lozano-Perez, T. (1986). On multiple moving objects. In *Proceedings of IEEE International Conference on Robotics and Automation*, (pp. 1419 – 1424). IEEE Press.

Fanucrobotics. (2011). *ARC Mate 50iC/5L*. Retrieved on November, 2011, from http://www.fanucrobotics.com/file-repository/DataSheets/Robots/ARC-Mate-50iC-5L.pdf

Farinelli, A., Iocchi, L., & Nardi, D. (2004). Multi-robot systems: A classification focused on coordination. *IEEE Transactions on Systems, Man, and Cybernetics. Part B, Cybernetics*, *34*(5), 2015–2028. doi:10.1109/TSMCB.2004.832155

Feil-Seifer, D., & Mataric, M. J. (2005). Defining socially assistive robotics. In *Proceedings of the 9th International Conference on Rehabilitation Robotics*, (pp. 465- 468). IEEE.

Ferguson, D., & Stentz, A. (2005). The delayed D* algorithm for efficient path replanning. In *Proceedings of the IEEE International Conference on Robotics and Automation*, (pp. 2045-2050). IEEE Press.

Fernandez, J. A., & Gonzalez, J. (1998). Hierarchical graph search for mobile robot path planning. In *Proceedings of the IEEE International Conference on Robotics and Automation,* (vol. 1), (pp. 656-661). IEEE Press.

Fierro, R. D., Spletzer, A., Esposito, J., Kumar, J., Ostrowski, V., & Pappas, J. P. (2002). A framework and architecture for multi-robot coordination. *The International Journal of Robotics Research*, *21*(10), 977–995. doi:10.1177/0278364902021010981

Florczyk, S. (2005). *Video-based indoor exploration with autonomous and mobile robots*. Berlin, Germany: Wiley. doi:10.1007/s10846-005-3508-y

Fogel, D. B. (1995). *Evolutionary computation: Toward a new philosophy of machine intelligence*. Los Alamitos, CA: IEEE Press.

Forrest, S., & Mitchell, M. (1993). What makes a problem hard for a genetic algorithm? Some anomalous results and their explanation. *Machine Learning*, *13*, 285–319. doi:10.1023/A:1022626114466

Fredslund, J., & Mataric, M. J. (2002). A general algorithm for robot formations using local sensing and minimal communication. *IEEE Transactions on Robotics and Automation*, *18*(5), 837–846. doi:10.1109/TRA.2002.803458

Fujita, M. (2000). Digital creatures for future entertainment robotics. In *Proceedings of the IEEE International Conference on Robotics and Automation,* (Vol. 1), (pp. 801-806). IEEE Press.

Furuhashi, T., Hasegawa, T., Horikawa, S., et al. (1993). An adaptive fuzzy controller using fuzzy neural networks. In *Proceedings of IEEE Fifth International Fuzzy Systems Association World Congress*, (pp. 769–772). IEEE Press.

Garrido, S., Moreno, L., & Blanco, D. (2006). Voronoi diagram and fast marching applied to path planning. In *Proceedings 2006 IEEE International Conference on Robotics and Automation*, (pp. 3049-3054). IEEE Press.

Gerke, M. (1999). Genetic path planning for mobile robots. In *Proceedings of the American Control Conference,* (Vol. 4), (pp. 2424–2429). IEEE.

Gerkey, B. P., & Mataric, M. J. (2004). A formal analysis and taxonomy of task allocation in multi-robot systems. *The International Journal of Robotics Research*, *23*(9), 939–954. doi:10.1177/0278364904045564

Ge, S. S., & Lewis, F. L. (2006). *Autonomous mobile robot*. Boca Raton, FL: CRC Press.

Goel, A. K., Ail, K. S., Donnellan, M. W., Gomez de Silva Garza, A., & Callantine, T. J. (1994). Multistrategy adaptive path planning. *IEEE Expert*, *9*(6), 57–65. doi:10.1109/64.363273

Goldberg, D. E. (1989). *Genetic algorithms in search, optimization, and machine learning*. Reading, MA: Addison-Wesley.

Goto, T., Kosaka, T., & Noborio, H. (2003). On the heuristics of A* or A algorithm in ITS and robot path-planning. In *Proceedings of the IEEE Conference on Intelligent Robots and Systems,* (vol. 2), (pp. 1159-1166). IEEE Press.

Graupe, D. (1999). *Principles of artificial neural networks*. Singapore, Singapore: World Scientific.

Grossberg, S. (1973). Contour enhancement, short term memory, and constancies in reverberating neural networks. *Studies in Applied Mathematics*, *52*, 217–257.

Guibas, L., Latombe, J.-C., LaValle, S., Lin, D., & Motwani, R. (1985). A visibility-based pursuit-evasion problem. *International Journal of Computational Geometry & Applications*, *4*(2), 74–123.

Gullapalli, V., Franklin, J. A., & Benbrahim, H. (1994). Acquiring robot skills via reinforcement learning. *IEEE Control Systems*, *14*(1), 13–24. doi:10.1109/37.257890

Guo, J., Liu, L., Liu, Q., & Qu, Y. (2009). An improvement of D* algorithm for mobile robot path planning in partial unknown environment. In *Proceedings of the IEEE Conference on Intelligent Computation Technology and Automation,* (vol. 3), (pp. 394-397). IEEE Press.

Han, W., Baek, S., & Kuc, T. (1997). GA based on-line path planning of mobile robots playing soccer games. In *Proceedings of the 40th Midwest Symposium on Circuits and Systems*, (Vol. 1), (pp. 522–525). IEEE.

Hansen, N., & Ostermeier, A. (1997). Convergence properties of evolution strategies with the derandomized covariance matrix adaptation: The (μ/μI,λ)-CMA-ES. In *Proceedings of the 5th European Congress on Intelligent Techniques and Soft Computing*, (pp. 650–654). Aachen, Germany: Verlag Mainz. Holland, J. H. (1975). *Adaptation in natural and artificial systems*. Ann Arbor, MI: University of Michigan Press.

Hazon, N., & Kaminka, G. (2008). On redundancy, efficiency, and robustness in coverage for multiple robots. *IEEE Transactions on Robotics and Automation*, *56*(12), 1102–1114.

Hedberg, S. (1995). Robots cleaning up hazardous waste. *AI Expert*, 20–24.

Hirose, S., & Fukushima, E. F. (2002). Development of mobile robots for rescue operations. *Advanced Robotics*, *16*(6), 509–512. doi:10.1163/156855302320535845

Hirzinger, G., Brunner, B., Dietrich, J., & Heindl, J. (1993). Sensor-based space robotics-ROTEX and its telerobotic features. *IEEE Transactions on Robotics and Automation*, *9*(5), 649–663. doi:10.1109/70.258056

Hohfeld, M., & Rudolph, G. (1997). Towards a theory of population-based incremental learning. In *Proceedings of the 4th IEEE Conference on Evolutionary Computation*, (pp. 1–5). IEEE Press.

Holland, J. H. (1975). *Adaptation in natural and artificial systems*. Ann Arbor, MI: University of Michigan Press.

Holland, J. M. (2004). *Designing autonomous mobile robots*. Boston, MA: Elsevier.

Horowitz, E., & Sahni, S. (1978). *Fundamentals of computer algorithms*. Baltimore, MD: Computer Science Press.

Hu, Y., & Yang, S. X. (2004). A knowledge based genetic algorithm for path planning of a mobile robot. In *Proceedings 2004 IEEE International Conference on Robotics and Automation,* (vol. 5), (pp. 4350-4355). IEEE Press.

Hu, Y., Yang, S. X., Xu, L., & Meng, Q. H. (2004). A knowledge based genetic algorithm for path planning in unstructured mobile robot environments. In *Proceedings of the IEEE Conference on Robotics and Biomimetic,* (pp. 767-772). IEEE Press.

Hui, N. B., & Pratihar, D. K. (2009). A comparative study on some navigation schemes of a real robot tackling moving obstacles. *Robotics and Computer-integrated Manufacturing*, *25*(4-5), 810–828. doi:10.1016/j.rcim.2008.12.003

Hui-Ying, D., Shuo, D., & Yu, Z. (2010). Delaunay graph based path planning method for mobile robot. In *Proceedings of the IEEE Conference on Communications and Mobile Computing,* (vol. 3), (pp. 528-531). IEEE Press.

Hwang, J. Y., Kim, J. S., Lim, S. S., & Park, K. H. (2003). A fast path planning by path graph optimization. *IEEE Transactions on Systems, Man, and Cybernetics. Part A, Systems and Humans*, *33*(1), 121–128. doi:10.1109/TSMCA.2003.812599

IBM. (2011). *Deep blue*. Retrieved on December, 2011, from http://www.research.ibm.com/deepblue/meet/html/d.3.html

Iocchi, L., Nardi, D., & Salerno, M. (2001). Reactivity and deliberation: A survey on multirobot systems. In *Proceedings of Balancing Reactivity and Social Deliberation in Multi-Agent Systems* (pp. 9–32). IEEE. doi:10.1007/3-540-44568-4_2

Isaacs, R. (1965). *Differential games*. New York, NY: Wiley.

Ishiguro, A., Watanabe, Y., Kondo, T., Shirai, Y., & Uchikawa, Y. (1997). A robot with a decentralized consensus-making mechanism based on the immune system. In *Proceedings of the Third International Symposium on Autonomous Decentralized Systems*, (pp. 231-237). IEEE.

Jan, G. E., Chang, K. Y., & Parberry, I. (2008). Optimal path planning for mobile robot navigation. *IEEE Transactions on Mechatronics*, *13*(4), 451–460. doi:10.1109/TMECH.2008.2000822

Jihong, L. (1995). A dynamic programming approach to near minimum-time trajectory planning for two robots. *IEEE Transactions on Robotics and Automation*, *11*(1), 160–164. doi:10.1109/70.345949

Jolly, K. G., Kumar, R. S., & Vijayakumar, R. (2009). A Bezier curve based path planning in a multi-agent robot soccer system without violating the acceleration limits. *Robotics and Autonomous Systems*, *57*(1), 23–33. doi:10.1016/j.robot.2008.03.009

JPL. (2012). In-situ exploration and sample return: Autonomous planetary mobility. *Jet Propulsion Laboratory*. Retrieved on January, 2012, from http://marsrover.nasa.gov/technology/is_autonomous_mobility.html

Juang, C. F. (2004). A hybrid of genetic algorithm and particle swarm optimization for recurrent network design. *IEEE Transactions on Systems, Man, and Cybernetics. Part B, Cybernetics*, *34*(2), 997–1008. doi:10.1109/TSMCB.2003.818557

Juidette, H., & Youlal, H. (2000). Fuzzy dynamic path planning using genetic algorithms. *IEEE Electronics Letters*, *36*(4), 374–376. doi:10.1049/el:20000314

Kaelbling, L. P. (1988). Artificial intelligence and robotics. In *Proceedings of the IEEE Conference on Computer Society International Conference*, (pp. 59-61). IEEE Press.

Kala, R., Shukla, A., & Tiwari, R. (2009a). Fusion of evolutionary algorithms and multi-neuron heuristic search for robotic path planning. In *Proceedings of the 2009 IEEE World Congress on Nature & Biologically Inspired Computing*, (pp. 684-689). Coimbatote, India: IEEE Press.

Kala, R., Shukla, A., & Tiwari, R. (2009b). Robotic path planning using multi neuron heuristic search. In *Proceedings of the ACM 2nd International Conference on Interaction Sciences: Information Technology, Culture and Human*, (pp. 1318-1323). Seoul, South Korea: ACM Press.

Kala, R., Shukla, A., & Tiwari, R. (2012). Robotic path planning using hybrid genetic algorithm particle swarm optimization. *International Journal of Information and Communication Technology*.

Kala, R., Shukla, A., Tiwari, R., Roongta, S., & Janghel, R. R. (2009). Mobile robot navigation control in moving obstacle environment using genetic algorithm, artificial neural networks and A* algorithm. In *Proceedings of the IEEE World Congress on Computer Science and Information Engineering, CSIE 2009*, (pp. 705-713). Los Angeles, CA: IEEE Press.

Kala, R., Shukla, A., Tiwari, R., Rungta, S., & Janghel, R. R. (2009b). Mobile robot navigation control in moving obstacle environment using genetic algorithm, artificial neural networks and A* algorithm. In *Proceedings of the IEEE World Congress on Computer Science and Information Engineering*, (vol. 4), (pp. 705-713). IEEE Press.

Kala, R., Shukla, A., & Tiwari, R. (2010). Dynamic environment robot path planning using hierarchical evolutionary algorithms. *Cybernetics and Systems*, *41*(6), 435–454. doi:10.1080/01969722.2010.500800

Kala, R., Shukla, A., & Tiwari, R. (2010). Fusion of probabilistic A* algorithm and fuzzy inference system for robotic path planning. *Artificial Intelligence Review*, *33*(4), 275–306. doi:10.1007/s10462-010-9157-y

Kala, R., Shukla, A., & Tiwari, R. (2010a). Fusion of probabilistic A* algorithm and fuzzy inference system for robotic path planning. *Artificial Intelligence Review*, *33*(4), 275–306. doi:10.1007/s10462-010-9157-y

Kala, R., Shukla, A., & Tiwari, R. (2010b). Evolving robotic path with genetically optimized fuzzy planner. *International Journal of Computational Vision and Robotics*, *1*(4), 415–429. doi:10.1504/IJCVR.2010.038196

Kala, R., Shukla, A., & Tiwari, R. (2011). Robotic path planning in static environment using hierarchical multi-neuron heuristic search and probability based fitness. *Neurocomputing*, *74*(14-15), 2314–2335. doi:10.1016/j.neucom.2011.03.006

Kala, R., Shukla, A., & Tiwari, R. (2011). Robotic path planning using evolutionary momentum-based exploration. *Journal of Experimental & Theoretical Artificial Intelligence*, *23*(4), 469–495. doi:10.1080/0952813X.2010.490963

Kala, R., Shukla, A., & Tiwari, R. (2012). *Robot path planning using dynamic programming with accelerating nodes*. Paladyn Journal of Behavioural Robotics. *3*(1), 23-34. doi:10.2478/s13230-012-0013-4

Kambhampati, S., & Davis, L. (1986). Multiresolution path planning for mobile robots. *IEEE Journal on Robotics and Automation*, *2*, 135–145. doi:10.1109/JRA.1986.1087051

Kandel, A. (1991). *Fuzzy expert systems*. Boca Raton, FL: CRC Press.

Karaboga, D. (2005). *An idea based on honey bee swarm for numerical optimization. Technical Report-Tr06*. Kayseri, Turkey: Erciyes University.

Karaboga, D., & Basturk, B. (2007). A powerful and efficient algorithm for numerical function optimization: Artificial bee colony (ABC) algorithm. *Journal of Global Optimization*, *39*(3), 459–471. doi:10.1007/s10898-007-9149-x

Kasabov, N. K. (1995). Hybrid connectionist fuzzy production systems: Toward building comprehensive AI. *Intelligent Automation and Soft Computing*, *1*(4), 351–360.

Kasabov, N. K. (1998). Foundations of neural networks. In *Fuzzy Systems, and Knowledge Engineering*. Cambridge, MA: MIT Press.

Kennedy, J., & Eberhart, R. C. (1995). Particle swarm optimization. In *Proceedings of IEEE International Conference on Neural Networks*, (pp. 1942-1948). Perth, Australia: IEEE Press.

Kim, J., & Ostrowski, J. P. (2003). Motion planning a aerial robot using rapidly-exploring random trees with dynamic constraints. In *Proceedings of the IEEE Conference on Robotics and Automation*, (vol. 2), (pp. 2200-2205). IEEE Press.

Kitano, H., Tadokoro, S., Noda, I., Matsubara, H., Takahashi, T., Shinjou, A., & Shimada, S. (1999). RoboCup rescue: Search and rescue in large-scale disasters as a domain for autonomous agents research. In *Proceedings of the 1999 IEEE International Conference on Systems, Man, and Cybernetics*, (vol. 6), (pp. 739-743). IEEE Press.

Kitano, H., Asada, M., Kuniyoshi, Y., Noda, I., & Osawa, E. (1997). Hitoshi matsubara robocup: A challenge problem for AI. *AI Magazine*, *18*(1), 73–85.

Klir, G. J., Clair, U. H., & Yuan, B. (1997). *Fuzzy set theory: Foundations and applications*. Upper Saddle River, NJ: Prentice-Hall.

Koenig, S., & Likhachev, M. (2002). Improved fast replanning for robot navigation in unknown terrain. In *Proceedings of the IEEE International Conference on Robotics and Automation*, (pp. 968 – 975). IEEE Press.

Konar, A. (1999). *Artificial intelligence and soft computing: Behavioral and cognitive modeling of the human brain*. Boca Raton, FL: CRC Press. doi:10.1201/9781420049138

Konar, A. (2000). *Artificial intelligence and soft computing: Behavioral and cognitive modeling of the human*. Boca Raton, FL: CRC Press.

Kuffner, J. J., & LaValle, S. M. (1999). Randomized kinodynamic planning. In *Proceedings of the IEEE International Conference on Robotics and Automation*, (pp. 473–479). IEEE Press.

Kuffner, J. J., & LaValle, S. M. (2000). RRT-connect: An efficient approach to single-query path planning. In *Proceedings of the IEEE International Conference on Robotics and Automation*, (vol. 2), (pp. 995-1001). IEEE Press.

Kuka Robotics. (2011). *Robocoaster*. Retrieved on November, 2011, from http://www.kuka-robotics.com/usa/en/products/systems/robocoaster/start.htm

Kuwata, Y., Karaman, S., Teo, J., Frazzoli, E., How, J. P., & Fiore, G. (2009). Real-time motion planning with applications to autonomous urban driving. *IEEE Transactions on Control Systems Technology*, *17*(5), 1105–1118. doi:10.1109/TCST.2008.2012116

Lai, X. C., Ge, S. S., & Al Mamun, A. (2007). Hierarchical incremental path planning and situation-dependent optimized dynamic motion planning considering accelerations. *IEEE Transactions on Systems, Man, and Cybernetics. Part B, Cybernetics*, *37*(6), 1541–1554. doi:10.1109/TSMCB.2007.906577

Langdon, W. B. (1998). *Genetic programming and data structures: Genetic programming + data structures = automatic programming!*Dordrecht, The Netherlands: Kluwer.

Laumanns, M., Thiele, L., Deb, K., & Zitzler, E. (2001). *On the convergence and diversitypreservation properties of multi-objective evolutionary algorithms*. TIK Report No. 108. Zurich, Switzerland: Swiss Federal Institute of Technology (ETH).

LaValle, S. M., Gonzalez-Banos, H. H., Becker, C., & Latombe, J. C. (1997). Motion strategies for maintaining visibility of a moving target. In *Proceedings of the IEEE International Conference on Robotics and Automation,* (vol. 1), (pp. 731-736). IEEE Press.

LaValle, S. M., & Hutchinson, S. A. (1998). Optimal motion planning for multiple mobile robots having independent goals. *IEEE Transactions on Robotics and Automation, 14*, 912–925. doi:10.1109/70.736775

Li, Y., Zonghai, C., & Feng, C. (2002). A case-based reinforcement learning for probe robot path planning. In *Proceedings of the IEEE Conference on Intelligent Control and Automation,* (vol. 2), (pp. 1161-1165). IEEE Press.

Lin, H., Xiao, J., & Michalewicz, Z. (1994). Evolutionary algorithm for path planning in mobile robot environment. In *Proceedings of the First IEEE Conference on Evolutionary Computation*, (pp. 211-216). IEEE Press.

Liu, Y., & Arimoto, S. (1990). A flexible algorithm for planning local shortest path of mobile robots based on reachability graph. In *Proceedings of the IEEE Conference on Intelligent Robots and Systems*, (vol. 2), (pp. 749-756). IEEE Press.

Lozano-Perez, T. (1987). A simple motion-planning algorithm for general robot manipulators. *IEEE Journal on Robotics and Automation, 3*(3), 224–238. doi:10.1109/JRA.1987.1087095

Luke, S., & Spector, L. (1997). A comparison of crossover and mutation in genetic programming. In *Proceedings of the Second Annual Conference on Genetic Programming*, (pp. 240–248). IEEE.

Luo, C., Yang, S. X., Stacey, D. A., & Jofriet, J. C. (2002). A solution to vicinity problem of obstacles in complete coverage path planning. In *Proceedings of the IEEE Conferences on Robotics and Automation,* (Vol. 1), (pp. 612-617). IEEE Press.

Maaref, H., & Barret, C. (2000). Sensor-based fuzzy navigation of an autonomous mobile robot in an indoor environment. *Control Engineering Practice*, *8*(7), 757–768. doi:10.1016/S0967-0661(99)00200-2

Mahadevan, S., & Connell, J. (1992). Automatic programming of behavior-based robots using reinforcement learning. *Artificial Intelligence*, *55*(2-3), 311–365. doi:10.1016/0004-3702(92)90058-6

Mahjoubi, H., Bahrami, F., & Lucas, C. (2006). Path planning in an environment with static and dynamic obstacles using genetic algorithm: A simplified search space approach. *IEEE Congress on Evolutionary Computation,* (pp. 2483-2489). IEEE Press.

Mamdani, E. (1977). Application of fuzzy logic to approximate reasoning using linguistic synthesis. *IEEE Transactions on Computers*, *26*, 1182–1191. doi:10.1109/TC.1977.1674779

Maniezzo, V., Gambardella, L. M., & Luigi, F. D. (2004). Ant colony optimization. In Onwubolu, G. C., & Babu, B. V. (Eds.), *Optimization Techniques in Engineering* (pp. 101–117). Heidelberg, Germany: Springer.

Manikas, T. W., Ashenayi, K., & Wainwright, R. L. (2007). Genetic algorithms for autonomous robot navigation. *IEEE Instrumentation & Measurement Magazine*, *10*(6), 26–31. doi:10.1109/MIM.2007.4428579

Martin, S. R., Wright, S. E., & Sheppard, J. W. (2007). Offline and online evolutionary bi-directional RRT algorithms for efficient re-planning in dynamic environments. In *Proceedings of the IEEE Conference on Automation Science and Engineering,* (pp. 1131-1136). IEEE Press.

Matthies, L., Kelly, A., Litwin, T., & Tharp, G. (1995). Obstacle detection for unmanned ground vehicles: A progress report. In *Proceedings of the IEEE Intelligent Vehicles Symposium*, (pp. 66-71). IEEE Press.

Maurya, R., & Shukla, A. (2010). Generalized and modified ant algorithm for solving robot path planning problem. In *Proceedings of the 3rd IEEE International Conference on* Computer *Science and Information Technology*, (vol. 1), (pp. 643-646). IEEE Press.

Michalewicz, Z. (1992). *Genetic algorithms + data structures = evolution programs*. Berlin, Germany: Springer.

Millan, J. R., Renkens, F., Mourino, J., & Gerstner, W. (2004). Noninvasive brain-actuated control of a mobile robot by human EEG. *IEEE Transactions on Bio-Medical Engineering, 51*(6), 1026–1033. doi:10.1109/TBME.2004.827086

Miranda, E. R., & Biles, J. A. (2007). *Evolutionary computer music*. Heidelberg, Germany: Springer. doi:10.1007/978-1-84628-600-1

Mital, D. P., Teoh, E. K., & Yong, I. N. (1988). A robotic vision system with artificial intelligence for automatic wafer inspection. In *Proceedings of the IEEE Conference on Intelligent Robots*, (pp. 289-295). IEEE Press.

Mitchell, M. (1999). *An introduction to genetic algorithms*. Cambridge, MA: MIT Press.

Mizumoto, M., & Zimmermann, H. (1982). Comparison of fuzzy reasoning methods. *Fuzzy Sets and Systems, 18*, 253–283. doi:10.1016/S0165-0114(82)80004-3

Murphy, R. (2000). *Introduction to AI robotics*. Cambridge, MA: MIT Press.

Naderan-Tahan, M., & Manzuri-Shalmani, M. T. (2009). Efficient and safe path planning for a mobile robot using genetic algorithm. In *Proceedings of the IEEE Congress on Evolutionary Computation*, (pp. 2091-2097). IEEE Press.

Noreils, F. R. (1990). Integrating multirobot coordination in a mobile-robot control system. In *Proceedings of the IEEE Conferences on Intelligent Robots and Systems 1990: Towards a New Frontier of Applications*, (Vol. 1), (pp. 43-49). IEEE Press.

O'Hara, K. J., Walker, D. B., & Balch, T. R. (2008). Physical path planning using a pervasive embedded network. *IEEE Transactions on Robotics, 24*(3), 741–746. doi:10.1109/TRO.2008.919303

O'Neill, M., & Brabazon, A. (2004). Grammatical swarm. *Lecture Notes in Computer Science, 3102*, 163–174. doi:10.1007/978-3-540-24854-5_15

O'Neill, M., & Ryan, C. (2001). Grammatical evolution. *IEEE Transactions on Evolutionary Computation, 5*, 349–358. doi:10.1109/4235.942529

O'Neill, M., & Ryan, C. (2003). *Grammatical evolution*. Boston, MA: Kluwer.

Olfati-Saber, R., Fax, J. A., & Murray, R. M. (2007). Consensus and cooperation in networked multi-agent systems. *Proceedings of the IEEE, 95*(1), 215–233. doi:10.1109/JPROC.2006.887293

Parker, L. E. (2003). Current research in multirobot systems. *Artificial Life and Robotics, 7*(1-2), 1–5. doi:10.1007/BF02480877

Parpinelli, R. S., Lopes, H. S., & Freitas, A. A. (2002). Data mining with an ant colony optimization algorithm. *IEEE Transactions on Evolutionary Computation, 6*(4), 321–332. doi:10.1109/TEVC.2002.802452

Patnaik, S., Jain, L. C., Tzafestas, S. G., Resconi, G., & Konar, A. (2005). *Innovations in robot mobility and control*. Berlin, Germany: Springer-Verlag.

Peasgood, M., Clark, C. M., & McPhee, J. (2008). A complete and scalable strategy for coordinating multiple robots within roadmaps. *IEEE Transactions on Robotics, 24*(2), 283–292. doi:10.1109/TRO.2008.918056

Peram, T., Veeramachaneni, K., & Mohan, C. K. (2003). Fitness-distance-ratio based particle swarm optimization. In *Proceedings of the 2003 IEEE Swarm Intelligence Symposium*, (pp. 174-181). IEEE Press.

Petrinec, K., & Kovacic, Z. (2005). The application of spline functions and Bezier curves to AGV path planning. In *Proceedings of the IEEE Conference on* Industrial *Electronics*, (vol. 4), (pp. 1453-1458). IEEE Press.

Pham, D. T., Eldukhri, E. E., & Soroka, A. J. (2006). The bees algorithm – A novel tool for complex optimisation problems. In *Proceedings of the Intelligent Production Machines and Systems Conference*, (pp. 454–459). IEEE.

Plonka, L., & Mrozek, A. (1995). Rule based stabilization of the inverted pendulum. *Computational Intelligence: An International Journal*, *11*(2), 348–356.

Pozna, C., Troester, F., Precup, R. E., Tar, J. K., & Preitl, S. (2009). On the design of an obstacle avoiding trajectory: Method and simulation. *Mathematics and Computers in Simulation*, *79*(7), 2211–2226. doi:10.1016/j.matcom.2008.12.015

Pradhan, S. K., Parhi, D., & Panda, A. K. (2009). Fuzzy logic techniques for navigation of several mobile robots. *Applied Soft Computing*, *9*(1), 290–304. doi:10.1016/j.asoc.2008.04.008

Reis, L. P., & Lau, N. (2001). FC Portugal team description: RoboCup 2000 simulation league champion. *Lecture Notes in Computer Science*, *2019*, 29–40. doi:10.1007/3-540-45324-5_2

Ren, W., Beard, R. W., & Atkins, E. M. (2005). A survey of consensus problems in multi-agent coordination. In *Proceedings of the 2005 American Control Conference*, (pp. 1859-1864). IEEE.

Rosca, J. P. (1997). Analysis of complexity drift in genetic programming. In *Proceedings of the Second Annual Conference on Genetic Programming*, (pp. 286–294). Morgan Kaufmann.

Rosenblatt, F. (1958). The perceptron, a probabilistic model for information storage and organization in the brain. *Psychological Review*, *65*, 386–408. doi:10.1037/h0042519

Rosenblatt, F. (1961). Principles of neurodynamics. In *Perceptrons and the Theory of Brain Mechanisms*. Washington, DC: Spartan Press.

Rumelhart, D. E., & McClelland, J. L. (1986). *Parallel distributed processing: Exploring in the microstructure of cognition*. Cambridge, MA: MIT Press.

Russell, S. J., & Norvig, P. (2003). *Artificial intelligence: A modern approach*. Upper Saddle River, NJ: Prentice Hall.

Sadati, N., & Taheri, J. (2002). Genetic algorithm in robot path planning problem in crisp and fuzzified environments. In *Proceedings of the IEEE International Conference on Industrial Technology*, (Vol. 1), (pp. 175–180). IEEE Press.

Sadik, A. M. J., Dhali, M. A., Farid, H. M. A. B., Rashid, T. U., & Syeed, A. (2010). A comprehensive and comparative study of maze-solving techniques by implementing graph theory. In *Proceedings of the IEEE Conference on Artificial Intelligence and Computational Intelligence*, (Vol. 1), (pp. 52-56). IEEE Press.

Sariff, N., & Buniyamin, N. (2006). An overview of autonomous mobile robot path planning algorithms. In *Proceedings of the IEEE Conference on Research and Development*, (pp. 183-188). IEEE Press.

Schwefel, H. P. (1995). *Evolution and optimum seeking*. New York, NY: Wiley.

Schwefel, H. P., & Rudolph, G. (1995). Contemporary evolution strategies. *Lecture Notes in Computer Science*, *929*, 893–907. doi:10.1007/3-540-59496-5_351

Sedighi, K. H., Ashenayi, K., Manikas, T. W., Wainwright, R. L., & Heng-Ming, T. (2004). Autonomous local path planning for a mobile robot using a genetic algorithm. In *Proceedings of the IEEE Conference on Evolutionary Computation*, (Vol. 2), (pp. 1338–1345). IEEE Press.

Seetharaman, G., Lakhotia, A., & Blasch, E. P. (2006). Unmanned vehicles come of age: The DARPA grand challenge. *Computer*, *39*(12), 26–29. doi:10.1109/MC.2006.447

Sheng, L., & Pan, Y. J. (2009). Distributed control for consensus of networked multi-agent robotic system with time delays. *International Journal of Information and Systems Science*, *5*(2), 161–178.

Shi, Y. (2004). Particle swarm optimization. *IEEE Neural Networks Society Bulletin*, 8–13.

Shibata, T., Fukuda, T., & Tanie, K. (1993). Fuzzy critic for robotic motion planning by genetic algorithm in hierarchical intelligent control. In *Proceedings of 1993 International Joint Conference on Neural Networks*, (pp. 77-773). IEEE.

Shimoda, S., Kuroda, Y., & Iagnemma, K. (2005). Potential field navigation of high speed unmanned ground vehicles on uneven terrain. In *Proceedings of the 2005 IEEE International Conference on Robotics and Automation*, (pp. 2828-2833). IEEE Press.

Shin, K., & McKay, N. (1986). A dynamic programming approach to trajectory planning of robotic manipulators. *IEEE Transactions on Automatic Control*, *31*(6), 491–500. doi:10.1109/TAC.1986.1104317

Shukla, A., Tiwari, R., & Kala, R. (2008). Mobile robot navigation control in moving obstacle environment using A* algorithm. In *Proceedings of the International Conference on Artificial Neural Networks in Engineering*, (pp. 113-120). St. Louis, MO: IEEE.

Shukla, A., & Kala, R. (2008). Multi neuron heuristic search. *International Journal of Computer Science and Network Security*, *8*(6), 344–350.

Shukla, A., Tiwari, R., & Kala, R. (2008). *Mobile robot navigation control in moving obstacle environment using A* algorithm. Intelligent Systems Engineering Systems through Artificial Neural Networks* (*Vol. 18*, pp. 113–120). ASME Publications.

Shukla, A., Tiwari, R., & Kala, R. (2010). *Real life applications of soft computing*. Boca Raton, FL: CRC Press. doi:10.1201/EBK1439822876

Shukla, A., Tiwari, R., & Kala, R. (2010a). *Real life applications of soft computing*. Boca Raton, FL: CRC Press. doi:10.1201/EBK1439822876

Shukla, A., Tiwari, R., & Kala, R. (2010b). *Towards hybrid and adaptive computing: A perspective*. Berlin, Germany: Springer-Verlag.

Sicard, P., & Levine, M. D. (1988). An approach to an expert robot welding system. *IEEE Transactions on Systems, Man, and Cybernetics*, *18*(2), 204–222. doi:10.1109/21.3461

Singh, S. (2002). *Reinforcement learning*. Reading, MA: Kluwer Academic Publishers.

Sitti, M., & Hashimoto, H. (1998). Tele-nanorobotics using atomic force microscope. In *Proceedings of the IEEE/RSJ International Conference on Intelligent Robots and Systems*, (vol. 3), (pp. 1739-1746). IEEE Press.

Sniedovich, M. (2011). *Dynamic programming: Foundation and principles*. Boca Raton, FL: CRC Press.

Southey, F., & Karray, F. (1999). Approaching evolutionary robotics through population-based incremental learning. In *Proceedings of the IEEE International Conference on Systems, Man, and Cybernetics*, (vol. 2), (pp. 710–715). IEEE Press.

Stentz, A. (1990). *The navlab system for mobile robot navigation*. (Ph.D. Thesis). Carnegie Mellon University School of Computer Science. Pittsburgh, PA.

Stentz, A. (1994). Optimal and efficient path planning for partially-known environments. In *Proceedings of the International Conference on Robotics and Automation*, (pp. 3310–3317). IEEE.

Stentz, A. (1995). The focussed D* algorithm for real-time replanning. In *Proceedings of the International Joint Conference on Artificial Intelligence*, (pp. 1652–1659). IEEE. Thornton, C., & du Boulay, B. (2005). *Artificial intelligence: Strategies, applications, and models through search*. New Delhi, India: New Age Publishers.

Stipanovic, D. M., Inalhan, G., Teo, R., & Tomlin, C. J. (2004). Decentralized overlapping control of a formation of unmanned aerial vehicles. *Automatica*, *40*(8), 1285–1296. doi:10.1016/j.automatica.2004.02.017

Stutzle, T., & Hoos, H. (1997). Improvements on the ant system: Introducing MAX–MIN ant system. In *Proceedings of the International Conference on Artificial Neural Networks and Genetic Algorithms*, (pp. 245–249). Springer.

Stutzle, T., & Dorigo, M. (1999). ACO algorithms for the traveling salesman problem. In Miettinen, K., & Neittaanmaki, P. (Eds.), *Evolutionary Algorithms in Engineering and Computer Science* (pp. 160–184). West Sussex, UK: Wiley.

Sugeno, M. (1974). *Theory of fuzzy integral and its applications*. (PhD Thesis). Tokyo Institute of Technology. Tokyo, Japan.

Sugeno, M. (1985). An introductory survey of fuzzy control. *Information Sciences*, *36*, 59–83. doi:10.1016/0020-0255(85)90026-X

Sutton, R. S., & Barto, A. G. (1998). *Reinforcement learning: An introduction*. Reading, MA: Kluwer Academic Publishers.

Tackett, W. A. (1993). Genetic programming for feature discovery and image discrimination. In *Proceedings of the Fifth International Conference on Genetic Algorithms*, (pp. 303–309). San Mateo, CA: Morgan Kaufmann.

Takagi, T., & Sugeno, M. (1985). Fuzzy identification of systems and its application to modeling and control. *IEEE Transactions on Systems, Man, and Cybernetics*, *15*(1), 116–132.

Thorisson, K. R. (1996). *Communicative humanoids: A computational model ofpsychosocial dialogue skills*. (PhD Thesis). Massachusetts Institute of Technology, Program in Media Arts & Sciences. Cambridge, MA.

Thornton, C., & du Boulay, B. (2005). *Artificial intelligence: Strategies, applications, and models through search*. New Delhi, India: New Age Publishers.

Tokuse, N., Sakahara, H., & Miyazaki, F. (2008). Motion planning for producing a give-way behavior using satio-temporal RRT. In *Proceedings of the IEEE Conference on System Integration,* (pp. 12-17). IEEE Press.

Tomono, M. (2003). Planning a path for finding targets under spatial uncertainties using a weighted Voronoi graph and visibility measure. In *Proceedings of the IEEE Conference on Intelligent Robots and Systems,* (vol. 1), (pp. 124-129). IEEE Press.

Toogood, R., Hong, H., & Chi, W. (1995). Robot path planning using genetic algorithms. In *Proceedings of the IEEE Conference on Systems, Man and Cybernetics,* (Vol. 1), (pp. 489–494). IEEE Press.

Tsai, C. H., Lee, J. S., & Chuang, J. H. (2001). Path planning of 3-D objects using a new workspace model. *IEEE Transactions on Systems, Man and Cybernetics. Part C, Applications and Reviews*, *31*(3), 405–410. doi:10.1109/5326.971669

Tsai, C., Lee, J., & Chuang, J. (2001). Path planning of 3-D objects using a new workspace model. *IEEE Transactions on Systems, Man and Cybernetics. Part C, Applications and Reviews*, *31*(3), 405–410. doi:10.1109/5326.971669

Tu, J., & Yang, S. (2003). Genetic algorithm based path planning for a mobile robot. In *Proceedings of the IEEE International Conference on Robotics and Automation*, (pp. 1221-1226). IEEE Press.

Tuci, E., Gross, R., Trianni, V., Mondana, F., Bonani, M., & Dorigo, M. (2006). Cooperation through self-assembly in multi-robot systems. *ACM Transactions on Autonomous and Adaptive Systems*, *1*(2), 115–150. doi:10.1145/1186778.1186779

Turban, E., & Schaeffer, D. M. (1988). Expert system-based robot technology: A systems management approach. In *Proceedings of the IEEE Conference on System Sciences,* (Vol. 4), (pp. 227-235). IEEE Press.

Tutte, W. T. (2001). *Graph theory*. Cambridge, UK: Cambridge University Press.

University of Reading. (2012). Biological interfaces with computer systems. *University of Reading*. Retrieved January, 2012, from http://www.prospectus.rdg.ac.uk/cirg/cirg-manmachine.aspx

Urdiales, C., Bantlera, A., Arrebola, F., & Sandoval, F. (1998). Multi-level path planning algorithm for autonomous robots. *IEEE Electronic Letters, 34*(2), 223-224.

Urdiales, C., Bandera, A., Arrebola, F., & Sandoval, F. (1998). Multi-level path planning algorithm for autonomous robots. *Electronics Letters*, *34*, 223–224. doi:10.1049/el:19980204

Veeramachaneni, K., Peram, T., Mohan, C., & Osadciw, L. A. (2003). Optimization using particle swarms with near neighbor interactions. In *Proceedings of Genetic and Evolutionary Computation Conference*, (pp. 110-121). IEEE.

Veloso, M. M. (2002). Robots: Intelligence, versatility, adaptivity. *Communications of the ACM*, *45*(3), 59–63.

Vidal, R., & Shakernia, O., Kim, Shim, D., & Sastry, S. (2002). Probabilistic pursuit-evasion games: Theory, implementation, and experimental evaluation. *IEEE Transactions on Robotics and Automation*, *18*, 662–669. doi:10.1109/TRA.2002.804040

Viguria, A., & Howard, A. M. (2009). An integrated approach for achieving multirobot task formations. *IEEE Trasactions on Mechatronics*, *14*, 176–186. doi:10.1109/TMECH.2009.2014056

Waldherr, S., Romero, R., & Thrun, S. (2000). A gesture based interface for human-robot interaction. *Autonomous Robots*, *9*(2), 151–173. doi:10.1023/A:1008918401478

Wang, C., Soh, Y. C., Wang, H., & Wang, H. (2002). A hierarchical genetic algorithm for path planning in a static environment with obstacles. In *Proceedings of IEEE Conference on Electrical and Computer Engineering,* (Vol. 3), (pp. 1652–1657). IEEE Press.

Wang, Y., Linnett, J. A., & Roberts, J. (1994). Kinematics, kinematic constraints and path planning for wheeled mobile robots. *Robotica*, *12*(5), 391–400. doi:10.1017/S026357470001794X

Warren, C. W. (1990). Multiple path coordination using artificial potential fields. In *Proceedings of the IEEE International Conference on Robotics and Automation*, (pp. 500-505). IEEE Press.

Washburn, A., & Kress, M. (2009). Unmanned aerial vehicles. In *Combat Modeling, International Series in Operations Research & Management Science* (pp. 185–210). Berlin, Germany: Springer.

Weisstein, E. W. (2010). *Bézier curve*. Retrieved from http://mathworld.wolfram.com/BezierCurve.html

Widrow, B., & Hoff, M. E. (1960). Adaptive switching circuits. In *Proceedings of the 1960 IREWESCON Convention,* (pp. 96–104). IREWESCON.

Widrow, B. (1962). Generalization and information storage in networks of ADALINE neurons. In Yovits, M. C., Jacobi, G. T., & Goldstein, G. D. (Eds.), *Self-Organizing Systems* (pp. 435–461). New York, NY: Pergamon.

Widrow, B., & Winter, R. (1988). Neural nets for adaptive filtering and adaptive pattern recognition. *Computer*, *21*, 25–39. doi:10.1109/2.29

Willms, A. R., & Yang, S. X. (2006). An efficient dynamic system for real-time robot-path planning. *IEEE Transactions on Systems, Man, and Cybernetics. Part B, Cybernetics*, *36*(4), 755–766. doi:10.1109/TSMCB.2005.862724

Wisegeek. (2011). *What is a SCARA robot?* Retrieved on November, 2011, from http://www.wisegeek.com/what-is-a-scara-robot.htm

Woong-Gie, H., Seung-Min, B., & Tae-Yong, K. (1997). Genetic algorithm based path planning and dynamic obstacle avoidance of mobile robots. In *Proceedings of the IEEE Conference on Systems, Man, and Cybernetics,* (Vol. 3), (pp. 2747–2751). IEEE Press.

Wyrobek, K. A., Berger, E. H., Van der Loos, H. F. M., & Salisbury, J. K. (2008). Towards a personal robotics development platform: Rationale and design of an intrinsically safe personal robot. In *Proceedings of the IEEE International Conference on Robotics and Automation*, (pp. 2165-2170). IEEE Press.

Xiao, J., Michalewicz, Z., Zhang, L., & Trojanowski, K. (1997). Adaptive evolutionary planner/navigator for mobile robots. *IEEE Transactions on Evolutionary Computation*, *1*(1), 18–28. doi:10.1109/4235.585889

Xing, X., Jia, Q., Ling, L., & Yuan, D. (2007). A novel genetic algorithm based on individual and gene diversity maintaining and its simulation. In *Proceedings of the IEEE Conference on Automation and Logistics,* (pp. 2754-2758). IEEE Press.

Yager, R., & Zadeh, L. (1992). *An introduction to fuzzy logic applications in intelligent systems*. Boston, MA: Kluwer Academic.

Yagnik, D., Ren, J., & Liscano, R. (2010). Motion planning for multi-link robots using artificial potential fields and modified simulated annealing. In *Proceedings of the IEEE Conference on Mechatronics and Embedded Systems and Applications,* (pp. 421-427). IEEE Press.

Yan, X., Wu, Q., Yan, J., & Kang, L. (2007). A fast evolutionary algorithm for robot path planning. In *Proceedings of the IEEE Conference on Control and Automation,* (pp. 84-87). IEEE Press.

Yang, S. (2005). Population-based incremental learning with memory scheme for changing environments. In *Proceedings of the 2005 Conference on Genetic and Evolutionary Computation*, (pp. 711–718). IEEE.

Yang, S., & Yao, X. (2003). Dual population-based incremental learning for problem optimization in dynamic environments. In *Proceedings of the 7th Asia Pacific Symposium on Intelligent and Evolutionary Systems*, (pp. 49–56). IEEE.

Yang, X., Moallen, M., & Patel, R. (2003). An improved fuzzy logic based navigation system for mobile robots. In *Proceedings of the IEEE Conference on Intelligent Robots and Systems*, (pp. 1709-2003). IEEE Press.

Yang, S. X., & Meng, M. (2000). An efficient neural network approach to dynamic robot motion planning. *Neural Networks*, *13*(2), 143–148. doi:10.1016/S0893-6080(99)00103-3

Yang, S. X., & Meng, M. Q.-H. (2003). Real-time collision-free motion planning of a mobile robot using a neural dynamics-based approach. *IEEE Transactions on Neural Networks*, *14*(6), 1541–1552. doi:10.1109/TNN.2003.820618

Yang, S., & Yao, X. (2005). Experimental study on population-based incremental learning algorithms for dynamic optimization problems. *Soft Computing*, *9*(11), 815–834. doi:10.1007/s00500-004-0422-3

Yang, S., & Yao, X. (2008). Population-based incremental learning with associative memory for dynamic environments. *IEEE Transactions on Evolutionary Computation*, *12*(5), 542–561. doi:10.1109/TEVC.2007.913070

Yanrong, H., & Yang, S. X. (2004). A knowledge based genetic algorithm for path planning of a mobile robot. In *Proceedings of the IEEE Conference on Robotics 845 and Automation*, (Vol. 5), (pp. 4350–4355). IEEE Press.

Yegenoglu, F., & Stephanou, H. E. (1989). Collision-free path planning for multirobot systems. In *Proceedings of the IEEE Symposium on Intelligent Control,* (pp. 537-542). IEEE Press.

Yogeswaran, M., & Ponnambalam, S. G. (2010). Swarm robotics: An extensive research review. In Fuerstner, I. (Ed.), *Advanced Knowledge Application in Practice*. New York, NY: InTech Publishers.

Yun, X., Alptekin, G., & Albayrak, O. (1997). Line and circle formation of distributed physical mobile robots. *Journal of Robotic Systems*, *14*(2), 63–76. doi:10.1002/(SICI)1097-4563(199702)14:2<63::AID-ROB2>3.0.CO;2-R

Zadeh, L. (1968). Probability measures of fuzzy events. *Journal of Mathematical Analysis and Applications*, *22*, 421–427. doi:10.1016/0022-247X(68)90078-4

Zadeh, L. (1979). A theory of approximate reasoning. In Hayes, M. M. (Ed.), *Machine Intelligence* (*Vol. 9*, pp. 149–194). New York, NY: Elsevier.

Zadeh, L. A. (1965). Fuzzy sets. *Information and Control*, *8*, 338–353. doi:10.1016/S0019-9958(65)90241-X

Zelinsky, A. (1994). Using path transforms to guide the search for findpath in 2D. *The International Journal of Robotics Research*, *13*(4), 315–325. doi:10.1177/027836499401300403

Zhang, J., & Bohner, P. (1993). A fuzzy control approach for executing subgoal guided motion of a mobile robot in a partially-known environment. In *Proceedings of the IEEE Conference on Robotics and Automation,* (vol 2), (pp. 545 -550). IEEE Press.

Zhang, Z., Ma, S., Lu, Z., & Cao, B. (2006). Communication mechanism study of a multi-robot planetary exploration system. In *Proceedings of the 2006 IEEE International Conference on Robotics and Biomimetics*, (pp. 49-54). Kunming, China: IEEE Press.

Zheng, T., Li, R., Guo, W., & Yang, L. (2008). Multi-robot cooperative task processing in great environment. In *Proceedings of the IEEE Conference on Robotics, Automation and Mechatronics,* (pp. 1113-1117). IEEE Press.

Zhu, D., & Latombe, J. (1991). New heuristic algorithms for efficient hierarchical path planning. *IEEE Transactions on Robotics and Automation*, *7*(1), 9–20. doi:10.1109/70.68066

About the Authors

Ritu Tiwari is serving as an Assistant Professor in the Indian Institute of Information Technology and Management Gwalior. Her field of research includes biometrics, artificial neural networks, speech signal processing, robotics, and soft computing. She has published over 75 research papers in various national and international journals/conferences. She has received the Young Scientist Award from Chhattisgarh Council of Science and Technology and also received a Gold Medal in her post graduation.

Anupam Shukla is serving as a Professor in the Indian Institute of Information Technology and Management Gwalior. He heads the Soft Computing and Expert System Laboratory at the Institute. He has 20 years of teaching experience. His research interest includes speech processing, artificial intelligence, soft computing, biometrics, and bioinformatics. He has published over 150 papers in various national and international journals/conferences. He is editor and reviewer in various journals. He received the Young Scientist Award from Madhya Pradesh Government and a Gold Medal from Jadavpur University.

Rahul Kala is pursuing a PhD from the School of Cybernetics at the University of Reading. He did his Bachelors and Masters in Information Technology at the Indian Institute of Information Technology and Management Gwalior. His areas of research are hybrid soft computing, robotic planning, autonomous vehicles, biometrics, artificial intelligence, and soft computing. He has published over 50 papers in various international journals and conferences and is the author of 2 books. He is the recipient of the Commonwealth Scholarship and Fellowship Program (2010), UK, and is the winner of the Lord of the Code Scholarship Contest organized by KReSIT, IIT Bombay, and Red Hat. He secured the all India 8th position in Graduates Aptitude Test in Engineering—2008 examinations.

Index

S

T

U